Microbial Technologies in Industrial Wastewater Treatment

Maulin P. Shah
Editor

Microbial Technologies in Industrial Wastewater Treatment

Springer

Editor
Maulin P. Shah
Environmental Microbiology Lab
Ankleshwar, Gujarat, India

ISBN 978-981-99-2434-9 ISBN 978-981-99-2435-6 (eBook)
https://doi.org/10.1007/978-981-99-2435-6

This Springer imprint is published by the registered company Springer Nature Singapore Pte Ltd.
The registered company address is: 152 Beach Road, #21-01/04 Gateway East, Singapore 189721, Singapore

Contents

Effect of Heavy Metals in Sewage Sludge

Simranjeet Singh, Harry Kaur, Daljeet Singh Dhanjal, Ruby Angurana, Dhriti Kapoor, Vaidehi Katoch, Dhriti Sharma, Praveen C. Ramamurthy, and Joginder Singh

1 Introduction

Heavy metal is a high-density metalloid or metal recognized for its potentially harmful effects, especially in the environment. Heavy metal toxic effects can be caused by an excess of desired levels or unwanted metals derived from natural sources, mostly on land that has been centered as a result of human activity (Gupta et al. 2021). These metals enter the trees, living creatures, and tissues within the body via ingestion, fluid intake, and handling, where they can attach to and disrupt the function of important cellular constituents (Ekere et al. 2020). Heavy metals

S. Singh · P. C. Ramamurthy (✉)
Interdisciplinary Centre for Water Research (ICWaR), Indian Institute of Science, Bangalore 560012, India
e-mail: praveen@iisc.ac.in

H. Kaur
Department of Biosciences and Bioengineering, Indian Institute of Sciences (IIT), Roorkee, Uttarakhand, India

D. S. Dhanjal · J. Singh (✉)
Department of Biotechnology, School of Bioengineering and Biosciences, Lovely Professional University, Phagwara, Punjab, India
e-mail: joginder.15005@lpu.co.in

R. Angurana
Department of Zoology, School of Bioengineering and Biosciences, Lovely Professional University, Phagwara, Punjab, India

D. Kapoor · D. Sharma
Department of Botany, School of Bioengineering and Biosciences, Lovely Professional University, Phagwara, Punjab, India

V. Katoch
Department of Forensic Science, School of Bioengineering and Biosciences, Lovely Professional University, Phagwara, Punjab, India

M. P. Shah (ed.), *Microbial Technologies in Industrial Wastewater Treatment*,
https://doi.org/10.1007/978-981-99-2435-6_1

were major environmental contaminants, as well as their toxic effects seem to be a growing concern for ecosystems, biological and social factors, health and nutrition, and geographic factors (Ali et al. 2019). In addition to cobalt, copper, zinc, manganese, chromium, lead, cadmium, silver, nickel, platinum, and iron, other metals and metalloids with densities larger than 4 g per cubic centimeter have atomic weights that are five times or higher than the density of water (Asati et al. 2016). A condition of the surroundings is described as all the elements that influence a species or group of species, particularly when it relates to the actual physiological conditions that physically impact the organism's development, growth, and sustenance (Awuchi et al. 2020). Usually, they are found scattered among rock formations. With increasing industrial development and growth, heavy metals have become more prevalent in the environment, reaching their highest concentrations in aquatic as well as soil ecosystems and having a comparatively smaller percentage in the particulate matter in the atmosphere or vapor (Wang et al. 2020). Because several heavy metals are thought toward being necessary for growth, the accumulation of heavy metals in plants depends largely on species of plants, individual metals, composition, chemical state, and soil pH (Hasan et al. 2019). These heavy metals, including copper and zinc, act as cofactors or activators of enzyme reactions. These heavy metals are readily accumulative and persistent in the environment, and they are not easily degraded or metabolized (Rehman et al. 2021). As these metals accumulate in the ecological food chain, they are taken up by primary producers and subsequently used up by consumers (Ghori et al. 2019). The plant's root is an important point of contact for heavy metal ions due to their stationary state. Plants in aquatic systems are exposed to these ions throughout their entire body. As a result of particles that have been accumulated on the foliar outer surface, heavy metals are also directly absorbed by the leaves. It is most common for heavy metals to enter the body via ingestion or inhalation since that is the predominant method of entry for the general public (Shahid et al. 2017).

Nevertheless, all of these factors, including urban development and transit, agricultural and industrial practices, industrial discharges, but also resource extraction, have all made a significant contribution to "human inhalation of heavy metals". Cd, Ni, As, and Cr are heavy metals that are hazardous to humans in a number of ways. As well as causing cancer, heavy metals may also cause other diseases (Karande et al. 2019). Cadmium consumption causes Itai-Itai disease, while mercury consumption causes Minamata disease (Raikwar et al. 2008). Poisoning from other heavy metals, such as kidney dysfunction, nervous system disorders, vascular damage, skin lesions, immune system dysfunction, cancer, and birth defects, is caused by contaminated drinking water (Baby et al. 2010). The haphazard disposal of industrial pollutants, human wastes, and other disposals often pollutes these environments, affecting beneficial microorganisms' survival and physical functions (Smith and Riddle 2009). Most metals gather and become magnified in the food chain (Vizzini et al. 2013). Heavy metals have an impact on all life forms and ecological functions, which include microbes (Englande et al. 2015). The bioaccumulation and biomagnification of trace elements in living organisms describe the procedures and these pollutants' pathways

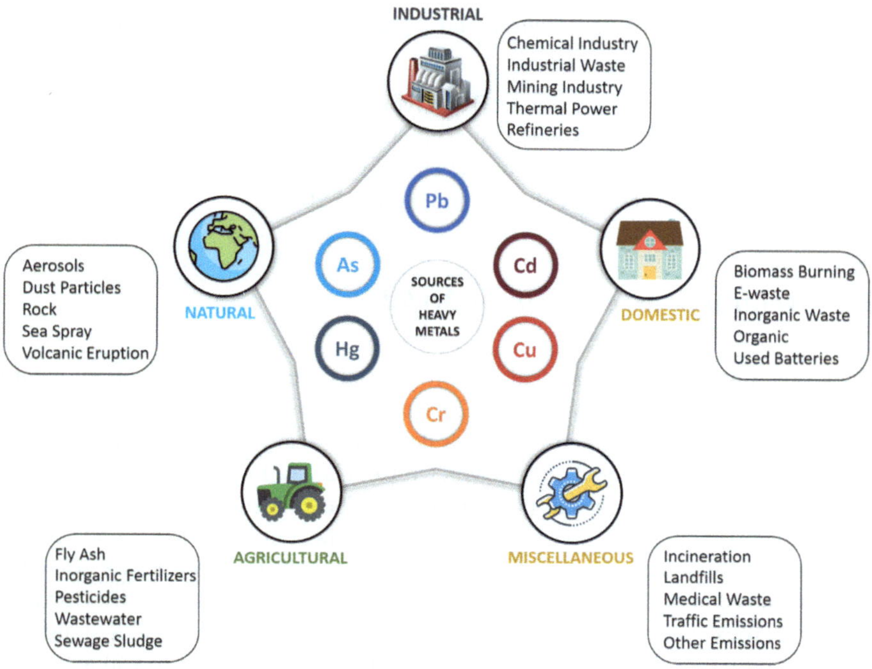

Fig. 1 Different sources of heavy metal in the environment

from one level to the next to the next, demonstrating the organisms' superior bioaccumulation capacity (Unrine et al. 2007). Humans consume some of these organisms, such as fish. However, they can spread serious illnesses such as Itai-Itai and Minamata (Ali et al. 2019). Based on the extent, complexity, and type of contamination, remediation methods such as digging and soil amendment, heating, electro reclamation, and soil maxing have been proposed, all of which are costly and dangerous to the environment (Szynkowska et al. 2017). This book chapter provides an overview of current research on the sources of heavy metals (Fig. 1) and their effects on plants. Heavy metals' effects on humans are also summarized.

2 Significance of Heavy Metals

Plants and animals require certain heavy metals (Cu, Zn, and Fe). The availability of heavy metals in the medium varies (Asati et al. 2016). The toxic limit of heavy metals and the safe intake of heavy metals for human health (Hu 2002). Heavy metals are also known as trace elements since they can be found in environmental matrices in trace or ultra-trace at different concentrations (Behbahani et al. 2013). Vital heavy metals (Mo, Cu, Mn, In, and Fe) perform physiological and biochemical roles in animals and plants (Sharma and Agrawal 2005). Heavy metals serve two major functions:

(a) participation in redox reactions (Nagajyoti et al. 2010) in addition
(b) direct participation as a fundamental component of numerous enzymes. Because heavy metals are formed inside the crust of Earth, their usual presence in soil results from weathering (Sharma and Agrawal 2005).

2.1 *Natural Sources of Heavy Metals*

Rock outcroppings and biological parent materials are the most important natural sources of heavy metals (Wen et al. 2020). Rock type and environmental conditions that activate the weathering process determine heavy metals' composition and concentration (Sandeep et al. 2019). Mn, Cr, Co, Cu, Ni, In, Sn, Pb, Cd, and Hg are commonly found in geological plant materials. The concentration of heavy metals also varies depending on the type of rock (Srivastava et al. 2017). In addition to being primarily composed of sedimentary rock, the soil is only a minor source of heavy metals due to its scarcity and ease of weathering (Khan et al. 2020). However, numerous rock types, such as hornblende, olivine, and augite, make a significant contribution in considerable quantities of Mn, Cu, Ni, Co, and ln to soil and water (Sharma et al. 2019). Shale has the highest concentrations of Mn, Cr, Co, Cu, Ni, In, Sn, Pb, Cd, and Hg among sediments, accompanied by limestone and sandstone (Sun et al. 2019).

2.2 *Heavy Metal Sources in Agricultural Soils*

Sources of heavy metals in the environment are sewage sludging, liming, irrigation waterways, and pesticide residues (Singh et al. 2019a, 2020a, 2020b, 2020c, 2021; Rebi et al. 2021). Other chemicals, particularly inorganic fertilizers, fungicides, especially the fertilizers mostly phosphate, contain varying amounts of Cr, Ni, Pb, and Cd in addition ln dependent on the source (Joimel et al. 2021). Cadmium is particularly dangerous in plants because it aggregates at an extremely high amount (mostly leaf part), which can be eaten by humans or animals (Yang et al. 2017). Cadmium enrichment can also be caused by sewage sludge, manures, and limes (Hamid et al. 2019). Although heavy metal levels in agricultural land are comparatively low, the frequent usage of phosphate fertilizer, combined with metals' elongated residence time, might result in a hazardously high buildup of some metals (Huang et al. 2018).

The use of soil application, including compost rejecting and nitrate herbicides and pesticides, can also contribute to heavy metals in the land (Sandeep et al. 2019). Liming effects heavy metal content in the soil more than nitrate fertilizers as well as compost waste (Masud et al. 2020). Among the most important sources of accumulation of heavy metals in the soil is sewage sludging (Muchuweti et al. 2006). Several pesticides based on heavy metals regulate illnesses in grains, fruit, and vegetable productions, but they also pollute the soil (Singh et al. 2014, 2016, 2017a, 2019b; Sidhu et al. 2019; Sharma and Agrawal 2005). As a result of rapid urbanization, the majority of urban sewage sludge has risen significantly in recent years (Han et al. 2021; Singh et al. 2020d). According to the most recent data, China's sewage sludge (80% water content) will reach 90 million tons by 2020. Several toxic and hazardous constituents in sewage sludge, such as eggs of the parasite, microbial pathogens, and heavy metals, can pollute water and soil forms if not handled appropriately (Djandja et al. 2020; Singh et al. 2017b, 2019c). Traditional sludge treatment procedures primarily comprise sanitary landfilling and incineration (Zhang et al. 2016). However, due to soil scarcity, waste disposal technology in China was already severely limited or even banned; e.g., due to the complex nature and unpredictability of the contaminant constituents in the sludge, urban effluents stabilized by lime can still be used for the non-food cultivated area (Hu et al. 2020; Odey et al. 2017). The higher cost of sludge incineration and treatment technology, as well as the hazard of secondary pollution, have hampered its development (Negrete-Bolagay et al. 2021). Heavy metals and pathogens are present in sewage sludge, classifying it as a hazardous waste (Geng et al. 2020). More advanced technologies for utilizing the nutrients derived from this unsafe sludge have recently been introduced. Successful carbon recovery from sludge can considerably reduce waste-to-energy conversion and promote sustainability (Liew et al. 2021). In the meantime, the restoration of nitrogen and trace minerals enables fertilizer production. The major risks associated with using sewage sludge for agricultural purposes are the accessibility and bioaccumulation of heavy metals (Agoro et al. 2020).

3 Disposal of Sewage Sludge

A cost-effective, environmentally friendly approach to dispose of this trash has been identified using sewage sludge as a fertilizer. As a result, sewage sludge treatment might be thought of as a two-in-one process (Madsen et al. 2011). Sewage sludge disposal on agricultural land is problematic due to hazardous compounds. They can be taken out utilizing an assortment of strategies (physical, substance, and natural). However, there is still no agreement on the best handling method (Fytili and Zabaniotou 2008). Chemical and physical treatments can result in secondary contamination and are sometimes quite costly. While natural methodologies have shown promising outcomes in evacuating metals and are less harmful to the climate, they have just been tried on a limited scale and take longer (Camargo et al. 2016). As a result, it is critical to encourage the development of innovative approaches that are both ecologically and economically beneficial. Alternatives for sewage sludge disposal include agricultural use, landfilling, and cremation, as well as additional nitrification and wet oxidation; all have delayed and non-linear effects (Convey and Peck 2019; Clarke et al. 2006).

A significant volume of sewage sludge is being burned throughout the world. Agricultural use has long-term repercussions on crop yield and groundwater, like a resource of intake water due to nitrogen and phosphorus molecules. Excessive fertilizer use and air pollution exacerbate these impacts (Lawniczak et al. 2016). Due to leaching and emissions into the atmosphere, sludge landfilling has an environmental impact. Co-disposal of (dewatered) sewage sludge in a landfill body may alter microbial decomposition (O'Kelly 2016). Heavy metals may be discharged instantly or slowly, depending on the shape of the dried sludge, the process parameters in a landfill body, the chemical makeup of the dried sewage sludge, and the moisture present, causing leachate treatment to be disrupted (Raheem et al. 2018). Because the process parameters in the landfill body are interdependent, the time route for decontamination cannot be determined. The leaching is invariably non-linear and time-delayed (Watson-Craik and Sinclair 2020). The major disposal option for sewage sludge is a land application, which takes advantage of the nutrients in the sludge (Zhang et al. 2017). It has become a significant environmental issue as the volume of sewage sludge has increased. A risk assessment is required to limit the possible danger of heavy metal contamination in soil. Before sewage sludge is put to land, proper treatment methods for extraction of chemicals and phytoremediation should be employed to limit this possible danger (Shukla et al. 2016). As the concentration of heavy metals remains high in sludge after using one of these treatment processes, it should be disposed of in a landfill or burned. In addition, for pollutants and heavy toxic metals in sludge from agricultural use, an effective guideline on sewage sludge land use should be implemented (Duan et al. 2017). Although it was not compulsory, it was the hardest requirement in relation to similar standards across the world. The standard should be changed to use sewage sludge, and the impediment on profound metals content in sewage overflow should be killed. Since long stretch use of sewage slop shorewards would spoil soil features and add the regular risk, soil biological cutoff should be fused, which isn't communicated in the standard. Before continuing to utilize sewage on the ground expecting the soil at this point contains a lot of profound metals, the soil ecological limit must be assessed (McBride 2003). As significant metals can bio-gather, attempts should be made to lessen the levels of profound metals in profluent and sludge delivered to the environment whatever amount could sensibly be anticipated, either directly or through reapplication for water framework and soil conditioner, to ensure the security of all purchasers of things got from such usage. Government agencies in charge of STP management must also look at new technology that has a higher capability for removing these pollutants of heavy metals (Muduli et al. 2021). Critical measures of labile weighty metals in debris (from sewage muck burning) elevate relocation to the dirt, less significantly than sewage slop discarded at the landfill with elective treatment methods. It implies numerous techniques for unloading sewage muck in a similar landfill (idea to be more secure, for example, geo-tubing and utilizing slag from sewage slime cremation) cause soil defilement here and there.

4 Immobilization of Heavy Metals by Sewage Sludge

Massive metal ejection and weighty metal immobilization are two strategies for reducing heavy metal in sewage (Fig. 2). During the first stage, heavy metals are removed from sewage sludge using compound draining, bioleaching, electrochemical techniques, or a combination of these cycles, completely reducing the contamination risk caused by heavy metals. Heavy metals are more expensive to remove from sludge than to immobilize in it.

Contaminants isolated from sewage sludge must be carefully examined in order to minimize secondary contamination. Heavy metals are converted into more stable fractions during the second step, limiting their mobility and bioavailability (Peng et al. 2009). Metals that have been immobilized can still be found in sewage sludge, and they have the ability to reactivate with the passage of moment in ecological circumstances. Sewage sludge composting and chemical immobilization are two popular immobilization procedures.

4.1 Composting of Sewage Sludge

Fertilizing the soil is a vigorous absorption process that goes through three stages: mesophilic, thermophilic, and cooling. Microorganisms digest the biodegradable division and turn it into stable humic components (Cui et al. 2020). Because of N, P, K, and other supplements, the finished product is rich in humic substances

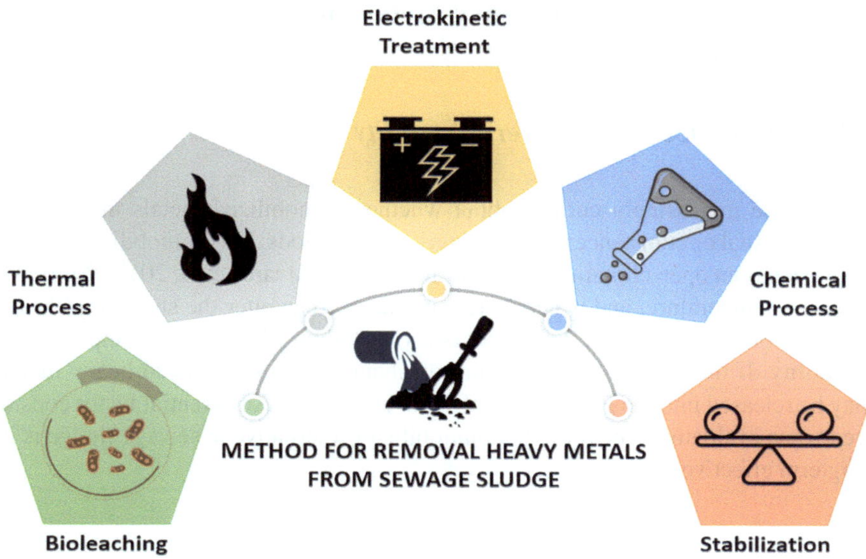

Fig. 2 Various processes for removal of heavy metals from sewage sludge

and may be used as a soil conditioner/compost. However, heavy metals in manure restricts its use as a soil conditioner or compost. Because of their bioaccumulation and biomagnification qualities, plants can ingest heavy metals, posing a hazard to human health. The dispersion and bioavailability of dangerous metals in the last fertilizer are influenced by metal speciation, slime characteristics, the fertilizing of the soil cycle, and physicochemical features of the last manure, such as how much natural carbon, humic matter substance, pH, and so on (Ahmed et al. 2007). While treating the soil is a potent technique for reducing the variety of heavy metals in sewage sludge, it comes with the drawbacks of requiring a large one-time investment, a large area, and a lengthy period to treat the soil. To improve fertilizer quality while lowering costs, a novel soil treatment innovation with minimal risk and high competency is required.

4.2 Chemical Immobilization

In order to slow down heavy metal migration, dispersion, ion exchange, adsorbent, and membrane filtration are examples of chemical treatments (Zhang et al. 2019). The volume of sludge created throughout the process remains constant or rises extremely slowly, minimizing future sludge transportation and storage costs. Immobilization of chemicals appealing strategy for immobilizing chemical toxin present in sewage as the agent is easy to use. Basic chemicals, aluminosilicate, phosphorus-bearing minerals, and sulphides are the most commonly used additives.

5 Problems During the Heavy Metal Immobilization Process in Sewage Sludge

5.1 Persistence or Long-Term Stability

Experts are increasingly curious about whether immobilized metals may remain steady in usual circumstances for time-consuming periods. The immobilized sewage slop reaches an open climatic framework using land (Alvarez et al. 2017). The environment, temperature, and water system in the area can alter the strength of heavy metals in the framework. Soil characteristics, including CEC, can also impact metal versatility. If the ecological circumstances change, the immobilized heavy metals may be released into the Phyto-available area once more. An essential factor considered when evaluating a heavy metal immobilising alternative in sewage sludge is its longterm effectiveness.

5.2 Environmental Sensitivity

Compost or chemical amendments would change environmental characteristics in the long run. Phosphoric corrosive may reduce the pH value of sewage slime, increasing the possibility of different metals filtering. Lime raises the pH of sewage muck, influencing its physicochemical properties (Samaras et al. 2008). Excessive pH elevation would be detrimental to plant growth. When deciding on a method to fix heavy metals, these factors must be taken into account.

5.3 Different Immobilizing Methods' Synergistic Effects

Weighty metal immobilization is improved by converting weighty metals into more stable speciation. Fixation aspects and effects on heavy metals vary according to obsession technique (Seneviratne et al. 2017). Integrating multiple strategies can amplify the immobilizing effect. Fertilizer revisions such as zeolite, lime, lime and sodium sulphide, and red mud are commonly used to improve manure quality by switching metals from portable to less portable or lingering structures (Singh and Kalamdhad 2013). Because of the presence of heavy metals, the administration of sewage muck as a result of the natural wastewater treatment process is becoming increasingly difficult. Both soil treatment and compound immobilization can effectively remove heavy metals from sewage muck, making land application a cost-effective option for sewage slop removal. Working on heavy metal addiction and ensuring long haul stability are two critical issues that must be addressed. Both fertilizing the soil and compound immobilization have advantages and disadvantages. A combination of techniques can have a greater impact than a single strategy alone. Microwave, bentonite pillared, and ultrasonic adjustment are commonly used to enhance the obsession impact of compound immobilizing specialists on heavy metals (Ninago et al. 2017).

6 Effect of Heavy Metals on Sewage Sludge

Sewage sludge contains organic matter and other biogenic materials in exceptionally high amounts, particularly primary nutrients like nitrogen and phosphorus, thus enhancing plant growth (Zhang et al. 2017; Turek et al. 2019). Despite this, it also invariably houses heavy metals like As, Cd, Cr, Cu, Hg, Ni, Pb, Zn, etc. (Tytła et al. 2015). Their nature turns toxic if present in elevated concentrations, which, when coupled with their high persistence and bioaccumulation potential, makes them fall into the category of priority or potentially toxic elements as per the US Environmental Protection Agency (Chen et al. 2015). Waste = water generated from both domestic and industrial sectors, sewerage pipes undergoing corrosion, runaway water from fields and roads are some principal sources of entry of heavy metals in the sludge (Duan et al. 2017; Milik et al. 2017).

Several pharma products, body hygiene materials, and illegal discharge of untreated wastewater also contribute to the addition of heavy metals (Rizzardini and Goi 2014). The measures of heavy metals are quite low in primary and mixed sludge is thickened state. Still, after subjecting the sludge to the processes of digestion by anaerobic means and dehydration, the metal content has been found to be the highest. Alvarez et al. (2002) linked this to loss of weight by the fresh sludge during the anaerobic digestion process followed by an increase in dry matter content due to dehydration (Alvarez et al. 2002). Moreover, modulations in different environment characteristics such as its organic matter amount, pH value, and Eh (Redox potential) value change the movement patterns of heavy metals, thus changing their availability from the sewage sludge to different parts of the ecosystem (Feizi et al. 2019).

7 Risk Assessment of Heavy Metal Presence in Sewage Sludge

Prior to applying sewage sludge to the soil to be amended, evaluation of the risks involved is a must. Suppose the land application of sewage sludge is continued for a longer time period without proper risk assessment. In that case, it might cause piling up of heavy metals in the soil beyond permissible levels, turning them highly toxic. Multiple factors as the likes of different soil attributes, absolute contents of heavy metals and their speciation, etc., affect the risk potential of heavy metal presence in the sewage sludge. The bioavailability of heavy metals in soil is also linked to their characteristics. Four alternative methods, namely (i) application of sewage sludge to the agricultural land, (ii) incinerating it with other wastes, (iii) incineration followed by recovery of phosphorus, and (iv) fractionation combined with the retrieval of phosphorus, have been compared by Lundin et al. (2004) out of which the first one gets the least preference from the point of ecological risks involved thereof while the others possess good sustainability potential (Lundin et al. 2004). Of late, considering the toxicity of heavy metal presence, Turunen et al. (2018) have contributed to developing a decision support tool based on multi-attribute value theory (MAVT). A simple scoring is performed to estimate the ecological risks of different sewage sludge disposal methods. This tool provides the highest ranking to pyrolysis of sewage sludge to composting and incineration (Turunen et al. 2018).

Besides this, there are two chief methods for evaluating the risk of the presence of heavy metals in the sludge, which are as follows:

(i) Geo-accumulation Index (Igeo): This quantitative value indicator helps estimate the extent of defiling caused by heavy metals in sewage sludge. German Scholar Muller developed it in 1969, and it quite efficiently assesses the degree of pollution caused by pollutants.

It is calculated as $I_{geo} = \log_2 (C_n/1.5B_n)$.

Here, C_n is the concentration of any heavy metal (n) from sewage sludge under examination; B_n denotes the geochemical background value of that very metal; 1.5 is a constant. Both C and B_n are taken in the same concentration unit.

(ii) Potential Ecological Risk Index (RI): Hakanson, a scientist of Swedish origin, utilized this indicator to assess the extent of heavy metal contamination.

It is estimated as $E_r^i = T_r^i C_n / B_n$

$$RI = \sum_{i=1}^{n} E_r^i$$

Here, RI is the ecological risk assessment value of heavy metals; E_r^i is the potential ecological risk value for a given metal; T_r^i depicts the toxic response coefficient for the metal, C_n is the examined concentration of the metal, and B_n represents the geochemical background reference value. Both C_n and B_n are taken in the same concentration unit.

These two indices aim at estimating the value of total content of heavy metals, but these toxic pollutants have different speciation forms owing to their percolation and interactivities with diverse elements of natural ecosystems, which in turn cast an effect on their mobility, availability, and perniciousness (Karwowska and Dąbrowska 2017). Therefore, speciation indices such as Individual Contamination Factor (ICF) and Risk Assessment Code (RAC) are also to be taken into account for the comprehensive and complete assessment of the likelihood of every possible threat imposed by the presence of heavy metals in the sewage sludge (Ikem et al. 2003; Zhao et al. 2012).

8 Influence of Heavy Metals Carrying Sludge on Environment

Being rich in organic matter and diverse nutrients, sewage sludge finds its important application as a fertilizer that enhances soil productivity. However, it has been well established over the years that the heavy metal content present in sewage sludge, upon exceeding its optimal concentrations, contaminates not only the soil but also defiles surface and groundwater resources which leads to negatively influencing the biotic communities of all types, i.e., plant, animal, and microbial. External supplementation of sewage sludge as fertilizer causes heavy metals to enter the soil from where their uptake, either by plants, percolation, or erosion by wind and water, etc., aggravates their harmful impacts. Reduction in yield, as well as the quality of plant products due to disruption of optimal functioning of plants, has been directly linked to heavy metal absorption. After the application of sludge, heavy metals get easily complexed with the soil particles and strike an equilibrium with them in their liquid phase. This leads to the retention of heavy metals mostly in the upper layers of soil.

However, soil attributes such as its pH, moisture, temperature, texture, type, organic matter, along cation exchange capacity have also been found to influence the movement of heavy metals in the soil layers, thereby affecting their accessibility to different life forms. Out of all of these, pH plays a key role in determining the availability of heavy metals to the biotic components in sludge treated soil. If the pH increases, heavy metals are adsorbed greatly on the surface of the clay and other organic constituents. The soils having pH in the range of 5–7 are able to precipitate and adsorb heavy metals, so they cause less harm. On similar lines, the sewage sludge's organic matter content regulates the bioavailability of heavy metals in soil, acting as an efficient agency for adsorbing the heavy metals in the soil. This organic matter in its soluble form possesses exchange sites for getting complexed with metals, thus reducing their availability. In order to lessen the toxicity caused by a labile fraction of heavy metals to the plants, fertilizers as the likes of Ca benovite, Na benovite (2% w/w) and Novaphos (0.05% w/w) have been reported to be quite effective by reducing bio-accessibility of heavy metals in wheat plant (Usman et al. 2006).

Further, deleterious effects of heavy metals get amplified upon entering the food chain, whereby human health is severely deteriorated (Kovacs and Szemmelveisz 2017). Various physiological and neurological malignancies are observed if levels of these heavy metals are elevated inside human bodies, as seen in the case of Pb in the blood of children all over the world (Luo et al. 2012). However, in soils with mineral deficiencies such as Cu, the presence of heavy metals in sewage sludge proves quite beneficial (Comission 2001). The activity of microorganisms helps in the upkeep of the quality and fertility of the soil. And the sludge having heavy metals in low concentrations prove advantageous for enhancing microbial activities, their biomass, and organic carbon content, but the converse of it happens if heavy metals are present in higher amounts (Knight et al. 1997).

9 Conclusion and Perspectives

Since sewage sludge has now gained a status of a renewable energy resource and is also used variously for material recovery, its post-production treatment thus holds an important place in the environment, which further helps in realizing the aim of ecological innovation. Different organic nutrients present in sewage sludge can be recovered safely only if applying it in raw form is discontinued. Further, the imminent threats posed to the biotic community by the potentially toxic pollutants as the likes of heavy metals present in sewage sludge can be estimated by biological tests. These tests can properly assess the safety of sewage sludge with respect to the environment and human health prior to its application for the purpose of soil reclamation or other agricultural operations. In the case of plants, the toxicity is mainly influenced by the physical composition of the soil, the nature of plant species to be grown, and the constituent made up of sewage sludge (Oleszczuk 2010). Various quality management programs of both voluntary and involuntary nature can address the safety concerns of sewage sludge application to the crop plants followed by their entry into the food chain. However, these management practices have to be formulated in recent times after taking into account the latest policies regarding climate change and greenhouse gas emissions.

Limiting the actions like landfilling and disposal along with achieving adequate recovery or recycling for reuse, legislations regarding adopting a common waste management system have also taken center stage. The other alternatives that can be opted for include incineration or anaerobic digestion of sewage sludge, leading to the destruction of toxic organic materials and the generation of energy in biofuel, heat, and electricity. Besides this, the leftover ash can be used for construction purposes as cementing material that causes heavy metals immobilization upon it, rendering them unable to cause any ecological harm. These methods also prove beneficial in the areas where land for recycling is unavailable due to public opposition or high population densities. This lowering of the content of heavy metals in sewage sludge ensures the mitigation of their harmful impacts on the ecosystem and its various components. Therefore it has to be adopted worldwide.

Acknowledgements Dr. Simranjeet Singh would like to acknowledge DBT HRD Project & Management Unit, Regional Center for Biotechnology, NCR Biotech Science Cluster, Faridabad, Haryana for Research Associateship (DBT-RA), funding under award letter No DBT-RA/2022/July/N/2044 dated January 12, 2023.

References

Agoro MA, Adeniji AO, Adefisoye MA, Okoh OO (2020) Heavy metals in wastewater and sewage sludge from selected municipal treatment plants in Eastern Cape Province, South Africa. Water 12(10):2746

Ahmed M, Idris A, Omar SRS (2007) Physicochemical characterization of compost of the industrial tannery sludge. J Eng Sci Technol 2(1):81–94

Ali H, Khan E, Ilahi I (2019) Environmental chemistry and ecotoxicology of hazardous heavy metals: environmental persistence, toxicity, and bioaccumulation. J Chem 2019:6730305

Alvarez EA, Mochón MC, Sanchez JCJ, Rodríguez MT (2002) Heavy metal extractable forms in sludge from wastewater treatment plants. Chemosphere 47(7):765–775

Alvarez A, Saez JM, Davila Costa JS, Colin VL, Fuentes MS, Cuozzo SA, Benimeli CS, Polti MA, Amoroso MJ (2017) Actinobacteria: current research and perspectives for bioremediation of pesticides and heavy metals. Chemosphere 166:41–62

Asati A, Pichhode M, Nikhil K (2016) Effect of heavy metals on plants: an overview. Int J Appl Innov Eng Manag 5(3):56–66

Awuchi CG, Igwe VS, Amagwula IO (2020) Nutritional diseases and nutrient toxicities: a systematic review of the diets and nutrition for prevention and treatment. Int J Adv Acad Res 6(1):1–46

Baby J, Raj JS, Biby ET, Sankarganesh P, Jeevitha MV, Ajisha SU, Rajan SS (2010) Toxic effect of heavy metals on aquatic environment. Int J Biol Chem Sci 4(4):939–952

Behbahani M, Najafi M, Amini MM, Sadeghi O, Bagheri A, Salarian M (2013) Dithizone-modified nanoporous fructose as a novel sorbent for solid-phase extraction of ultra-trace levels of heavy metals. Microchim Acta 180:911–920

Camargo FP, Sérgio Tonello P, dos Santos ACA, Duarte ICS (2016) Removal of toxic metals from sewage sludge through chemical, physical, and biological treatments—A Review. Water Air Soil Pollut 227(12):1–11

Chen H, Teng Y, Lu S, Wang Y, Wang J (2015) Contamination features and health risk of soil heavy metals in China. Sci Total Environ 512–513:143–153

Clarke A, Johnston NM, Murphy EJ, Rogers AD (2006) Introduction. Antarctic ecology from genes to ecosystems: the impact of climate change and the importance of scale. Philos Trans R Soc B Biol Sci 362(1477):5–9

European Comission (2001) SEWAGE SLUDGE 2/1/02 17:27 Pagina 3 Compuesta C M Y CM MY CY CMY K Disposal and recycling routes for sewage sludge Part 3-Scientific and technical report

Convey P, Peck LS (2019) Antarctic environmental change and biological responses. Sci Adv **5**(11):eaaz0888

Cui H, Ou Y, Wang L, Yan B, Li Y, Ding D (2020) The passivation effect of heavy metals during biochar-amended composting: emphasize on bacterial communities. Waste Manag 118:360–368

Djandja OS, Wang ZC, Wang F, Xu YP, Duan PG (2020) Pyrolysis of municipal sewage sludge for biofuel production: a review. Ind Eng Chem Res 59(39):16939–16956

Duan B, Zhang W, Zheng H, Wu C, Zhang Q, Bu Y (2017) Disposal situation of sewage sludge from municipal wastewater treatment plants (WWTPs) and assessment of the ecological risk of heavy metals for its land use in Shanxi, China. Int J Environ Res Public Health 14(7):823

Ekere NR, Ugbor MCJ, Ihedioha JN, Ukwueze NN, Abugu HO (2020) Ecological and potential health risk assessment of heavy metals in soils and food crops grown in abandoned urban open waste dumpsite. J Environ Heal Sci Eng 18(2):711–721

Englande AJ, Jr, Krenkel P, Shamas J (2015) Wastewater Treatment &Water Reclamation. Ref Modul Earth Syst Environ Sci

Feizi M, Jalali M, Renella G (2019) Assessment of nutrient and heavy metal content and speciation in sewage sludge from different locations in Iran Nat Hazards J Int Soc Prev Mitig Nat Hazards 95(3):657–675

Fytili D, Zabaniotou A (2008) Utilization of sewage sludge in EU application of old and new methods—a review. Renew Sustain Energy Rev 12(1):116–140

Geng H, Xu Y, Zheng L, Gong H, Dai L, Dai X (2020) An overview of removing heavy metals from sewage sludge: achievements and perspectives. Environ Pollut 266:115375

Ghori NH, Ghori T, Hayat MQ, Imadi SR, Gul A, Altay V, Ozturk M (2019) Heavy metal stress and responses in plants. Int J Environ Sci Technol 16(3):1807–1828

Gupta K, Joshi P, Gusain R, Khatri OP (2021) Recent advances in adsorptive removal of heavy metal and metalloid ions by metal oxide-based nanomaterials. Coord Chem Rev 445:214100

Hamid Y, Tang L, Yaseen M, Hussain B, Zehra A, Aziz MZ, He ZL, Yang X (2019) Comparative efficacy of organic and inorganic amendments for cadmium and lead immobilization in contaminated soil under rice-wheat cropping system. Chemosphere 214:259–268

Han W, Jin P, Chen D, Liu X, Jin H, Wang R, Liu Y (2021) Resource reclamation of municipal sewage sludge based on local conditions: a case study in Xi'an, China. J Clean Prod 316:128189

Hasan MM, Uddin MN, Ara-Sharmeen I, Alharby HF, Alzahrani Y, Hakeem KR, Zhang L (2019) Assisting phytoremediation of heavy metals using chemical amendments. Plants 8(9):295

Hu H (2002) Human health and heavy metals exposure. In: Life support environ & human health. MIT Press. Cambridge, pp 65–82

Hu G, Feng H, He P, Li J, Hewage K, Sadiq R (2020) Comparative life-cycle assessment of traditional and emerging oily sludge treatment approaches. J Clean Prod 251:119594

Huang Y, Chen Q, Deng M, Japenga J, Li T, Yang X, He Z (2018) Heavy metal pollution and health risk assessment of agricultural soils in a typical peri-urban area in southeast China. J Environ Manage 207:159–168

Ikem A, Egiebor NO, Nyavor K (2003) Trace elements in water, fish and sediment from Tuskegee Lake, Southeastern USA. Water Air Soil Pollut 149(1):51–75

Joimel S, Cortet J, Consalès JN, Branchu P, Haudin CS, Morel JL, Schwartz C (2021) Contribution of chemical inputs on the trace elements concentrations of surface soils in urban allotment gardens. J Soils Sediments 21(1):328–337

Karande UB, Kadam A, Umrikar BN, Wagh V, Sankhua RN, Pawar NJ (2019) Environmental modelling of soil quality, heavy-metal enrichment and human health risk in sub-urbanized semiarid watershed of western India. Model Earth Syst Environ 6(1):545–556

Karwowska B, Dąbrowska L (2017) Bioavailability of heavy metals in the municipal sewage sludge. Ecol Chem Eng A 24(1):75–86

Khan FSA, Mubarak NM, Khalid M, Walvekar R, Abdullah EC, Mazari SA, Nizamuddin S, Karri RR (2020) Magnetic nanoadsorbents' potential route for heavy metals removal—a review. Environ Sci Pollut Res 27(19):24342–24356

Knight BP, McGrath SP, Chaudri AM (1997) Biomass carbon measurements and substrate utilization patterns of microbial populations from soils amended with cadmium, copper, or zinc. Appl Environ Microbiol 63(1):39–43

Kovacs H, Szemmelveisz K (2017) Disposal options for polluted plants grown on heavy metal contaminated brownfield lands—a review. Chemosphere 166:8–20

Lawniczak AE, Zbierska J, Nowak B, Achtenberg K, Grześkowiak A, Kanas K (2016) Impact of agriculture and land use on nitrate contamination in groundwater and running waters in central-west Poland. Environ Monit Assess 188(3):1–17

Liew CS, Kiatkittipong W, Lim JW, Lam MK, Ho YC, Ho CD, Ntwampe SKO, Mohamad M, Usman A (2021) Stabilization of heavy metals loaded sewage sludge: reviewing conventional to state-of-the-art thermal treatments in achieving energy sustainability. Chemosphere 277:130310

Lundin M, Olofsson M, Pettersson GJ, Zetterlund H (2004) Environmental and economic assessment of sewage sludge handling options. Resour Conserv Recycl 41(4):255–278

Luo XS, Yu S, Zhu YG, Li XD (2012) Trace metal contamination in urban soils of China. Sci Total Environ 421–422:17–30

Madsen M, Holm-Nielsen JB, Esbensen KH (2011) Monitoring of anaerobic digestion processes: a review perspective. Renew Sustain Energy Rev 15(6):3141–3155

Masud MM, Baquy MAA, Akhter S, Sen R, Barman A, Khatun MR (2020) Liming effects of poultry litter derived biochar on soil acidity amelioration and maize growth. Ecotoxicol Environ Saf 202:110865

McBride MB (2003) Toxic metals in sewage sludge-amended soils: has promotion of beneficial use discounted the risks? Adv Environ Res 8(1):5–19

Milik JK, Pasela R, Lachowicz M, Chalamoński M (2017) The concentration of trace elements in sewage sludge from wastewater treatment plant in Gniewino. J. Ecol. Eng. 18(5):118–124

Muchuweti M, Birkett JW, Chinyanga E, Zvauya R, Scrimshaw MD, Lester JN (2006) Heavy metal content of vegetables irrigated with mixtures of wastewater and sewage sludge in Zimbabwe: Implications for human health. Agric Ecosyst Environ 112(1):41–48

Muduli M, Sonpal V, Trivedi K, Haldar S, Kumar MA, Ray S (2021) Enhanced biological phosphate removal process for wastewater treatment: a sustainable approach. In: Wastewater treatment reactors, 1st edn, Elsevier, pp 273–287

Nagajyoti PC, Lee KD, Sreekanth TVM (2010) Heavy metals, occurrence and toxicity for plants: a review. Environ Chem Lett 8(3):199–216

Negrete-Bolagay D, Zamora-Ledezma C, Chuya-Sumba C, De Sousa FB, Whitehead D, Alexis F, Guerrero VH (2021) Persistent organic pollutants: the trade-off between potential risks and sustainable remediation methods. J Environ Manage 300:113737

Ninago MD, López OV, Gabriela Passaretti M, Fernanda Horst M, Lassalle VL, Ramos IC, Di Santo R, Ciolino AE, Villar MA (2017) Mild microwave-assisted synthesis of aluminum-pillared bentonites: Thermal behavior and potential applications. J Therm Anal Calorim 129:1517–1531

O'Kelly BC (2016) Geotechnics of municipal sludges and residues for landfilling. Geotech Res 3(4):148–179

Odey EA, Li Z, Zhou X, Kalakodio L (2017) Fecal sludge management in developing urban centers: a review on the collection, treatment, and composting. Environ Sci Pollut Res 24:23441–23452

Oleszczuk P (2010) Testing of different plants to determine influence of physico–chemical properties and contaminants content on municipal sewage sludges phytotoxicity. Environ Toxicol 25(1):38–47

Peng, JF, Song, YH, Yuan P, Cui XY, Qiu GL (2009) The remediation of heavy metals contaminated sediment. J Hazard Mater161(2–3):633–640
Raheem A, Sikarwar VS, He J, Dastyar W, Dionysiou DD, Wang W, Zhao M (2018) Opportunities and challenges in sustainable treatment and resource reuse of sewage sludge: a review. Chem Eng J 337:616–641
Raikwar MK, Kumar P, Singh M, Singh A (2008) Toxic effect of heavy metals in livestock health. Vet World 1(1):28–30
Rebi A, Rehman T, Wahaj M, Raza HA, Usman M, Gul N, Bano M, Khan Z, Ghazanfar S, Irfan M, Naz M, Ahmed MI (2021) Impact of sewage water on human health and agricultral land: a review. Ann Rom Soc Cell Biol 25(7):1366–1376
Rehman AU, Nazir S, Irshad R, Tahir K, Rehman K, Islam RU, Wahab Z (2021) Toxicity of heavy metals in plants and animals and their uptake by magnetic iron oxide nanoparticles. J Mol Liq 321:114455
Rizzardini CB, Goi D (2014) Sustainability of domestic sewage sludge disposal. Sustain 6(5):2424–2434
Samaras P, Papadimitriou CA, Haritou I, Zouboulis AI (2008) Investigation of sewage sludge stabilization potential by the addition of fly ash and lime. J Hazard Mater 154(1–3):1052–1059
Sandeep G, Vijayalatha KR, Anitha T (2019) Heavy metals and its impact in vegetable crops. Int J Chem Stud 7(1):1612–1621
Seneviratne M, Weerasundara L, Ok YS, Rinklebe J, Vithanage M (2017) Phytotoxicity attenuation in Vigna radiata under heavy metal stress at the presence of biochar and N fixing bacteria. J Environ Manage 186:293–300
Shahid M, Dumat C, Khalid S, Schreck E, Xiong T, Niazi NK (2017) Foliar heavy metal uptake, toxicity and detoxification in plants: a comparison of foliar and root metal uptake. J Hazard Mater 325:36–58
Sharma A, Giri RK, Chalapathi Rao NV, Rahaman W, Pandit D, Sahoo S (2019) Arc-related pyroxenites derived from a long-lived Neoarchean subduction system at the Southwestern margin of the Cuddapah Basin: geodynamic implications for the evolution of the Eastern Dharwar Craton, Southern India. J Geol 127(5):567–591
Sharma RK, Agrawal M (2005) Biological Effects of Heavy Metals: An Overview. J Environ Biol26(2 suppl):301–313
Shukla SK, Mishra RK, Pandey M, Mishra V, Pathak A, Pandey A, Kumar R, Dikshit A (2016) Land reformation using plant growth–promoting rhizobacteria in the context of heavy metal contamination. In: Plant metal interaction: emerging remediation techniques, 1st edn, Elsevier, pp 499–529
Sidhu GK, Singh S, Kumar V, Dhanjal DS, Datta S, Singh J (2019) Toxicity, monitoring and biodegradation of organophosphate pesticides: a review. Crit Rev Environ Sci Technol 49(13):1135–1187
Singh J, Kalamdhad AS (2013) Chemical speciation of heavy metals in compost and compost amended soil—a review. Int J Environ Eng Res 2(2):27–37
Singh J, Singh S, Kumar V, Upadhyay N, Kumar V, Kaur S, Datta S (2014) Environmental exposure and health risks of the insecticide monocrotophos-a review. J Biodivers Environ Sci 5(1):111–120
Singh S, Singh N, Kumar V, Datta S, Wani AB, Singh D, Singh K, Singh J (2016) Toxicity, monitoring and biodegradation of the fungicide carbendazim. Environ Chem Lett 14:317–329
Singh S, Kumar V, Kanwar R, Wani AB, Gill JPK, Garg VK, Singh J, Ramamurthy PC (2021) Toxicity and detoxification of monocrotophos from ecosystem using different approaches: A review. Chemosphere 275:130051
Singh S, Kumar V, Chauhan A, Datta S, Wani AB, Singh N, Singh J. (2017a) Toxicity, degradation and analysis of the herbicide atrazine. Environ Chem Lett 16:211–237
Singh S, Kumar V, Upadhyay N, Singh J, Singla S, Datta S. (2017b) Efficient biodegradation of acephate by Pseudomonas pseudoalcaligenes PS-5 in the presence and absence of heavy metal ions [Cu(II) and Fe(III)], and humic acid. 3 Biotech 7(4):1–10

Singh S, Kumar V, Singh S, Singh J (2019a) Influence of humic acid, iron and copper on microbial degradation of fungicide Carbendazim. Biocatal Agric Biotechnol20:101196

Singh S, Kumar V, Singh J (2019b) Kinetic study of the biodegradation of glyphosate by indigenous soil bacterial isolates in presence of humic acid, Fe(III) and Cu(II) ions. J Environ Chem Eng**7**(3):103098

Singh S, Kumar V, Sidhu GK, Datta S, Dhanjal DS, Koul B, Janeja HS, Singh J (2019c) Plant growth promoting rhizobacteria from heavy metal contaminated soil promote growth attributes of Pisum sativum L. Biocatal Agric Biotechnol 17:665–671

Singh S, Kumar V, Gill JPK, Datta S, Singh S, Dhaka V, Kapoor D, Wani AB, Dhanjal DS, Kumar M, Harikumar SL, Singh J (2020a) Herbicide glyphosate: toxicity and microbial degradation. Int J Environ Res Public Health 17(20):7519

Singh S, Kumar V, Upadhyay N, Singh J (2020b) The effects of Fe(II), Cu(II) and humic acid on biodegradation of atrazine. J Environ Chem Eng8(2):103539

Singh S, Kumar V, Datta S, Wani AB, Dhanjal DS, Romero R, Singh J (2020c) Glyphosate uptake, translocation, resistance emergence in crops, analytical monitoring, toxicity and degradation: a review. Environ Chem Lett 18(3):663–702

Singh S, Kumar V, Dhanjal DS, Datta S, Bhatia D, Dhiman J, Samuel J, Prasad R, Singh J (2020d) A sustainable paradigm of sewage sludge biochar: Valorization, opportunities, challenges and future prospects. J Clean Prod269:122259

Smith JJ, Riddle MJ (2009) Sewage disposal and wildlife health in Antarctica. Heal Antarct Wildl, 271–315

Srivastava V, Sarkar A, Singh S, Singh P, de Araujo ASF, Singh RP (2017) Agroecological responses of heavy metal pollution with special emphasis on soil health and plant performances. Front Environ Sci 5:64

Sun L, Guo D, Liu K, Meng H, Zheng Y, Yuan F, Zhu G (2019) Levels, sources, and spatial distribution of heavy metals in soils from a typical coal industrial city of Tangshan, China. CATENA 175:101–109

Szynkowska MI, Pawlaczyk A, Maćkiewicz E (2017) Bioaccumulation and biomagnification of trace elements in the environment. In: Recent advances in trace elements, 1st edn. John Wiley & Sons, Ltd, Hoboken, New Jersey, pp 251–276

Turek A, Wieczorek K, Wolf WM (2019) Digestion procedure and determination of heavy metals in sewage sludge—an analytical problem. Sustain 11(6):1753

Turunen V, Sorvari J, Mikola A (2018) A decision support tool for selecting the optimal sewage sludge treatment. Chemosphere 193:521–529

Tytła M, Widziewicz K, Zielewicz E (2015) Heavy metals and its chemical speciation in sewage sludge at different stages of processing. Environ Technol 37(7):899–908

Unrine JM, Hopkins WA, Romanek CS, Jackson BP (2007) Bioaccumulation of trace elements in omnivorous amphibian larvae: Implications for amphibian health and contaminant transport. Environ Pollut 149(2):182–192

Usman ARA, Kuzyakov Y, Lorenz K, Stahr K (2006) Remediation of a soil contaminated with heavy metals by immobilizing compounds. J Plant Nutr Soil Sci 169(2):205–212

Vizzini S, Costa V, Tramati C, Gianguzza P, Mazzola A (2013) Trophic transfer of trace elements in an isotopically constructed food chain from a semi-enclosed marine coastal area (Stagnone di Marsala, Sicily, Mediterranean). Arch Environ Contam Toxicol 65(4):642–653

Wang Y, Duan X, Wang L (2020) Spatial distribution and source analysis of heavy metals in soils influenced by industrial enterprise distribution: case study in Jiangsu Province. Sci Total Environ 710:134953

Watson-Craik IA, Sinclair KJ (2020) Co-Disposal of industrial wastewaters and sludges. In: Microbiology of landfill sites, 2nd ed. CRC Press, Boca Raton, Florida, United States, pp 91–130

Wen Y, Li W, Yang Z, Zhang Q, Ji J (2020) Enrichment and source identification of Cd and other heavy metals in soils with high geochemical background in the karst region, Southwestern China. Chemosphere 245:125620

Yang Y, Ge Y, Zeng H, Zhou X, Peng L, Zeng Q (2017) Phytoextraction of cadmium-contaminated soil and potential of regenerated tobacco biomass for recovery of cadmium. Sci Reports 7(1):1–10
Zhang QH, Yang WN, Ngo HH, Guo WS, Jin PK, Dzakpasu M, Yang SJ, Wang Q, Wang XC, Ao D (2016) Current status of urban wastewater treatment plants in China. Environ Int 92–93:11–22
Zhang X, Wang XQ, Wang DF (2017) Immobilization of heavy metals in sewage sludge during land application process in China: a review. Sustain 9(11):2020
Zhang L, Shang Z, Guo K, Chang Z, Liu H, Li D (2019) Speciation analysis and speciation transformation of heavy metal ions in passivation process with thiol-functionalized nano-silica. Chem Eng J 369:979–987
Zhao S, Feng C, Yang Y, Niu J, Shen Z (2012) Risk assessment of sedimentary metals in the Yangtze Estuary: new evidence of the relationships between two typical index methods. J Hazard Mater 241–242:164–172

Nanotechnology for Bioremediation of Heavy Metals

Abel Inobeme, Charles Oluwaseun Adetunji, Mathew John Tsado, Alexander Ikechukwu Ajai, Jonathan Inobeme, Mutiat Oyedolapo Bamigboye, Sandra Onyeaku, Munirat Maliki, Chinenye Eziukwu, and Kelani Tawa

1 Introduction

Owing to the skyrocketing advancements in industrialization, urbanization, and current agricultural practices, environmental pollution has become inherent. Hence the most remarkable challenge for researchers is the elimination of contaminants from the environment. Various inorganic and organic pollutants constitute a serious threat to humans and the environment at large. Of more concern is the fact that in recent times their concentrations and persistence in various environmental matrices have increased tremendously (Ali et al. 2019). Amongst the various contaminants, heavy metals constitute a class of serious concern due to their persistence and tendency to bioaccumulate in tissues. Heavy metals are components of the environment and are

A. Inobeme (✉) · M. Maliki · C. Eziukwu · K. Tawa
Department of Chemistry, Edo State University Uzairue, Edo State, Nigeria
e-mail: inobeme.abel@edouniversity.edu.ng

C. O. Adetunji
Applied Microbiology, Biotechnology and Nanotechnology Laboratory, Department of Microbiology, Edo State University Uzairue, PMB 04, Edo State, Nigeria

M. J. Tsado
Department of Chemistry, Ibrahim Badamasi Babangida University Lapai, Lapai 911101, Nigeria

A. I. Ajai · S. Onyeaku
Department of Chemistry, Federal University of Technology Minna, Minna, Nigeria

J. Inobeme
Department of Geography, Ahmadu Bello University Zaria, Zaria, Nigeria

M. O. Bamigboye
Department of Chemical Sciences, Kings University, Odeomu, Nigeria

M. P. Shah (ed.), *Microbial Technologies in Industrial Wastewater Treatment*,
https://doi.org/10.1007/978-981-99-2435-6_2

of high natural abundance with a very high density and atomic number. Their densities are known to be five times that of water. These metals constitute one of the major environmental contaminants. Heavy metal contaminants such as nickel, chromium (VI), zinc, cadmium, lead, and mercury are highly toxic and non-biodegradable in nature and have been reported to be associated with severe health-related challenges in humans and animals (Cimboláková et al. 2019).

Various technologies have been employed for the remediation of heavy metals from soil, water, and other environmental resources. Some of the approaches that have been employed for the removal of heavy metals from the environment include solvent extraction, chemical oxidation, biological degradation, and adsorption using activated carbon, amongst others. Most of the various conventional techniques employed were shown to possess their inherent limitations. Some of which are expensive and not very efficient (Cimboláková et al. 2019; Shah 2020). These limitations of the traditional approaches gave rise to bioremediation which is relatively more efficient and greener in nature.

Nanotechnology proffers promising outcomes in the remediation of heavy metals. However chemically synthesized nanomaterials are of concern with respect to cost of production, and the fact that toxic chemicals which have negative effects on the environment and humans are used. In recent years, bio nanotechnology has emerged as an indispensable tool for the removal of heavy metals. This makes it possible to eliminate the effect of high cost, and the introduction of a clean and environmentally friendly approach to nanoparticle synthesis and applications. The biogenic production of nanoparticles is achievable with the aid of various groups of microorganisms and plants and highly remarkable for the removal of heavy metals from the environment (Adekeye et al. 2019).

Bioremediation technology has been documented to be of high efficiency for the management and treatment of various classes of contaminants and also suitable for broad applicability. The bioremediation of various heavy metals contaminants was found to be enhanced through biostimulation in which the local microorganisms are stimulated thereby speeding up the bioremediation processes. The utilization of bioremediation alone was however proven to have its own limitation, one of which is its inefficiency in the treatment of heavily polluted sites. On this premise, the integration of nanotechnology with bioremediation becomes a remarkable solace for the efficient removal of heavy metals as well as various refractory compounds (Rahman and Singh 2020; Yap et al. 2021; Shah 2021). Nanotechnology involves the use of various categories of nanomaterials. Nanomaterials are particles that have sizes that are within 100 nm or in some cases less than in at least one of its dimensions. This combinational approach that involves two remarkable areas moves bioremediation beyond its restrictions. This makes it feasible for broader range and scope of applications with highly reduced cost and eco-friendly impact on the environment. This has also been documented to be of high efficiency for the removal of heavy metals. Nanobioremediation technology has shown that plants and microorganisms have the potential of inducing the immobilization of different heavy metals thereby transforming them into less harmful substances. This has also proven to be effective for other inorganic and organic pollutants. In the recent years, numerous positive outcomes have been reported for the incorporation of nanomaterials and biotechnology for the elimination of heavy metals from the environment (Sayqal and Ahmed 2021).

The combination of bioremediation and nanotechnology for the treatment and management of heavy metal pollution has proven to be more promising. Since there are positive and negative consequences that may come up, it becomes paramount to understand comprehensively the various interactions between the contaminants (heavy metals), the microorganisms employed, and the nanomaterials involved. While some of the nanomaterials employed in bioremediation could act as suitable stimulants for the microbes, some others have been reported to be of high toxicity. The proper choice of the microorganism and the nanoparticles of interest become paramount. The use of microorganisms for the extracellular and intracellular production of nanoparticles with varying chemical compositions, shapes, sizes, and adapted monodispersity could be a novel, promising, economically favorable, and green approach that can be utilized for the reduction of various environmental contaminants including heavy metals (Nunez et al. 2020). This chapter presents a general overview of heavy metals, the role of nanotechnology in bioremediation, types of nanomaterials used in the bioremediation of heavy metals, mechanisms involved, and limitations of nanotechnology in bioremediation. Finally an attempt is made in highlighting the most recent reports and future trends in this regard (Fig. 1).

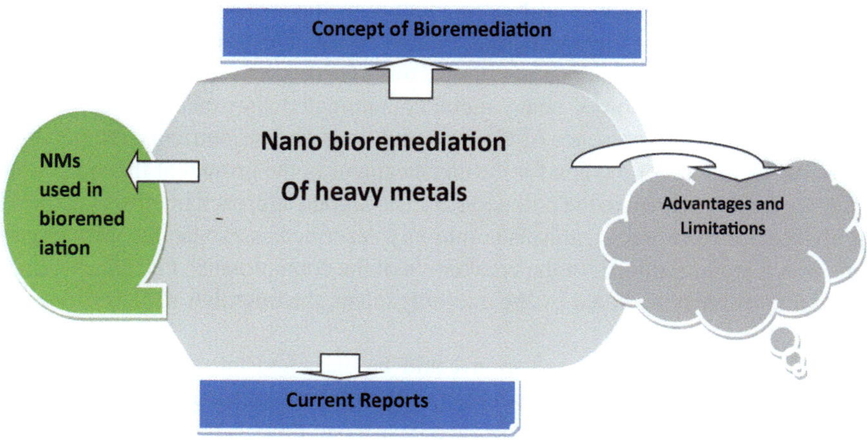

Fig. 1 Nanotechnology for bioremediation of heavy metals

2 Principles Heavy Metal Bioremediation

Heavy metals are ubiquitous hence their concentrations in the environment continue to rise due to anthropogenic contributions. Apart from their tendencies to bioaccumulate and persist within the environment, some of these metals are also known to be carcinogenic and harmful to various living cells and organs. The various groups of heavy metals show high levels of variability in some of their chemical properties and cytotoxic effects alongside their varying interactions with various abiotic and biotic factors within the environment. This has further limited the successful application of bioremediation alone. The combination of nanomaterials with bioremediation technology could therefore proffer a promising step for the efficient remediation and enhancement of the speed and efficiency of degradation (Zand and Tabrizi 2021). Besides the utilization of physical and chemical processes for the remediation of various sites contaminated by heavy metals, the use of biological treatment approach becomes of immense relevance as a result of its economical feasibility and wider applicability. Bioremediation technology involves various processes such as biosorption, bio-stabilization, biotransformation, and bioaccumulation. This technology basically involves the use of microorganisms and plants as well as the combination of both (Sarkar et al. 2020).

Bioremediation could be used on its own in which case it is called intrinsic bioremediation (natural attenuation), while in other cases when it is induced to occur through the addition of various stimulating agents for the improvement of bioavailability within the medium then it is known as biostimulated bioremediation. There are various kinds of bioremediation technologies the prominent which include bioventing, bioaugmentation, rhizofiltration, biostimulation, bioleaching, composting, phytoremediation, and bioreactors. There are two major paths through which the bioremediation of heavy metal contaminated sites takes place. The first approach requires the presence of various conditions such as nutrients, temperature, and suitable amount of oxygen for the enhancement of the growth of the indigenous microorganisms present in the polluted site. The second approach involves the role of highly specialized microorganisms commonly described as exogenous microorganisms which are introduced for the breakdown of the contaminants. Once the targeted contaminants are remediated by the microorganism, the microbes die (Kebede et al. 2021).

The application of bioremediation could be classified into in situ and ex situ bioremediation. In in situ bioremediation, the toxic contaminants are treated within their location hence it is relatively less expensive and there is no discharge of much amount of the contaminants since they are managed within the polluted site. This approach of application may however be relatively slower and may be difficult for effective management. Ex situ bioremediation on the other hand needs the complete removal or excavation of the contaminated materials prior to their treatment. This technique could however be faster, easier to moderate, and suitable for the treatment of a broad class of contaminants and the nature of the medium when compared to in situ bioremediation (Verma 2021) (Fig. 2).

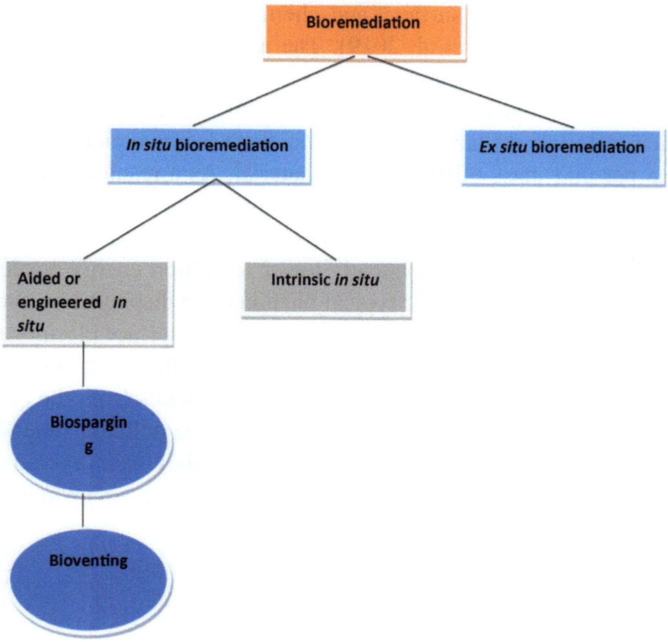

Fig. 2 Types of bioremediation technologies

3 Nanobioremediation of Heavy Metals

Over time, nanomaterials have been introduced into various biological phenomena for the purpose of accelerating and enhancing the removal of heavy metals from the environment. Various terms have therefore been employed in this regard. The term nanobioremediation refers to processes in which nanoparticles together with plants or microorganisms are employed for the removal of heavy metals and other environmental contaminants. In some other studies, such practices were grouped on the basis of the nature of microorganisms that were introduced during nanobioremediation. On this premise, various related terms such as microbial nanobioremediation, zoo nanoremediation, and phytonanoremediation which involve the use of microorganisms, animals, and plants respectively together with the nanoparticles. Basically, since bioremediation technology makes use of various groups of living organisms for the remediation of the polluted environment there is need for proper interaction of the living organisms and the nanoparticles. Various factors such as sizes of the nanoparticles, nanonutrition, and nanotoxicity are paramount for effective nanobioremediation of heavy metals (Yan et al. 2022).

Various researchers have identified several factors which affect the interactions between the nanoparticles and the contaminants. Some of these factors include pH of the media, temperature, and nature of organism, shape and size of the nanoparticle, presence of coatings on the surface, and chemical compositions of the nanoparticle, amongst others (Khan et al. 2019). The above listed factors can also affect the overall stability of the nanoparticles produced. There are numerous reasons why nanoparticles are employed in the bioremediation of various environmental contaminants. When materials are taken to nanoscale, there is a remarkable increase in the surface area per unit mass of the original material. As a result of this, it becomes easier for a larger component of the material to come in close contact with the materials targeted which also affects its reactivity. Nanoparticles are capable of showing quantum effects; hence there is relatively lower activation energy needed to make the reaction to be feasible. Another typical phenomenon displayed by nanoparticles which makes them suitable for nanobioremediation is the surface Plasmon resonance. With respect to the varying sizes and shapes, various nanomaterials, both metallic and non-metallic could be used for the cleanup of heavy metals from the environment. The nanoparticles have the potential of diffusing into the polluted zone where other larger macro and micro particles cannot reach; also, they have a relatively higher reactivity to contaminants that are vulnerable to redox processes (Wang et al. 2017).

4 Types of Nanomaterials Employed for the Bioremediation of Heavy Metals

There are various classes of nanomaterials that have been utilized alongside biotechnologies for the removal of heavy metals from the environment. These nanomaterials were also documented to present varying efficiencies in the removal of heavy metals and other diverse contaminants. Diverse groups of microorganisms were also combined alongside the nanoparticles for the removal of specific heavy metal of interest. Some of the nanomaterials that have been successfully applied include biogenic uranite, dendrimers, single enzyme nanomaterials, carbon-derived nanoparticles, engineered polymeric nanomaterials, and iron-based nanoparticles (Naseer et al. 2018).

The nature of the organism utilized is also vital. Phytonanoremediation tends to show certain advantages over the microbial nanoremediation. This is because plants are capable of producing different classes of molecules which are capable of inducing the transformation of heavy metals and other contaminants. Some of such active compounds present in plants include reactive oxygen intermediates, flavonoids, glutathione, and other bioactive compounds that are capable of acting during stress. Furthermore it is relatively easier to cultivate and handle plants when compared to microorganisms during nanobioremediation (Navya and Daima 2016).

5 Advantages of Using Nanotechnology in the Bioremediation of Heavy Metals

The application of nanotechnology in the bioremediation of heavy metals has several advantages over other existing approaches. Some of the advantages include its economy in terms of cost, high efficiency in the removal of various contaminants including heavy metals, minimization of chemical usage, high selectivity for metal removal, lack of special requirements for nutrient supply, potential for regeneration of nanoparticles-based adsorbent for heavy metals removal, and the remarkable effectiveness for metal recovery. Besides the fact that nanoparticles effectively remove the heavy metals contaminants, they are also able to react with various abiotic and biotic elements. Several studies have therefore been carried out with a view to evaluating the synergistic impact of the combination of the duo processes as well as investigating their physicochemical and biological connections in various environmental samples (Khan et al. 2019; Nunez et al. 2020).

6 Regulations of Nanobioremediation for Heavy Metals Removal

The incorporation of nanotechnology into the area of environmental bioremediation has the potential to drive the technology for the enhancement of environmental safety in various countries of the world. On this premise studies have been carried out with a view to understanding the processes involved in the decontamination. For example, it has been reported that nanoparticles could bring about a decrease in the extent of biodiversity and abundance of the communities of microorganisms in various ecosystems. Also nanomaterials were found to bring about a reduction in the concentrations of various enzymes that were involved in the ecological processes, however this was found to increase again after some time. The need for various regulating agencies becomes paramount as a result of the biosafety linked to the use of various kinds of nanoparticles (Babatunde et al. 2019). There is serious need for standard protocols for the assessment of the nanoparticles and their utilization on human safety, biodiversity preservation, and overall impact on the ecosystems. Various international agencies have therefore put forward some regulatory standards in this regard. Some of these include the United State Environmental Protection Agencies (USEPA), International Standard Organization (ISO) Technical Committee, and European Observatory for Nanoparticles (EON), amongst others. It has been predicted that the global market for the nanobioremediation technology could skyrocket above US 125 billion dollars by the year 2024. USEPA has further stated that phytoremediation and bioremediation market on a global scale may have up to a yearly growth of US 1.5 billion dollars annually (Pathakoti et al. 2018; Ahmed et al. 2021).

7 Current Reports on Studies in Nanobioremediation of Heavy Metals

Wang et al. (2014) from their studies documented that gold nanoparticles showed higher stability in buffer and aqueous media, however as the pH was increased from 4 to 10, there was a loss in the stability.

Abuhatab et al. (2020) investigated the removal of copper, zinc, and chromium (III) ions from contaminated water. Through the utilization of nickel oxide and magnesium oxide nanoparticles are synthesized through green approach induced by microorganisms. The process was found to be efficient and sustainable for the removal of the metals.

Similarly, in a study carried out by Ranjan et al. (2019) the recombinant cyanate hydratase enzyme was immobilized on the surface of iron oxide nanoparticle filled magnetic carbon nanotube. This was found to be efficient in the removal of lead, chromium copper, and iron from synthetic waste water.

In another investigation, San Keskin et al. (2018) adopted bionanoremediation technology for the removal of nickel and chromium ion from environmental matrices. The microorganism used for the induction of the production of the nanoparticle was Lysinibaciullus spp. This gave rise to the Electrospuncyclodextrinfibers which was shown to be highly efficient for the removal of metals.

Indraratne et al. (2021) opined that nanomaterials having a high reactivity could be employed as soil amendments for the remediation of heavy metal pollution through the immobilization of the elements. In their work, they compared the potential of iron oxide and aluminum oxide nanoparticles for the stabilization of lead, cadmium, and zinc in contaminated site. They reported a remarkable decrease in the concentrations of the heavy metals on treatment with the nanoparticles.

In a similar investigation, Tang et al. (2016) also reported that varying methods of synthesis affect the stability of copper nanoparticles to heat. Hence they opined for the need for a thorough experimental design with a view to evaluating the impact of pH and temperature on the interaction between the microorganisms and the nanoparticles.

Mohsenzadeh and Rad (2012) in a related study documented the bioaccumulation potential of nanoparticles produced from N. mucronanta in their study using simulated heavy metal solutions. The findings from their study revealed that the concentration of the heavy metals decreased several times during a period of three days experimentation.

Carbon nanotubes-based polymeric nanomaterials, thiacalixarenes, and Calixarenenes have been produced and characterized for the elimination of cadmium and lead ions from contaminated waste water. Similarly the adsorption potential for copper ions was also investigated. The findings from the study showed that the efficiency for the removal of Cu (II) ions reached up to 67.9% at a pH of 2.1 (Li et al. 2010).

Kumari and Singh (2016) in their work reported that nanomaterials are vital as facilitating agents during bioremediation of heavy metals and they do this through the enhancement of the growth of microorganisms, agents of remediation, or through the stimulation of microbial enzymes involved in remediation.

In another study, it was demonstrated that the nanoparticles enhanced the production of microbial biosurfactants, thereby increasing the solubilities of the various heavy metals contaminants and creating an environment that is conducive to the bioremediation of the heavy metals (Decesaro et al. 2017).

Hoseinian et al. (2020) in a related work fabricated an amino functionalized graphene oxide nanocollector for the efficient removal of nickel ions through the use of ion floatation. Using this approach they were able to achieve up to 100% nickel removal from the effluents.

8 Conclusion and Future Trends

This review work has presented comprehensively the application of nanotechnology in the bioremediation of heavy metals. It has also presented the various types of nanoparticles that have been employed in this regard as well as various factors affecting the effective utilization of these duo technologies. Nanotechnology has stimulated significant interest amongst scientists as a result of the beneficial effects, and potential for diverse applications. The integration of this with the bioremediation technology has proffered a green technology toward the efficient management of heavy metals. Bionanomaterials have also been seen to present various advantages over metallic nanoparticles in that they are environmental friendly due to their biodegradability. There is however need for further studies for effective understanding of the synergetic relationship between the nanoparticles and biotechnologies during the process of bioremediation and the combined impact of the processes. Furthermore, it is worth knowing that up to present times, there is dearth of studies which document the long-term impacts on health of the utilization of nanoparticles and microorganisms for the removal of heavy metals. Furthermore, another area of much concern is the commercialization of these bionanotechnological products since only about 1% has been commercialized so far. The full potential of nanobioremediation when pursued on a commercial scale would be a significant breakthrough for industries. In conclusion, the application of bionanotechnology for the decontamination of heavy metal polluted areas could be of immense relevance due to its sustainability, cost effectiveness, and green nature when compared to other existing technologies.

Table 1 Advantages and limitations of bioremediation for treatment of heavy metals

Bioremediation	Advantages	Limitations
	Ease of application	Not efficient for the treatment of highly polluted areas
	It is a green approach to remediation and suitable for various heavy metals	For some contaminants the intermediate formed could be more toxic when compared to the starting contaminant
	Could be applied for various organic and inorganic pollutants	May not be efficient for certain compounds especially some high molecular weight organic compounds
In situ bioremediation	Treatment is at the site of contamination Structural component of the polluted area is kept intact	There is need to install equipments for the process which is additional cost
Engineered in situ bioremediation	Does not take much time	There are high nutrient requirements
Ex situ bioremediation	Recommendable for small polluted sites	Cost for conveying and treatment away from the contaminated sites

References

Abuhatab S, El-Qanni A, Al-Qalaq H, Hmoudah M, Al-Zerei W (2020) Effective adsorptive removal of Zn^{2+}, Cu^{2+}, and Cr^{3+} heavy metals from aqueous solutions using silica-based embedded with NiO and MgO nanoparticles. J Environ Manag 268:110713

Adekeye D, Popoola O, Asaolu S, Adebawore O, Aremu K (2019) Adsorption and conventional technologies for environmental remediation and decontamination of heavy metals: an overview. E-ISSN: 2349-9788; P-ISSN: 2454-2237

Ahmed T, Noman M, Ijaz M, Rizwan M, Li B (2021) Current trends and future prospective in nanoremediation of heavy metals contaminated soils: a way forward towards sustainable agriculture. Ecotoxicol Environ Saf

Ali H, Khan E, Ilahi I (2019) Environmental chemistry and ecotoxicology of hazardous heavy metals: environmental persistence, toxicity, and bioaccumulation. J Chem. https://doi.org/10.1155/2019/6730305

Babantude D, Denwigwe I, Babantunde O, Agboola O (2019) Environmental and societal impact of nanotechnology. IEEE Access 99:1–1. https://doi.org/10.1109/ACCESS.2019.2961513

Cimbolakova I, Uher I, Lakticova K, Vargova M, Kimakova T, Papajova I (2019) Heavy metals and the environment. Open access peer-reviewed chapter. Doi:https://doi.org/10.5772/intechopen.86876

Decesaro A, Machado TS, Cappellaro ÂC, Reinehr CO, Thomé A, Colla LM (2017) Biosurfactants during in situ bioremediation: factors that influence the production and challenges in evaluation. Environ Sci Pollut Res 24:20831–20843

Hoseinian FS, Rezai B, Kowsari E, Chinnappan A, Ramakrishna S (2020) Synthesis and characterization of a novel nanocollector for the removal of nickel ions from synthetic wastewater using ion flotation. Sep Purif Technol 240:116639. https://doi.org/10.1016/j.seppur.2020.116639

Indraratne S, Pierzynski G, Baker L, Prasad V (2021) Nano-oxides immobilize cadmium, lead, and zinc in mine spoils and contaminated soils facilitating plant growth. Can J Soil Sci 101(3):543–554. https://doi.org/10.1139/cjss-2020-0127

Kebede G, Tafese T, Abda E, Kamaraj M, Assefa F (2021) Factors influencing the bacterial bioremediation of hydrocarbon contaminants in the soil: mechanisms and impacts. Open Access. Article ID 9823362 | https://doi.org/10.1155/2021/9823362

Keskin NO, Celebioglu A, Sarioglu OF, Uyar T, Tekinay T (2018) Encapsulation of living bacteria in electrospuncyclodextrin ultrathin fibers for bioremediation of heavy metals and reactive dye from wastewater. Colloid Surface B 161:169–176. https://doi.org/10.1016/j.colsurfb.2017.10.047

Khan I., Saeed K., Khan I. (2019). Nanoparticles: properties, applications and toxicities. Arab J Chem 12(7):908–931. https://doi.org/10.1016/j.arabjc.2017.05.011

Kumari B, Singh DP (2016) A review on multifaceted application of nanoparticles in the field of bioremediation of petroleum hydrocarbons. Ecol Eng 97:98–105

Li Y, Liu F, Xia B (2010) Removal of copper from aqueous solution by carbon nanotube/calcium alginate composites. J Hazard Mater 177(1–3):876–880

Mohsenzadeh F, Rad AC (2012) Bioremediation of heavy metal pollution by nano-particles of noaea mucronate. Int J Biosci, Biochem Bioinform 2:85–89

Naseer B, Srivastava G, Qadri OS, Faridi SA, Islam RU, Younis K (2018) Importance and health hazards of nanoparticles used in the food industry. Nanotechnol Rev 7(6):623–641. https://doi.org/10.1515/ntrev-2018-0076

Navya PN, Daima H (2016) Rational engineering of physicochemical properties of nanomaterials for biomedical applications with nanotoxicological perspectives. Nano Converg 3:1. https://doi.org/10.1186/s40580-016-0064-z

Nunez EV, Molina CE, Julain M, Rosa G (2020) Use of nanotechnology for the bioremediation of contaminants: a review. Processes 1(7):1–18. https://doi.org/10.3390/pr8070826

Pathakoti K, Manjunath M, Hwang H (2018) Nanotechnology applications for environmental industry. Doi:https://doi.org/10.1016/b978-0-12-813351-4.00050-x

Rahman Z, Singh VP (2020) Bioremediation of toxic heavy metals (THMs) contaminated sites: concepts, applications and challenges. Environ Sci Pollut Res Int 27(22):27563–27581

Ranjan B, Pillai S, Permaul K, Singh S (2019) Simultaneous removal of heavy metals and cyanate in a wastewater sample using immobilized cyanate hydratase on magnetic-multiwall carbon nanotubes. J Hazard Mater 363:73–80

Sarkar S, Enamala M, Chavali M, Sarma S, Chandrasekhar K (2020) Nanophytoremediation: an overview of novel and sustainable biological advancement. Open access peer-reviewed chapter. Doi:https://doi.org/10.5772/intechopen.93300

Sayqal A, Ahmed O (2021) Advances in heavy metal bioremediation: an overview. Appl Bionics Biomech 1609149. Doi:https://doi.org/10.1155/2021/1609149

Shah MP (2020) Microbial bioremediation & biodegradation. Springer

Shah MP (2021) Removal of refractory pollutants from wastewater treatment plants. CRC Press

Tang H, Chen X, Niu YW, Luo X, Wang Z, Chen M, Shi G (2016) Thermal stability characteristics of in situ nano-particles formed in metal melt. Mater Lett 162:261–264

Verma A (2021) Bioremediation techniques for soil pollution: an introduction. Open access peer-reviewed chapter. DOI:https://doi.org/10.5772/intechopen.99028

Wang L, Hu C, Shao L (2017) The antimicrobial activity of nanoparticles: present situation and prospects for the future. Int J Nanomedicine 12:1227–1249. https://doi.org/10.2147/IJN.S121956

Wang A, Ng HP, Xu Y, Li Y, Zheng Y, Yu J, Han F, Peng F, Fu L (2014) Gold nanoparticles: synthesis, stability test, and application for the rice growth. J Nanomater 1:1–6

Yan C, Qu Z, Wang J, Cao L, Han Q (2022) Microalgal bioremediation of heavy metal pollution in water: recent advances, challenges, and prospects. Chemosphere 286(3):131870. https://doi.org/10.1016/j.chemosphere.2021.131870

Yap H, Zakaria N, Zulkharnain A, Sabri S, Fluentes C, Ahmad S (2021) Bibliometric analysis of hydrocarbon bioremediation in cold regions and a review on enhanced soil bioremediation. Biology (basel) 10(5):354. https://doi.org/10.3390/biology10050354

Zand A, Tabrizi A (2021) Effect of zero-valent iron nanoparticles on the phytoextraction ability of *Kochia scoparia* and its response in Pb contaminated soil. Environ Eng Res 26(4):200227. Doi:https://doi.org/10.4491/eer.2020.227

Diversity, Biodegradation and Bioremediation of Polycyclic Aromatic Hydrocarbons

Abel Inobeme, Charles Oluwaseun Adetunji, John Tsado Mathew, Alexander Ikechukwu Ajai, Abdullahi Mann, Jonathan Inobeme, Bamigboye Oyedolapo, Mathew Adefusika Adekoya, and Sandra Onyeaku

1 Introduction

The advancement in the area of industrialization within the last few decades has led to an increase in various anthropogenic substances in the environment. Most of the prevalent contaminants that have been identified include petroleum-based hydrocarbons, PAHs, pesticides, halogenated hydrocarbons and various salts. PAHs are one of the classes of contaminants of utmost concern. They have been reported to be mutagenic, toxic and carcinogenic hence their presence within the environment is of serious threat. USEPA has enlisted them as one of the priority pollutants which need to be continuously monitored in different effluents discharged from industries. Sixteen priority PAHs have also been identified by USEPA (Fig. 1). The physiological impacts

A. Inobeme (✉)
Department of Chemistry, Edo State University Uzairue, Edo State, Nigeria
e-mail: inobeme.abel@edouniversity.edu.ng

C. O. Adetunji
Applied Microbiology, Biotechnology and Nanotechnology Laboratory, Department of Microbiology, Edo State University Uzairue, PMB 04, Edo State, Nigeria

J. T. Mathew
Department of Chemistry, Ibrahim Badamasi Babangida University Lapai, Lapai 911101, Nigeria

A. I. Ajai · A. Mann · S. Onyeaku
Department of Chemistry, Federal University of Technology Minna, Minna, Nigeria

J. Inobeme
Department of Geography, Ahmadu Bello University Zaria, Zaria, Nigeria

B. Oyedolapo
Department of Chemical Sciences, Kings University, Odeomu, Nigeria

M. A. Adekoya
Department of Physics, Edo State University Uzairue, Edo State, Nigeria

M. P. Shah (ed.), *Microbial Technologies in Industrial Wastewater Treatment*,
https://doi.org/10.1007/978-981-99-2435-6_3

of PAHs on human and organisms have also been well documented. Various States and Agencies have identified PAHs to be highly toxic. PAHs are well-documented environmental contaminants that are distributed ubiquitously in the environment. PAHs are sometimes known as polynuclear aromatic hydrocarbons, abbreviated as PNAs, fused ring aromatics or condensed aromatic rings (Bisht et al. 2015). They are made up of aromatic rings that are fused together. The most commonly found within the environment include those with two member rings such as naphthalene and the seven member ring such as coronene, although there are also high member ring PAHs in large abundance. Their formation occurs during the incomplete combustion of oil, wood, coal and gasoline. There is a presence of high concentrations of these compounds in asphalt, crude oil and coal tar. More recently there is much interest in the diversity and distribution of this class of compounds as a result of their deleterious effects on human and the ecosystem at large. This issue of concern has therefore prompted various scientists on the need to proffer various feasible and reliable approaches for their remediation from the environment. Bioremediation is one of the effective approaches that has been employed for the remediation of polluted waters and lands and also aids the promotion of various natural attenuations of these contaminants through the use of in situ community of various microbes present in the site. There are various species of bacteria and fungi that have found relevance for the bioremediation of PAHs due to their unique potential of transforming these contaminants into simpler inorganic materials that are eco-friendly (Marcon et al. 2021). Various anaerobic and aerobic pathways are involved in the biodegradation of PAHs. Environmental concern has continually risen and various techniques have been put forward accordingly to remove them from the environment. Most of the traditional approaches have their inherent limitations such as incomplete removal, cost of execution and impact on the environment among others. One of the most promising approaches for the treatment of PAHs is bioremediation. Bioremediation is an approach in which various microorganisms such as algae, bacteria and fungi are utilized for the degradation and transformation of the various molecules of PAHs and other contaminants into simple inorganic materials such as water, carbon dioxide and inorganic salts. Various studies have documented the use of various classes of these organisms for the biodegradation of PAHs (Kebede et al. 2021). This chapter reviews the diversity, biodegradation and bioremediation of PAHs. It discusses the sources, classifications and physicochemical properties of PAHs, the principles and mechanisms involved in the biodegradation and bioremediation of PAHs and finally attempts is made in highlighting the recent reports on techniques involved in the bioremediation and biodegradation of PAHs.

2 Physicochemical Properties of PAHs

PAHs are colorless compounds, with high boiling and melting points. They have limited solubility in water and low vapor pressure. In their pure forms, they exist as solids that have low volatility at normal temperature and their appearances vary from

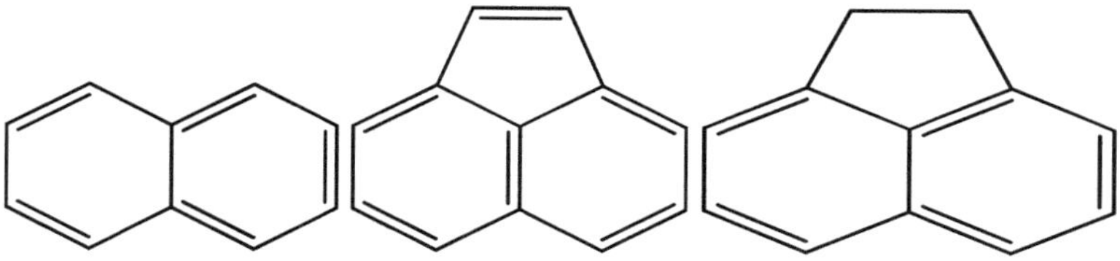

Naphthalene Acenaphthylene Acenaphthene

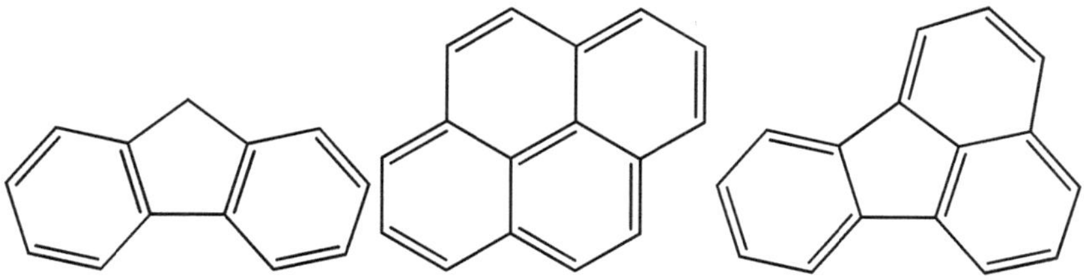

Fluorene Pyrene Fuoranthene

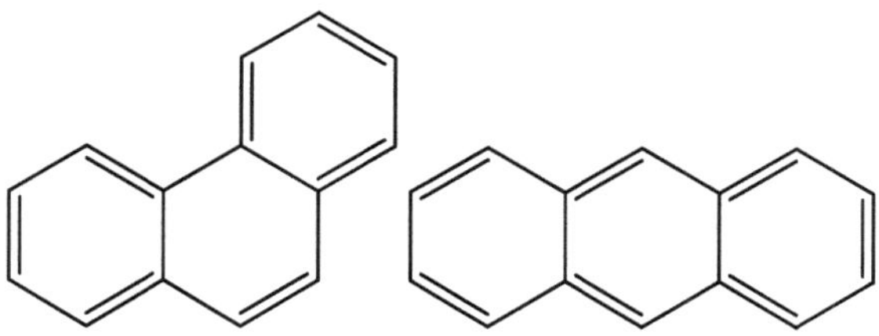

Phenanthrene Anthracene

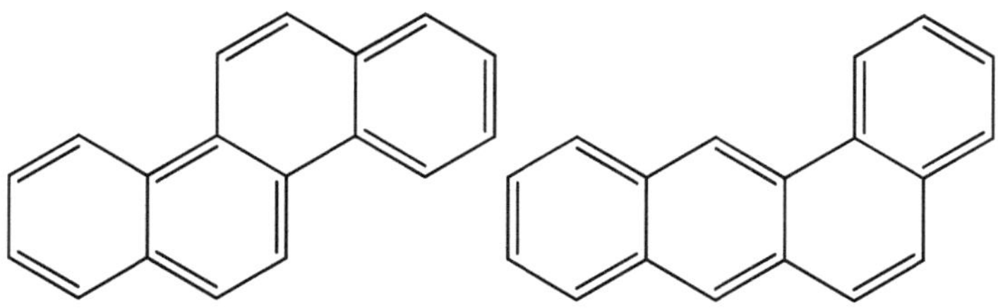

Chrysene Benz[a]anthracene

Fig. 1 Structures of the sixteen USEPA priority PAHs

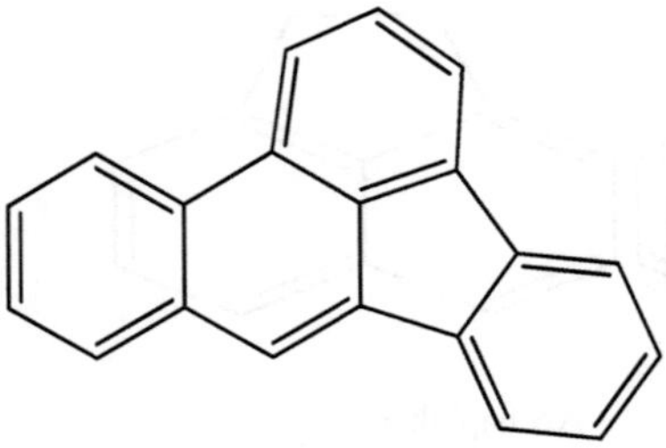

Benzo[b]flouranthene

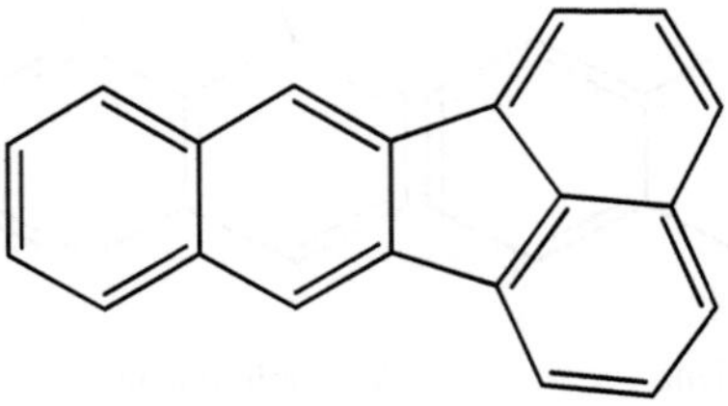

Benzo[kflouranthene

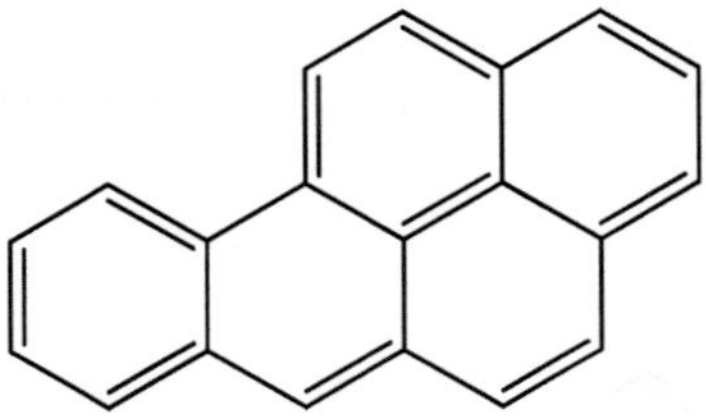

Benzo[a]pyrene

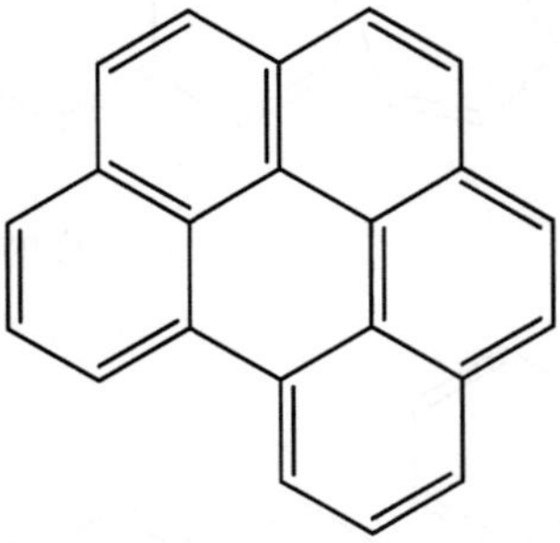

Benzo[ghi]perylene

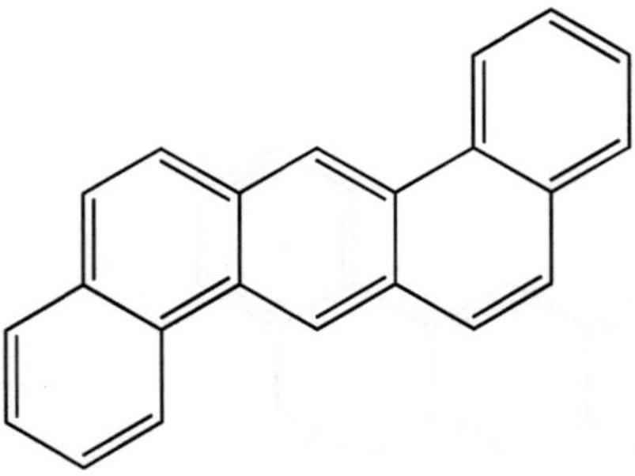

Diben[a,h]anthracene

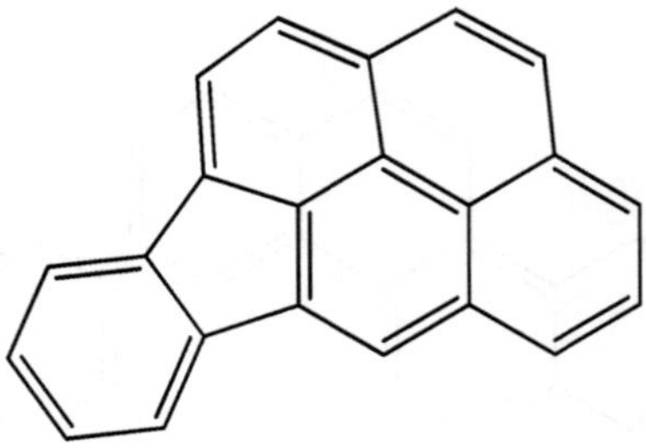

Indeno[1,2,3-cd]pyrene

Fig. 1 (continued)

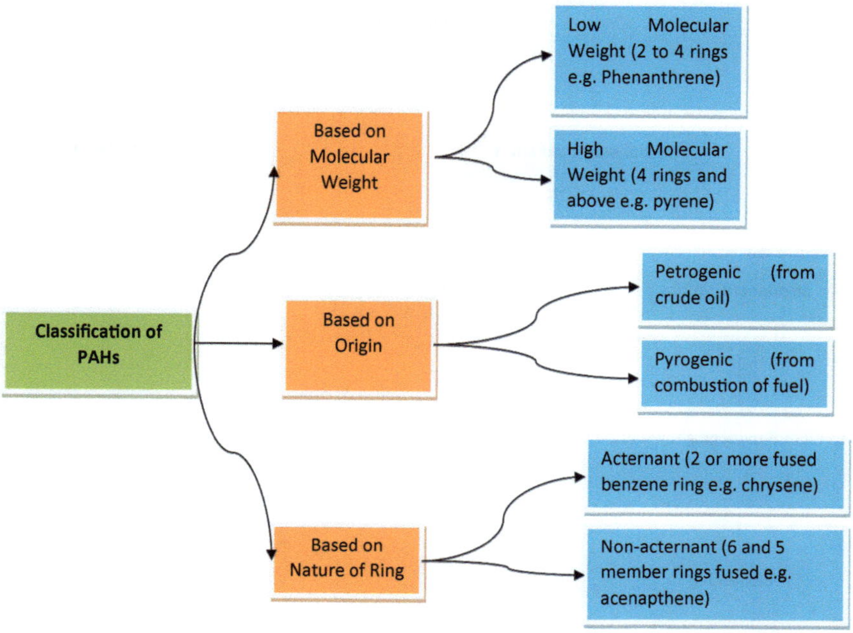

Fig. 2 Classification of PAHs

white, colorless to light yellow. They are non-polar organic molecules and made of hydrogen and carbon atoms. They have a faint pleasant smell and are readily photo-oxidized and degraded into smaller substances. PAHs from anthropogenic sources are released into the atmosphere where they are either adsorbed into particulate matter or in their gaseous state (Patel et al. 2020). Some of the properties of the sixteen USEPA priority PAHs are presented in Table 1.

3 Sources and Distribution of PAHs

Various kinds of combustion processes give rise to different types of PAHs in varying quantities as well as their isomers. Some of the natural sources of the petrogenic PAHs include mainly the seepages of oil and the erosion of various petroliferous shales, while the natural origins of PAHs from pyrolysis or combustion include those generated during the incomplete burning of biomass and woods through grass and forest fires. Anthropogenic sources of PAHs may bring about the formation of similar though not identical PAHs. The most prominent and ubiquitous input of anthropogenic PAHs is those connected with pyrogenic origin. PAHs are also generated through natural processes such as volcanic eruption, forest fire among others. The release of the residual amount of PAHs into the environment brings about serious

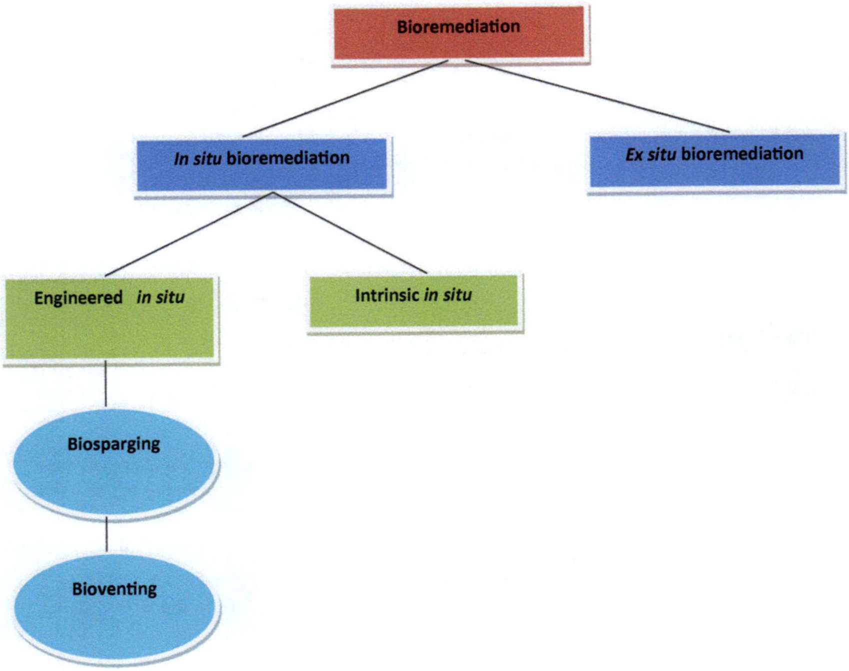

Fig. 3 Types of bioremediation technology

Table 1 Priority PAHs based on USEPA and some of their properties

Names of PAHs	Molecular weight (g/mol)	Number of rings	Log Kw	Solubility (mg/dm^3)
Naphthalene	128.17	2	3.37	31
Fluorene	166.22	3	4.18	1.9
Phenanthrene	178.23	3	4.57	1.1
Acenapthylene	152.2	3	3.92	16.1
Anthrancene	178.23	3	4.54	0.045
Acenaphthene	154.21	3	3.92	3.8
Fluranthene	202.26	4	5.22	0.26
Belz[a]anthracene	228.29	4	5.91	0.011
Chrysene	228.29	4	5.91	0.0015
Pyrene	202.26	4	5.18	0.132
Benzo[a]pyrene	252.32	5	5.91	0.0038
Benzo[b]fluoranthene	252.32	5	5.80	0.0015
Benzo[k]fluoranthene	252.32	5	6.00	0.0008
Dibenzo[a,h]anthracene	278.35	6	6.75	0.0005
Benzo[g,h,j]perylene	276.34	6	6.50	0.00026
Indeno[1,2,3-cd]pyrene	276.34	6	6.50	0.062

health hazard. A typical example is Phenanthrene, which has been documented to affect human skin since it has the potential of acting as a photosensitizer and moderate allergen. When present in soil, these compounds are also able to absorb sediment that is rich in organic matter and also accumulate within the bodies of various aquatic organisms and are eventually passed to humans through the food chain (Gao et al. 2021). The pollution of the marine environment by PAHs can also occur through accidental discharge of oil into sea water. Currently one of the major sources of PAHs includes hydrocarbon fuel such as diesel which has excessive quantity of PAHs in it. It is possible to determine the source of the PAHs using the ratio of the parent PAHs and the alkylated ones. This tells if the source is petroleum-based or combustive origin. High content of alkylated PAHs indicates the pollution that emerged from petroleum-based products such as leakages of fuel resulting in spillages. PAHs are then subjected to various mechanisms of removal such as photolytic reactions, oxidative reactions and dry–wet depositions. Prevailing winds and air current bring about the dispersion and movement of PAHs through long distances after which they are deposited on the surfaces of soil and water (Yang et al. 2015). Formation of PAHs basically involves the Diels-Alder rearrangement processes. One of the vital factors that affect both the diversity and structure of the PAH formed is temperature. The large PAHs are formed at relatively lesser level when compared to the small PAHs due to kinetic limitation during the introduction of more rings. Aside their biodegradation, their fate within the environment varies based on the nature of the environment, for instance, in the atmosphere, PAHs are capable of undergoing photo oxidation process, but when present in water and soil, they are able to undergo chemical oxidation as well as photo oxidation. Some PAHs like alkyl naphthalene and naphthalene are lost through volatilization. The occurrence of structures that are highly specific is lower due to the possibility of different isomeric compounds for the larger PAHs (Hamidi et al. 2016).

4 Classification of PAHs

PAHs are a broad class of diverse organic molecules which have two or more aromatic rings that are fused together ranging from those with just two members such as naphthalene and its derivatives to the complex ones which have about 10 rings fused together. PAHs that have up to 6 ring members fused together are commonly described as the "small" PAHs, while those that have up to six aromatic rings are known as the large PAHs. PAHs have also been grouped into nonalternant and alternant PAHs. The alternant PAHs are those molecules that are made up mainly of benzene rings of six-membered fused together. The nonalternant PAHs on the other hand are made of six member and five member rings of carbon that are fused together. The variation in the configuration of the rings contributes significantly to the differences in their physicochemical properties (Abdel-Shafy and Mansour 2015).

On the basis of origin, PAHs are grouped into pyrogenic PAHs which are formed from the combustion of fossil fuels and petrogenic PAHs which are linked to crude oil and water contamination after the spillage of oil and oil products. The structural composition of the pyrogenic and petrogenic PAHs also differ. The petrogenic PAHs are either alkylated extensively or undergo complete oxygenation resulting to the formation of PAHs quinones. PAHs existing as complex mixtures made of more than 100 chemical compounds include various derivative compounds such as hydroxyl and nitro-PAHs together with heterocyclic PAHs. The USEPA outlined sixteen PAHs which they named the priority PAHs of high pollution concern. These PAHs of greater priority include fluorene, fluoranthene, benzo[*k*]fluoranthene, dibenz[*a,h*]anthracene, acenapthene, benzo[*a*]pyrene (B[*a*]P), benzo[*g,h,i*]perylene, acenapthylene, chrysene, pyrene, indeno[1,2,3-*cd*]pyrene, benz[*a*]anthracene and benzo[*b*]fluoranthene (presented in Table 1). Out of the identified priority PAHs, B[a] P is considered as human carcinogen hence commonly employed as indicator for exposure to PAHs. PAHs can also be grouped based on their molecular weights in which case we have the low molecular weight PAHs (LMW) which are made of two and three member rings. These are soluble and more readily degraded. They also have higher volatility. The high molecular weight (HMW) PAHs have rings up to four or more and are readily adsorbed into sediments and soils and also show higher resistance to the process of degradation by microorganisms due to their high hydrophobicity and high molecular mass (Adeniji et al. 2017).

5 Exposure of Humans to PAHs

PAHs are found in air, food, water, soil and other plant materials. The primary route of exposure to PAHs includes dermal contact, ingestion through the mouth and inhalation. In some cases, some of the exposure could involve more than a single route at ones thereby affecting the total absorbed dose. The occupational routes of exposure primarily include persons that work in coal gasification sites, municipal incineration plants, aluminum production plants and smokehouses. The non-occupation means of exposure to these toxic compounds could include burning of woods and coals, smoking of tobacco and diet. Among which eating of contaminated food is the major route for non-occupational exposure to PAHs (Liu et al. 2019). PAHs are found in soil, sediments, air, ground water and surface water. They are distributed to various vegetations from the atmosphere resulting in their accumulation in various food chains, through which they eventually get to man.

6 Bioremediation of PAHs

Bioremediation technology is one of the most reliable for the treatment of various environmental matrices such as water, sediments and soil that are contaminated by PAHs. This approach to the cleanup of contaminants however has its inherent limitations, such as activity and diversity of the indigenous hydrocarbon degrading microorganisms, low abundance of associated bacteria, the slow growth and the restricted bioavailability of PAHs in the aqueous medium. It is a fast-emerging technological process that can be utilized together with various chemical and physical processes of treatment for the effective removal of various classes of organic and inorganic contaminants in the environment. It is a remarkable and promising technique for efficient environmental management of contaminants. Bioremediation is a remediation technology that involves the utilization of various microorganisms, fungi or plants for the degradation of contaminants thereby enhancing the transformation of the substances. There are various groups of microorganisms that play a role in the process of bioremediation with bacteria being one of the most prominent. Bacteria are ubiquitous and are found in various environmental media such as soil, water and air. The introduction of nutrients, co-substrates and electron acceptors so as to improve the activity of the indigenous microbes is however expensive to achieve, also the chemicals added could easily spread away from the targeted molecules which clearly indicates the limitation associated with the use of bioremediation for the treatment of PAHs contaminated site. Bioremediation is applied commonly for the treatment of PAHs contaminated sites such as sediments and soil (Yan et al. 2020).

7 Concept of Bioremediation of PAHs

Bioremediation is one of the most reliable approaches for the removal of PAHs from the environment. There are various processes and microorganisms involved in the bioremediation of PAHs. Processes that involve the utilization of localized bacteria for the degradation of contaminants under existing conditions of subsurface are known as the natural attenuation or passive bioremediation process (Yan et al. 2020). Natural attenuation processes are most common within the sub-surfaces where the population of the bacteria responsible for the degradation is high. The enhanced process of bioremediation is the type in which the indigenous bacteria community is stimulated through the addition of electron donors or substrates so as to bring about an increase in the growth of the bacteria thereby enhancing faster rates of biodegradation. The choice of substrate introduced is depended on the nature of the bacteria that are being stimulated for the degradation of the contaminants (Xu et al. 2018). Various agencies, industries and researchers have utilized some organic substrates for the promotion of anaerobic reductive removal of numerous PAHs into their final innocuous end results. Various remarkable and promising outcomes have been documented in the field of anaerobic bioremediation applications. Phytoremediation and bioremediation have advanced progressively and proved efficient most especially in the area of treatment of contaminated areas. Sites polluted with recalcitrant and PAHs have been shown to prove more resistance to these techniques, but there are however remarkable progress in the field and laboratory. Some of the most recent breakthroughs in the area of bioremediation of PAHs include advancement in various anaerobic and aerobic processes, bioreductive dechlorination processes, biomonitoring, bioaugmentation and phytoremediation. Various advanced processes of bioremediation which involve destructive and non-destructive approaches are being employed in the process of bioremediation. Some of the approaches include phytoremediation, biodegradation, thermal incineration, reductive and advanced oxidation processes (Sharma et al. 2018).

8 Types of Bioremediation Techniques for PAHs

There are various types of bioremediation techniques that could be employed for the removal of PAHs from the environment. One of the techniques in bioremediation is Land farming. This involves the stimulation of the local biodegrading microbes through the provision of nutrients, oxygen and water so as to enhance the aerobic breakdown of the contaminants. Composting involves the breakdown of the PAHs together with various agricultural wastes like manures. Hence it can be deduced that irrespective of the technique selected, an ideal approach to bioremediation of PAHs could be chosen on the basis of understanding the nature of the contaminant and the microorganisms needed, together with their responses to various environmental conditions (Azuibuike et al. 2016). There are various strategies involved in the process of bioremediation of PAHs. Two major approaches for the bioremediation of PAHs present in ground water have been identified. In the first approach, the water is pumped to the surface and treatment is carried out above the ground inside a bioreactor. In another approach, the aquifer is remediated within (in situ).

8.1 In Situ Bioremediation of PAHs

In situ approach of bioremediation is a better and more reliable approach when compared to traditionally employed methods. In situ bioremediation was first utilized as a natural approach for petroleum-based compounds in contaminated underground aquifers. Further studies later reported the potential of microbial population in the degradation of PAHs as well as other organic and inorganic contaminants. The in situ bioremediation approach is further classified into two. The first is intrinsic in situ bioremediation while the other is engineered in situ bioremediation. In intrinsic in situ bioremediation, the primary interest is monitoring the process of degradation which is already on going so as to ensure that the contamination plume does not expand. For the engineered in situ bioremediation, natural degradation is not taking place in the environment and in some other cases its occurrence is very slow, hence the subsurface environment is first manipulated so as to stimulate the process of biodegradation and the rate of the process is enhanced. In the engineered approach, the primary strategies involve the supply of nutrients such as electron acceptors, phosphorus and nitrogen to the subsurface. One of the most prominent acceptor of electron employed is oxygen. However, as a result of the low solubility of oxygen gas in water and the accompanied high biomass production, anaerobic processes have been applied recently (Philp 2015). This approach involves the treatment of contaminated substances in the area of contamination. Hence it does not need the process of excavation; therefore, it is usually followed by small or no disturbance to the structural composition of the soil. Normally this approach is supposed to be relatively cheaper in comparison to the ex situ technique of bioremediation since there is no extra cost required for the excavation of the soil around the site. However, the economic constrains of this approach is the additional cost that is required for the onsite installation and design of some of the complex equipment required for the technique. Some of the in situ approaches could be aided (phytoremediation, biosparging and bioventing) while some others could proceed without the need for any form of enhancement (natural attenuation and intrinsic bioremediation). Intrinsic bioremediation technique has been efficiently employed for the treatment of site contaminated by PAHs. However, an effective and successful application of various conditions such as pH, moisture content, temperature and availability of nutrients must be suitable for the process (Gao et al. 2021).

8.2 Ex situ Bioremediation of PAHs

In this approach, the contaminants are removed through excavation from the site of pollution and then conveyed to another area for the purpose of treatment. This approach is commonly considered on the basis of extent of pollution, economic implication of treatment, performance criteria as well as geographical location.

9 Biodegradation of PAHs

Though PAHs are capable of undergoing various processes such as photolysis, volatilization, adsorption and chemical degradation, the major route of degradation is the microbial breakdown of PAHs. The degradation of PAHs is affected by various environmental factors such as the type and population of microorganism, the chemical structure and nature of the PAH being degraded among others. The PAHs are biodegraded and transformed into simpler metabolites, and finally mineralized into simple inorganic substances such as water and carbon dioxide. The rate of degradation is further affected by presence of oxygen, temperature, pH, population of microorganisms, presence of nutrients, chemical properties of the PAHs, structure of the compounds and the chemical partitioning in the growth medium. There are varying species of bacteria that are associated with the degradation of PAHs and most of these organisms are isolated from polluted sediment or soils. Some of the PAHs degrading bacteria that have been isolated include: Paenibacillus spp., Mycobacterium spp., Aeruginosa spp., Haemophilus spp., Pseudomonas spp. among others (Yan et al. 2020). There are also some groups of fungi such as the lignolytic fungi that have the potential of degrading PAHs. Typical species of fungi that have been linked with the potential of degrading PAHs include Pleurotus ostreatus, Phanerochaete chrysosporium and Bjerkandera adusta. There are various kinds of enzymes that are also associated with the process of microbial degradation of PAHs. Some of the enzymes are lignolytic enzymes, dehydrogenase and oxygenase. The fungi lignolytic enzymes include laccase, lignin peroxidase and manganese peroxidase. These enzymes speed up the formation of radicals through the process of oxidation thereby bringing about the destabilization of the bonds within the molecule. The microbial degradation of PAHs has been studied under anaerobic and aerobic conditions and was reported that the speed of degradation could be aided through chemical and physical preliminary treatment of the polluted soil. The bioavailability of the PAHs to the process of degradation could be enhanced through the addition of light oils and biosurfactant-producing microbes. The process of biodegradation could also be improved through the addition of various supplements to the polluted soil, such as compost. It has also been reported that Wetlands are also relevant for the removal of PAHs from polluted water. The high biological processes within such ecological system bring about a high rate of heterotrophic and autotrophic processes. Similarly, aquatic weeds such as Scirpus lacustris and Typha spp. have been employed in vertical-horizontal macrophyte derived Wetland for the treatment of PAHs. The elimination of PAHs depends on the ionization potential of the system. An intensive assessment of the kinetic of the process has not been fully comprehended hence becomes of more concern when fungi are employed for the bioremediation of PAHs. The processes of degradation in soils are more complex when compared to those of liquid cultures due to the non-homogeneity of soils. Hence several factors such as the bioavailability of PAHs and the rate of desorption must be taken into consideration when studying bioremediation in soil matrix. Different routes of degradation have been put forward for PAHs. The peroxidases are enzymes associated with the production of heme which has similar cycles of catalysis. The resting enzyme is oxidized by a mole of hydrogen peroxide thereby withdrawing electrons. Hence the reduction of peroxidase takes place in two steps. Laccases are oxidases that are rich in copper. They are capable of oxidizing phenolic derivates and also bring about the reduction of molecular oxygen to water (Sharma et al. 2018).

10 Factors Affecting the PAHs Biodegradation

10.1 Temperature

Temperature has a significant impact on the biodegradation of various classes of PAHs since most of the contaminated places are usually not at ambient temperature required for the indigenous microorganisms. Basically, the solubility of PAHs usually increases with an increase in temperature; this in turn brings about a rise in the bioavailability of the PAHs. However, the increase in temperature also brings about a decrease in the dissolved oxygen content, which in turn results in a decrease in metabolic activities. Transformation of some of the PAHs also occurs into different compounds which in most cases tend to be more toxic than the starting compounds. For most of the investigation, the assessment of the temperature effect was done under moderate temperature (Abdel-Shafy and Mansour 2015).

10.2 PH

This is another vital parameter that affects the degradation of PAHs. Generally, most microbes are sensitive to pH and require a near-neutral pH range of 6.5 to 7.5 for their optimum performance. It has however been documented that for most PAHs polluted sites, the pH is far from the optimum condition. For example, in some relegated sites for gas works, where there are large quantities of waste emerging from demolition, the materials are easily leached bringing about a rise in pH of the soil. Acidic conditions can also be generated around coal spoils as a result of the oxidation of sulfides. Such acidic or alkaline condition usually slows down the degradation of PAHs. It is therefore recommended that prior to the commencement of the bioremediation process; the pH should be adjusted to near neutral through the addition of some chemicals.

10.3 Nutrients

Another rate-determining factor in the degradation of PAHs is the presence of nutrients in the contaminated site. Aside from a reliable source of carbon, most PAHs degrading microbes also require other materials such as phosphorus, iron, nitrogen and potassium for growth and effective metabolism. Thus, the stimulation of bioremediation in a contaminated site that has nutrient deficiency requires the addition of amendments containing the desired nutrients. In the marine environment, the poor degradation of PAHs is usually due to the deficiency in phosphorus and nitrogen which are required nutrients for biodegradation (Gao et al. 2021).

10.4 Oxygen Availability

Bioremediation and degradation of PAHs can take place under anaerobic and aerobic conditions. Most studies have reported more on aerobic environment where oxygen functions as a rate-limiting factor and as a co-substrate. During the process of biodegradation of PAHs, oxygen is a major requirement for the effective action of both dioxygenase enzymes and monooxygenase during the initial oxidation of the aromatic compounds. For in situ bioremediation of PAHs, oxygen is usually added in some cases. The addition of oxygen from external source is usually achieved through tilling, drainage and introduction of chemicals that are capable of producing oxygen.

10.5 Bioavailability

This is another prominent factor that influences the process of biodegradation of PAHs. The low aqueous solubility of PAHs might affect their biodegradation; also these compounds are readily adsorbed on the surface of various minerals and sediments thereby reducing their bioavailability. Adsorption of PAHs onto various surfaces therefore brings about a reduction in their rate of bioremediation due to the fall in their overall bioavailability. The bioavailability of PAHs is also affected by their molecular weights. Basically lower molecular mass PAHs have lower bioavailability when compared to the high molecular weight. PAHs easily get entrapped into black carbon soot and coals and this can bring about a poor remediation as a result of the decrease in the bioavailability of the PAHs. This is one of the primary challenges associated with the remediation of PAHs in contaminated areas (Sharma et al. 2018).

10.6 Presence of Novel and Modified Microbes

The biodegradation of several xenobiotic compounds and PAHs is affected by the potential of the microorganisms to mineralize these compounds into water and carbon dioxide. The development and utilization of organisms that have been genetically modified are highly promising and cost-effective for the decontamination of PAHs from polluted sites. The development and application of various genetically modified microorganisms promote an eco-friendly approach. However, the introduction of such organisms is sometimes questionable. The use of microbes for the degradation of PAHs has been considered as a green technique for the cleanup of PAHs (Yan et al. 2020).

10.7 Availability of Organic and Inorganic Amendments Substances

Also, some microorganisms are capable of excreting out biosurfactants which aid the bioavailability of various organic contaminants. Several microorganisms also exhibit chemotaxis toward the contaminants. Such strategies generally bring about an improvement in the degradation of the contaminants. The biodegradation of PAHs is also enhanced by the addition of a trace amount of biosurfactant thereby increasing the bioavailability of the PAH, small amount of other chemicals such as salicylic acid also induces the formulation of catabolic operons. The presence of various organic amendments affects the population of the indigenous microorganism together with the efficiency of the bioremediation of these compounds in polluted soil. Also, instead of treatment using a single microbe, a well-defined consortium of microorganisms could give a more efficient result in some cases for the effective remediation of the polluted site. There is a need for more investigations in the area of catabolic capabilities of PAHs degrading bacteria consortia, which would further improve the current knowledge of the microbial consortia enhanced remediation of polluted environment. Also, the contamination area could be treated using algae, fungi, bacteria or their combinations for more efficient removal of pollutants (Amodu et al. 2012).

11 Mechanisms and Techniques of Degradation of PAHs

11.1 Chemical Degradation

The determination of the extent of persistence of PAHs after the process of degradation in various environments is a tedious task. The presence of PAHs in an anaerobic environment is dependent on pH, substrate interaction and redox conditions. The degradation of PAHs in soil takes place through biotic processes. After undergoing chemical processes, the PAHs are converted into other derivative compounds. Since large amount of energy is needed for the conversion of an aromatic ring into a non-aromatic compound, PAHs usually retain their aromaticity. The efficiency of the chemical degradation of PAHs is restricted due to their limited solubility in aqueous medium as well as low vapor pressure. It has been documented that various surfactants have the potential of overcoming the challenges connected with the low solubility of PAHs in aqueous medium. The surfactant enhances the solubility of PAHs in aqueous medium by inducing the decrease in the surface tension within interfacial regions (Bisht et al. 2015).

11.2 Microbial Degradation

For the process of bioremediation to be successful the microbes and their enzymes must be in close physical contact with the PAHs, since the property of the matrix and the nature of the PAHs determine their bioaccessibility and bioavailability. Bioavailability generally refers to the component or fraction of the contaminant that is absorbed by the cells thereby bringing about toxicity. Bioaccessibility refers to the fraction that is available for the biota present in the soil and is also known as environmental availability. Microbial biodegradation of PAHs in soil occurs in two steps. The first is the uptake of the PAHs by the microorganisms, while the second step involves the metabolism of the PAH. The bioavailability of the PAHs in the environment influences their degradation, distribution and sequestration. The lower degradation of some PAHs is due to their poor bioavailability which is due to their low solubility in water, low degree of dissolution and high adsorption to soil particles (Jaiswal and Shukla 2020).

The bioavailability of PAHs is affected by several factors such as soil properties, physical characteristics as well as the receptor organisms. The lower molecular weight PAHs are more volatile, soluble and could be used as a primary source of carbon by microbes and can be eliminated faster through biological means. The bioavailability of these compounds within the environment varies with the stage of weathering and time. The bioavailability of PAHs in the soil is affected by factors such as aging and aggregation, soil texture and organic matter content of the soil. Higher molecular weight pure PAHs show higher resistance to biotransformation while those with lower molecular mass tend to be easily transformed. However, the presence of particular PAHs could affect the biodegradation of the other PAHs. In a study, it was observed that the presence of pyrene, fluorene, phenanthrene and anthracene brought about a decrease in the biodegradation of acenapthene and phenanthrene. There are numerous compounds of PAHs in the environment, however the determinations are usually restricted to 6–16 PAHs. Each PAH differs significantly in its physicochemical properties (Vásquez-Murrieta et al. 2016).

12 Recent Reports

Bisht et al. (2015) in their review work opined on the significance of rhizoremediation of PAHs when compared to other approaches to remediation. They also discussed various environmental factors that may affect the efficiency of the rhizoremediation approach. Their review work further identified various groups of microorganisms that have the potential of biodegrading PAHs such as Rhodococcus spp., Pseudomonas aeruginosa, Mycorbacterium spp. among others.

Also, Moghadam et al. (2014) isolated and identified various species of PAHs degrading bacteria present on the surface of sediment and evaluated their efficiency based on various experiments. The PAHs degrading bacteria were isolated from the surface of the sediments. The purified strains were then identified after isolation through the use of 16S rDNA sequence analysis. The optimum biodegradation of fluorene and phenanthrene was also determined. It was observed that the enriched consortium had the highest degradation potential when compared to the mixture and single strain culture. The maximum efficiency of degradation was recorded at 35 °C, at a pH of 8. The findings from their study showed that indigenous microbes from the surface sediments possess a high potential of breaking down phenanthrene and fluorine.

Similarly, Clemente et al. (2001) carried out an investigation in which a total of thirteen deuteromycete fungal species were cultured in a medium that contain PAHs for a duration of 6 to 10 days. The addition of the PAHs was done directly through the use of inocula. The best strains were then selected on the basis of the extent of degradation of the PAHs as well as the ligninolytic activities that were generated by the fungi. It was observed that the highest degradation of naphthalene was 69%. The highest degradation of phenanthrene was observed in the strain containing laccase and Mn peroxidase activities.

Ghosal et al. (2016) in a related review work, made a provision for in-depth knowledge on various fungi, bacteria, algae and halophilic archaea that are associated with the degradation of PAHs. They discussed the various factors which affect the degradation of PAHs within the environment. They also highlighted some of the recent advancements in the aspect of genomics, genetics, metabolomics and proteomics that are relevant in the area of bioremediation of PAHs.

Mandree et al. (2021) in their study identified potential strain of bacillus that could be employed effectively for the bioremediation of PAHs of high and low molecular masses from soil matrix. They were able to identify a total of six strains which were then formulated into two different prototypes and checked for their capacity of removing PAHs from contaminated soils around the industry. It was observed that the soil samples that were dosed using prototype 2 were able to degrade all the PAHs (100%) which include pyrene, phenanthrene and fluoranthene. In the case of prototype 1, the accumulation of phenanthrene and pyrene was observed. They concluded that the prototype 2 which was made of Bacillus subtilis and Bacillus cereus strains was more efficient for the biodegradation of PAHs and their intermediate derivatives. They were able to demonstrate from their investigation that Bacillus spp. has a remarkable potential of bioremediating polluted sites.

Subashchandrabose et al. (2018) reported from their work that the bacterium Rhodococcus wratislaviensis was able to degrade completely 40% 50 μM pyrene, 280 μM of phenanthrene, and 28% 40 μM B[a]P during supplementation of the culture medium for a period of 7 days. The various groups of gene-specific primers that were identified to be linked with the degradation of the PAHs which code for naphthalene dioxygenase, 2,3-dihydroxybiphenyl, 4-hydroxybenzoate 3-monooxygenase and extradiol dioxygenase. It was also observed from the orthogonal experimental design that B[a]P degradation was significantly enhanced through the addition of surfactant like linoleic acid, Tween 80 and Triton X which suggested the possible bioavailability enhancement.

Kosnar et al. (2019) in their experiment compared the effect of vermicomposting and composting in the degradation of PAHs present in environmental biomass. A comparison was also done between the elimination of PAHs from organic waste mixture with the same mixture after spiking with PAHs. They observed a higher dissipation of the total PAH concentration up to about 84.5% at the end of the composting when compared to 61.6% recorded for vermicomposting. They concluded that the vermicompost and compost generated from the bioremediation treatments were useable as amendments for soil.

Revathy et al. (2015) reported the isolation and characterization of Burkholderia spp. with the potential of degrading PAHs obtained from lagoon sediments from India. On the basis of phylogenic details, the species was identified as Burkholderia and labeled as VITRSB1. An assessment was done on the initial PAHs degrading potential of the specie through the use of basal salt medium enhanced with kerosene, aniline, phenol, naphthalene and diesel. The strain was observed to be highly effective in the degradation of aniline, diesel and toluene at higher concentrations. Further experiments were carried out with a view to determining the composition of the final product of degradation using FTIR. Results from the study also revealed that none of the final products of the biodegradation is toxic.

13 Conclusion

In more recent times greater emphasis has been put on research focusing bioremediation of PAHs. On this note, various classes of microorganisms have been isolated and also characterized and documented to possess the potential for the degradation of PAHs. Also, several enzymes having varied catabolic efficiencies connected with the degradation of PAHs have been obtained in their pure forms and various novel biochemical routes for the degradation have been put forward. The recent breakthroughs in the field of metabolomics, genetics, proteomics and genomics which are useful in the studies of catabolism of various classes of PAHs have made remarkable contributions toward the comprehension of the ecology, physiology and various processes occurring at the contaminated sites. However, there are various areas of bioremediation of PAHs that require more knowledge and have not been fully exploited. There are limited studies on the aspect of enzymes, genes together with the molecular mechanisms involved in the degradation of PAHs in highly saline environments or anaerobic conditions. There is also limited information on the aspect of transmembrane trafficking of various PAHs together with their metabolites. There are several transporters that have been assumed to be involved in the transportation of the PAHs into the microorganism; however none has been characterized up to date. Additionally, further studies in the area of genetic engineering could be used for boosting the catabolic potential of microorganisms employed during the process of bioremediation. Currently, it is possible for scientists to create a unique PAHs metabolic route through the combination of various catabolic genes from different groups of organisms in a particular host cell. However, prior to the introduction of such genetically modified organisms into the environment, it is paramount to check comprehensively the possibility of any undesired side effects. Also, authorities must ensure that the proposed GMOs are more efficient, cheaper and safer.

References

Abdel-shafy H, Mansour M (2015) Review a review on polycyclic aromatic hydrocarbons: source, environmental impact, effect on human health and remediation. Egypt J Pet. Doi:https://doi.org/10.1016/j.ejpe.2015.03.011

Adeniji A, Okoh O, Oko A (2017) Analytical methods for polycyclic aromatic hydrocarbons and their global trend of distribution in water and sediment: a review. Open access peer-reviewed chapter. Doi:https://doi.org/10.5772/intechopen.71163

Amodu O, Ojumu T, Ntwampe S (2012) Bioavailability of high molecular weight polycyclic aromatic hydrocarbons using renewable resources. Open access peer-reviewed chapter. Doi:https://doi.org/10.5772/54727

Azubuike C, Chikere C, Gideon C (2016) Bioremediation techniques–classification based on site of application: principles, advantages, limitations and prospects. World J Microbiol Biotechnol 32(11):180. https://doi.org/10.1007/s11274-016-2137-x

Bisht S, Pandey P, Bhargava B, Sharma S, Kumar V, Sharma K (2015) Bioremediation of polyaromatic hydrocarbons (PAHs) using rhizosphere technology. Braz J Microbiol 46(1):7–21. Doi:https://doi.org/10.1590/S1517-838246120131354

Clemente A, Anazawa T, Durrant L (2001) Biodegradation of polycyclic aromatic hydrocarbons by soil fungi. Braz J Microbiol 32(4). Doi:https://doi.org/10.1590/S1517-83822001000400001

DOI:http://dx.doi.org/https://doi.org/10.7314/APJCP.2016.17.1.15 Polycyclic Aromatic Hydrocarbons and Bioaccessibility in. Egyptian Journal of Petroleum, 25, 1, 107–123

Gao Y, Han Y, Xia J, Tan J, Wang Y, Wang S (2021) Composition and distribution of aliphatic hydrocarbon compounds and biomarkers in seafloor sediments from offshore of the Leizhou Peninsula (South China). ACS Omega 6(50):34286–34293. Doi:https://doi.org/10.1021/acsomega.1c03529

Ghosal D, Ghosh S, Dutta TK, Ahn Y (2016) Current state of knowledge in microbial degradation of polycyclic aromatic hydrocarbons (PAHs): a review. Front Microbiol 7:1369. https://doi.org/10.3389/fmicb.2016.01369

Hamidi E, Jajeb P, Selamat J, Faizal A, Razis A (2016) Polycyclic aromatic hydrocarbons (PAHs) and their bioaccessibility in meat: a tool for assessing human cancer risk. Asian Pac J Cancer Prev 17(1):15–23

Jaiswal S, Shukla P (2020) Alternative strategies for microbial remediation of pollutants via synthetic biology. Front Microbiol. https://doi.org/10.3389/fmicb.2020.00808

Kebede G, Tafese T, Abda E, Kamaraj M, Assefa F (2021) Factors influencing the bacterial bioremediation of hydrocarbon contaminants in the soil: mechanisms and impacts. ID 9823362 | https://doi.org/10.1155/2021/9823362

Košnář Z, Wiesnerová L, Částková T, Kroulíková S, Bouček J, Tlustoš F (2019) Bioremediation of polycyclic aromatic hydrocarbons (PAHs) present in biomass fly ash by co-composting and co-vermicomposting. J Hazard Mater 369(5):79–86. https://doi.org/10.1016/j.jhazmat.2019.02.037

Liu J, Zhang J, Zhan C, Liu H, Hu T, Xing X, Qu C (2019) Polycyclic aromatic hydrocarbons (PAHs) in urban street dust of Huanggang, central China: status, sources and human health risk assessment. Aerosol Air Qual Res 19:221–233

Mandree P, Masika W, Naicker J, Moonamy G, Ramchuran S, Lalloo R (2021) Bioremediation of Polycyclic Aromatic Hydrocarbons from Industry Contaminated Soil Using Indigenous *Bacillus* spp. Processes 9(9):1606. Doi:https://doi.org/10.3390/pr9091606

Marcon L, Oliveras J, Puntes V (2021) *In situ* nanoremediation of soils and groundwaters from the nanoparticle's standpoint: a review. Sci Total Environ. https://doi.org/10.1016/j.scitotenv.2021.148324

Moghadam M, Ebrahimipour G, Abtahi B, Ghassempour A, Hastroudi M (2014) Biodegradation of polycyclic aromatic hydrocarbons by a bacterial consortium enriched from mangrove sediments. J Environ Health Sci Eng 12:114. https://doi.org/10.1186/s40201-014-0114-6

Patel A, Shaikh S, Jain K, Desai C, Madamwar D (2020) Polycyclic aromatic hydrocarbons: sources, toxicity, and remediation approaches. Front. Microbiol, 5 November 2020 | https://doi.org/10.3389/fmicb.2020.562813

Philp R (2015) Bioremediation: the pollution solution? https://microbiologysociety.org/blog/bioremediation-the-pollution-solution.html

Revathy T, Jayasri M Suthindhiran K (2015). Biodegradation of PAHs by *Burkholderia* sp. VITRSB1 Isolated from Marine Sediments. Research Article | Open Access |Article ID 867586 | https://doi.org/10.1155/2015/867A586

Sharma I (2020) Bioremediation techniques for polluted environment: concept, advantages, limitations, and prospects. Open access peer-reviewed chapter. DOI:https://doi.org/10.5772/intechopen.90453

Sharma B, Dangi AK, Shukla P (2018) Contemporary enzyme based technologies for bioremediation: a review. J Environ Manage 210:10–22. https://doi.org/10.1016/j.jenvman.2017.12.075

Subashchandrabose S, Venkateswarlu K, Naidu R, Megharaj M (2018) Biodegradation of high-molecular weight PAHs by *Rhodococcus wratislaviensis* strain 9: Overexpression of amidohydrolase induced by pyrene and BaP. Sci Total Environ 651(1):813–821. https://doi.org/10.1016/j.scitotenv.2018.09.192

Vasquez-Murriet M, Hernandez O, Cruz-Maya J, Cancino-Dia, Jan-Roblero J (2016) Approaches for removal of PAHs in soils: bioaugmentation, biostimulation and bioattenuation. Open access peer-reviewed chapter. Doi:https://doi.org/10.5772/64682

Xu X, Liu W, Tian S, Wang W, Qi Q, Jiang P, Gao X, Li F, Li H, Yu H (2018) Petroleum hydrocarbon-degrading bacteria for the remediation of oil pollution under aerobic conditions: a perspective analysis. Front Microbiol. Doi:https://doi.org/10.3389/fmicb.2018.02885

Yan A, Wang Y, Tan S, Yusof M, Ghosh S, Chen Z (2020). Phytoremediation: a promising approach for revegetation of heavy metal-polluted land. Front Plant Sci. https://doi.org/10.3389/fpls.2020.00359

Yang Q, Chen H, Li B (2015) Polycyclic aromatic hydrocarbons (PAHs) in indoor dusts of Guizhou, Southwest of China: status, sources and potential human health risk. Doi:https://doi.org/10.1371/journal.pone.0118141

Integrated Omics Approaches for Structural and Functional Characterization of Environmental Microorganisms

Anurag Singh, Prachi Srivastava, and Vinod P. Sharma

1 Introduction

In context to the biological systems, the suffix **"-ome"** is generally an indication of a collection like **genome** (collection of total genes in an organism)**; proteome** (total protein present within an organism) and so on and so forth, and when this **-ome** is changed to **-omics**, it becomes the study of this collection like **genomics** (study of genome)**; proteomics** (study of proteome)**,** etc. In earlier times it was only DNA that scholars and scientists were concerned about but as the world has witnessed a paradigm shift, it is not only DNA that they are looking into but RNA, protein, metabolite and everything else that may contribute in giving them a clear picture of the whole functionality and inside machinery of an organism. Thus, at this level it is not just -**omics** but an **omics cascade** that is into play and the integration of data received from different omics approaches is a way to structurally and functionally characterize any biological organism (Fig. 1) (Hollister et al. 2015).

The developments in bioinformatics and in silico tools have been a driving force behind omics data generation, integration and analyses. The data received from different omics approaches are very large and not easy to comprehend thus, computational and statistical algorithms are the only way to analyze those datasets and present a clear picture for us. Now, in recent times there are a lot of new omics methodologies coming to light viz. lipidomics, glycomics, interactomics, etc., and

A. Singh
Department of Biochemistry, University of Lucknow, Lucknow 226007, India

P. Srivastava
Amity Institute of Biotechnology, Amity University Uttar Pradesh, Lucknow 226028, India

V. P. Sharma (✉)
Regulatory Toxicology Group, CSIR-Indian Institute of Toxicology Research, Lucknow 226001, India
e-mail: vpsitrc1@rediff.com

M. P. Shah (ed.), *Microbial Technologies in Industrial Wastewater Treatment*,
https://doi.org/10.1007/978-981-99-2435-6_4

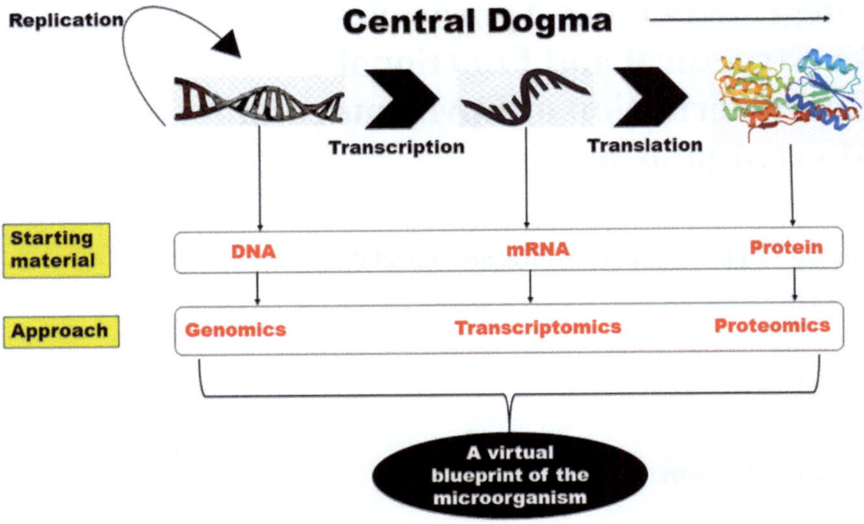

Fig. 1 A schematic illustration showing the overview of different mainstream omics approaches used in getting the structural and functional insight of a microorganism of interest

along with these there are still a few waiting to share the limelight and be accepted as a part of the mainstream scientific disciplines (Hollister et al. 2015).

The well-established omics approaches are being utilized for the specific characterization of microorganisms effectively from contaminated environments. Thus, in the chapter we have attempted to focus on the various mainstream omics approaches that are being used for structural and functional characterization of various environmental microbes. We will also very briefly be discussing how the integration of these omics dataset is the new way of efficiently and economically obtaining insights of the microbial colonies and anthropogenic threats.

2 Genomics

Genomics, as we discussed earlier, is the study of the collection of all the hereditary information of an organism. The genetic information is well stored in the Deoxyribonucleic Acid and it can be used for the diagnosis as well as prognosis of futuristic diseases in an organism. The genomic sequencing may help us to deal with better understanding of our evolution and biology with simultaneous disclosure of medical discoveries in critical pediatric as well as geriatric diseases namely aging, neurological disorders and transformations at cellular/molecular level. Although, DNA may be the hereditary molecule for most organisms but it isn't for all. For many viruses, the genome is not DNA but RNA only. The advances in sequencing methodologies and consistent fall in the cost of generating whole genome sequences have now made it possible to get a "Genome Map" of the microorganism of interest (Fleischmann et al. 1995). As per Genome Online Database (GOLD), as of January 2022, there are total 25,367 sequencing projects complete and available to the scientific community ready to use (Markowitz et al. 2010). With the availability of such high number of reference genome one specific field of genomics comes to light and that is "Comparative Genomics". With comparative genomic studies, we may make comparison of the sequence of our microorganism of interest with already available reference genomes and then we can pinpoint as to which gene is causing what kind of expression and identify regulatory regions of the genome in question, interpret unprecedented toxic impacts and also trace back their phylogenetic evolution.

Genomics and comparative genomics together might help us to understand how an organism is responding at the genetic level and provides us with the blueprint of the microorganism and we can point out exactly what genes are responsible for a particular function in an organism or natural ecosystem.

3 Transcriptomics

The subset of genes transcribed inside of a living system is known as its transcriptome and the field which involves the study and analysis of this transcriptome is transcriptomics (Schena et al. 1998). Modern techniques of genomics and comparative genomics give a vast amount of data which mainly deals with the DNA of a microorganism but what it is not able to tell is the expressed phenotype. In order to know that, one needs to turn to check the RNA or protein expression levels of a microorganism. RNA, or more specifically mRNA, is the molecule that contains the actual information stored in the genes and this mRNA determines what kind of a protein is going to be synthesized and what function it will bear. The environmental conditions are not static they keep on changing and the only way to adapt to those changes is to alter the gene expression regulation. Now transcriptomics is one such field that describes this phenomenon on a genome-wide scale (Golyshin et al. 2003; Diaz 2004).

Transcriptomics also has its limitations like any other technology. The major issue with transcriptomics is the isolation of mRNA because the total RNA extract of an organism is composed mostly of rRNA and tRNA and only a little amount of mRNA is present which makes it hard to isolate. The eukaryotic mRNA has a 3' poly-A tail which could be used to purify eukaryotic mRNA, however prokaryotic mRNA lacks the poly-A tails (Nakazato et al. 1975). Once mRNA is successfully isolated and enriched, it is then converted to cDNA as cDNA is the preferred choice of template and it is subjected to sequencing in order to identify which genes were expressed and what kind of phenotype they must result in. Using the bioinformatics tools and algorithms, the cDNA sequence reads are trimmed and quality checked and then mapped with reference genomes to assign gene function and their correct identification. There are various assemblers available since when we deal with transcriptomics data, assemblers are needed like Velvet (Zerbino and Birney 2008).

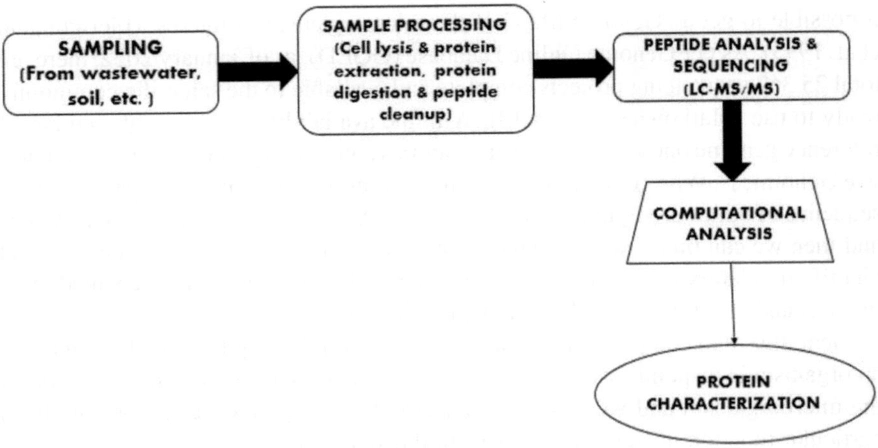

Fig. 2 A schematic proteomics workflow from sampling to protein characterization

4 Proteomics

The total amount of protein produced by any microorganism at a given point of time is known as its **proteome**. Now although DNA and RNA can give a very vivid picture of the environmental role of a particular microorganism but it is protein that actually determines the function of the microbes. The final gene product is protein (i.e., enzymes) that regulate microbial function. The proteome, unlike genome is dynamic in nature and it represents all the proteins expressed in a microbe at a particular time.

Analyzing the proteome of a microorganism has its own benefits like proteome analysis could lead us to identify different proteins and their function in relation to microbes' response to external stimuli, it can also tell which proteins are differently expressed and present a clear picture as to how the environmental microbes regulate their microbial activity and function (Fig. 2). There are studies that have shown that these high-throughput proteomics-based approaches could be used to elucidate and discover the biodegradative pathways in various microorganisms (Kim et al. 2004).

5 Metabolomics

Metabolomics is an upcoming field with applications in health and disease, agriculture, nutraceutical, nutritional sciences and exposomics. Moreover, the investigation of low atomic weight metabolites arises from the interaction between microorganisms and energizers. It explains the function of a microorganism; nonetheless, proteomics and metabolomics uncover data connected with the "final" genome product. Metabolomics significantly helps to quantify specific metabolites and proteins so as to target therapeutic interventions for understanding the reversal pattern of degenerative processes in an organism. The new technologies and better tools for monitoring and management of diseases are anticipated to pave the way for precision medicine and comprehensive metabolic signatures. It is expected that the narrow range of chemical analysis will be replaced by focused interventions, better differential diagnosis, predictive prognostic, diagnostic and surrogate markers for critical understanding of mechanism of diseases and subsequent medications. Thus, pharmaco-metabolomics and pharmaco-genomics may play a complementary role for precision medicine (Dunn 2008; Zhang et al. 2010; Beger et al. 2016).

6 Interactomics

Genome-wide mRNA profiling can't give any data about the movement, game plan, or last objective of the gene products, the proteins. Different proteomic approaches, then again, can effectively give straight responses. It is exceptionally uncommon that any protein atom goes about as a novel point of support during the physiological reaction in bioremediation interaction of any foreign substance when cell proteins and different other related cell articulations are on peak. In general, interactomics deals with interaction network of multiple proteins involved in a multicomponent aggregate system. A protein–protein interaction network tells us exactly how closely or distinctly one protein is related to another and how the expression of a particular protein affects the expression of another one. This methodology is also helpful in identifying the major protein responsible for a particular function and could be used to develop a targeted approach to manipulate the expression (Kuhner et al. 2005; Gao et al. 2004; Eyers et al. 2004; Segura et al. 2005).

7 Integration of Omics Data

Integration of omics datasets is important from the established framework. Endeavors to incorporate multiple omics technologies with each other are still generally not many, particularly in blended microbial networks, and they frequently depend on the layering of omics information onto reference pathways or the relationship of one omics informational index with another. When we integrate various omics data we can easily get information regarding various structural and functional regulation taking place in a microorganism relating to different environmental stimulus (Stenton et al. 2020).

8 Conclusion

The use of genomics, proteomics, transcriptomics, metabolomics and so on and so forth for structural and functional characterization of microbes is the new approach toward understanding environmental degradation. The omics approach creates datasets that are too large, otherwise known as "Big data" and the bioinformatic tools are one way of handling and understanding this big dataset. There are a number of reference databases that help in comparing and layering the omics dataset and help in its integration to better understand the structural complexity of the microbes' genomes, dynamism of their proteome, its expression and how it contributes to the functionality of the microbe. Thus, compared to the conventional methodologies of microbe characterization, integrated omics approach has presented a paradigm shift and is a new way forward.

Acknowledgements Authors are thankful to Knowledge Resource Centre of CSIR-Indian Institute of Toxicology Research for providing the details through available journals, books and databases. Moreover, the support of all the heads of co-partner academic institutions viz. Department of Biochemistry, University of Lucknow and Amity University Uttar Pradesh, for support in the preparation of this manuscript is acknowledged.

References

Beger RD, Dunn W, Schmidt MA, Gross SS, Kirwan JA, Cascante M et al (2016) Metabolomics enables precision medicine: "a white paper, community perspective." Metabolomics 12(9):1–15

Diaz E (2004) Bacterial degradation of aromatic pollutants: a paradigm of metabolic versatility. Int Microbiol 7:173–180

Dunn WB (2008) Current trends and future requirements for the mass spectrometric investigation of microbial, mammalian and plant metabolomes. Phys Biol 5(1):011001

Eyers L, George I, Schuler L et al (2004) Environmental genomics: exploring the unmined richness of microbes to degrade xenobiotics. Appl Microbiol Biotechnol 66:123–130

Fleischmann RD et al (1995) Whole-genome random sequencing and assembly of Haemophilus influenzae Rd. Science 269(5223):496–512

Gao H, Wang Y, Liu X et al (2004) Global transcriptome analysis of the heat shock response of Shewanellaoneidensis. J Bacteriol 186:7796–7803

Golyshin PN, Martins Dos Santos VA, Kaiser O et al (2003) Genome sequence completed of Alcanivorax borkumensis, a hydrocarbon-degrading bacterium that plays a global role in oil removal from marine systems. J Biotechnol 106:215–220

Hollister EB, Brooks JP, Gentry TJ (2015) Bioinformation and 'omic approaches for characterization of environmental microorganisms. In: Environmental microbiology. Academic Press, San Diego, pp 483–505

Kim SI, Kim JY, Yun SH, Kim JH, Leem SH, Lee C (2004) Proteome analysis of Pseudomonas sp. K82 biodegradation pathways. Proteomics 4(11):3610–3621

Kuhner S, Wohlbrand L, Fritz I et al (2005) Substrate-dependent regulation of anaerobic degradation pathways for toluene and ethylbenzene in a denitrifying bacterium, strain EbN1. J Bacteriol 187:1493–1503

Markowitz VM et al (2010) The integrated microbial genomes system: an expanding comparative analysis resource. Nucleic Acids Research 38(suppl_1):D382–D390

Nakazato H, Venkatesan S, Edmonds M (1975) Polyadenylic acid sequences in E. coli messenger RNA. Nature 256(5513):144–146

Schena M, Heller RA, Theriault TP et al (1998) Microarrays: biotechnology's discovery platform for functional genomics. Trends Biotechnol 16:301–306

Segura A, Godoy P, van Dillewijn P et al (2005) Proteomic analysis reveals the participation of energy- and stress-related proteins in the response of Pseudomonas putida DOT-T1E to toluene. J Bacteriol 187:5937–5945

Stenton SL, Kremer LS, Kopajtich R, Ludwig C, Prokisch H (2020) The diagnosis of inborn errors of metabolism by an integrative "multi-omics" approach: a perspective encompassing genomics, transcriptomics, and proteomics. J Inherit Metab Dis 43(1):25–35, Jan. https://doi.org/10.1002/jimd.12130. Epub June 25, 2019

Zerbino DR, Birney E (2008) Velvet: algorithms for de novo short read assembly using de Bruijn graphs. Genome Res 18(5):821–829

Zhang W, Li F, Nie L (2010) Integrating multiple 'omics' analysis for microbial biology: application and methodologies. Microbiology 156(2):287–301

Microbial Fuel Cell Assisted Wastewater Treatment: A Review on Current Trends

Archika Dutta, Samir Kumar Mukherjee, and Sk Tofajjen Hossain

1 Introduction

All components and organisms in nature work together in a highly harmonized manner and maintain their sustainable balanced existence in the ecosystem. Technological advancement and modernization have rapidly increased urban and industrial development that negatively impacts on such synchronization of the environmental network. As a need for socio-economic development and over-exploitation, natural resources are in increasing jeopardy. In the name of unplanned development, the water bodies are being affected vastly because of such anthropogenic activities. As a direct consequence of increasing domestic and industrial waste discharge onto the water bodies, the key parameters of such are being abruptly altered. The hazards are mainly coming up from organic, inorganic and metal toxic contaminants which are persistently being discharged into aquatic domains and thus affecting the quality of water used for drinking and agriculture purposes. Eutrophication and scarcity of fresh water have become the most crucial concern of environmental crisis. For combating such qualitative deterioration of water, environmentalists and researchers worldwide are taking the challenge to address the catastrophe to aware the general people for preventing unplanned disposal of wastewater, and are also trying their level best to develop superior management techniques for such mitigation.

Nature has its own, minimal potentially by which wastewater can be managed, but the wastewater generated from anthropogenic activities containing high load of non-biodegradable contaminants are beyond such natural attenuation process, thus warrant special attention to ameliorate the aquatic environmental condition. Due to the increasing scarcity of potable water, distinct measures should be considered for the conversion of wastewater into its reusable form. Domestic sewage wastewater contains detergent, organic waste products and fecal materials whereas, industrial

A. Dutta · S. K. Mukherjee · S. T. Hossain (✉)
Department of Microbiology, University of Kalyani, Kalyani, West Bengal 741235, India
e-mail: sktofajjen.hossain@gmail.com

M. P. Shah (ed.), *Microbial Technologies in Industrial Wastewater Treatment*,
https://doi.org/10.1007/978-981-99-2435-6_5

effluents contain substantial amounts of poisonous heavy metals and other persistent organic contaminants. Agricultural runoff is polluted with urea, nitrogenous waste, pesticides, phosphorous and organic nutrients. Different approaches are required to treat wastewater originated from different sources. Conventional wastewater management techniques include a series of site-specific physical, chemical and biological treatments. However, these established purification methods are less efficient, labor-intensive, time-consuming as well as expensive too.

Alternatively, a more efficient and sustainable implementation of microorganisms for removal of undesirable contaminated substances from the environment is termed as bioremediation. During bioremediation, contaminants may be completely or partially degraded into their less-toxic form by an enzymatic process. Microorganism-based bioremediation was first instigated by George M. Robinson during the 1960s for decontamination of oil-split media and municipal sewage treatment plant. Since then, bioremediation has become a cost-effective and eco-friendly decontamination tool for wastewater management. Bioremediation may be classified as biosorption and bioaccumulation processes. The former being a fast, revocable, inert adsorption process while, the latter involves both intra- and extracellular processes (Chojnacka 2010).

A more advanced emerging technology for wastewater treatment is the microbial fuel cell (MFC) system which can not only decontaminate the wastewater but also produce electricity. It incorporates microbial and electrochemical processes to generate electricity by oxidizing organic substances (Logan et al. 2006). MFCs can degrade and decrease contaminants under optimum reaction conditions and use less energy in contrast to other technologies.

2 Definition of Wastewater

Wastewater is defined as polluted water that has been generated on a daily basis from domestic, agricultural and industrial sources across the world and that must be decontaminated before discharge into aquatic bodies to avoid further pollution. The type of wastewater can be classified based on different anthropogenic activities and its various sources like rainwater, stormwater flooding, regular domestic activities and from various commercial origins like industrial plants (Fig. 1).

2.1 Municipal Wastewater

Municipal wastewater comprises of the daily discharge from various household activities. The quality and quantity of domestic effluents depend upon the people's standard of living. The municipal effluents constitute discharges from kitchens, bathrooms, car wash stations, drain water, etc. Physiological and domestic activities, determine the composition of domestic runoff ranging from soluble, colloidal, or

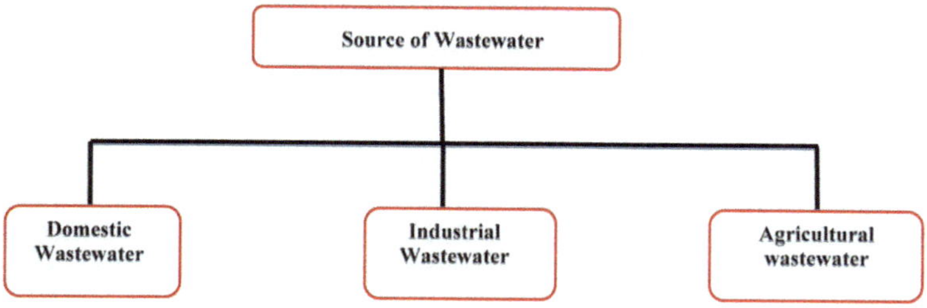

Fig. 1 Sources of wastewater in aquatic systems

suspended elements. Fecal waste and toilet runoff contain pathogenic microorganisms which can spread various kinds of contagions that may lead to mass infection. Oxygen depletion due to the discharge of biodegradable organic matter affects freshwater lives in aquatic bodies. Finally, toxic chemicals, organic materials, detergents and lubricants accumulate and enter the food chain and persist as a hazardous clash on living beings.

2.2 *Industrial Wastewater*

Industrial discharge comprises about one-third of total water pollution, which includes wastewater effluents, solid organic or inorganic waste, heavy metals and various other hazardous waste (Lokhande et al. 2011; Kansal et al. 2013; Shah 2020). The increasing quantity of effluents from the industrial sectors adds on to the aquatic systems and poses a substantial hazard to all life lives in aquatic bodies and also to outward user. The major issue of the industrial waste discharge is that the decontamination procedure is not uniform throughout the treatment plant, and thus various byproducts are generated and mixed into the water bodies.

2.3 *Agricultural Wastewater*

Agricultural waste consists of crop residues, pesticides, insecticides, fungicides, fertilizer wastes, nitrogenous wastes and nutrients, which are regularly discharged into the aquatic structure. The manure and wastes produced by livestock and various other agricultural runoffs decomposed and become hazardous and contaminant to the freshwater bodies. The soil textures, climate and socioeconomic condition of the country determine the composition and quality of the waste runoff.

3 Contaminants of Wastewater

An immense concern is related to tenacious, bioaccumulative and noxious pollutants present wastewater in the form of inorganic, organic or organometallic compounds. These wastewater pollutants show great capacity and resistance toward decomposition under the treatment of abiotic and biotic agents, which gives them a huge movability into the environment. Contaminants in wastewater can be divided into four main classes depending on the nature of the waste created.

3.1 *Radioactive Contaminants*

Chemical elements that have an unbalanced number of protons and neutrons and thus can emit ionizing radiation are termed as radioactive wastes. Many of the radioactive contaminants observe in public drinking water sources occur naturally due to the presence of some radioisotopes in rock and soil. Various radioisotopes are often generated while discharging and dismantling nuclear reactors. Some of the common radiological contaminants are radium and uranium, cesium and plutonium.

3.2 *Biological Contaminants*

Biological contaminants that account a significant portion of drinking water contamination can be classified into three major groups, which are microorganisms like bacteria, viruses, protozoa, etc. Natural organic substances, that act as a source of carbon, exacerbate the growth of microorganisms. Biological toxins are found in the treatment plant due to the presence of a particular microorganism that is lysed during the treatment procedure and releases specific toxins (Upadhyayula et al. 2009; Shah 2021). The presence of coliform bacteria in wastewater indicates fecal pollution as a result of human and animal waste.

3.3 *Organic Contaminants*

The carbon-based chemical compounds such as organic solvents, pesticides and volatile compounds even in minute quantities present in the soil, water and air, causing pernicious impacts on animal health, due to toxicity, tenacity and long-range transport capacity, are defined as organic contaminants. Most of these contaminants are chemically active and are frequently discharged along with industrial sewage. The toxicity of these organic contaminants is determined by the structure of the respective compounds. During the degeneration process of organic contaminants, the dissolved oxygen in the given water bodies may be depleted at a substantial rate, causing oxygen-deficient situations and having dreadful repercussions for the stream biota. Wastewater with organic contaminants also contains huge amounts of suspended solids which diminish the light accessible to photosynthetic organisms and consequently change the feature of the river bed, creating an inappropriate environment for many invertebrates.

3.4 Inorganic Contaminants

Inorganic contaminants are released into the environment due to various anthropogenic activities like mining, transportation and other metropolitan behavior. Alkali metals react with the different physical and chemical components of both soil and water and thereby interfering with the rate of nutrient uptake and its cycling processes. Most of the heavy metals and metalloids also strongly interact with water and soil constituents, but the rate of interaction varies depending on the element and its oxidation state. Major heavy metal pollutants from industrial wastes include As, Pb, Cr, Cd, Zn, Ni, Cu, Hg which are discharged into water as potential hazards.

Arsenic (As): It is a hazardous metalloid found naturally in the environment and also often liberate from different anthropogenic activities such as mining, fossil fuels combustion, fertilizers, etc. Several countries worldwide are threatened and affected by As-contamination and almost 150 million people from 70 countries worldwide drink As-contaminated water, beyond the permissible limit of arsenic concentration, as stated by the World Health Organization (WHO). Extensive use of this As-contaminated ground water for drinking and irrigation purposes may lead to various diseases in human (Basu et al. 2021). However, some indigenous microorganisms have adapted themselves to As-induced toxicity and contribute to the geochemical cycle of arsenic (Mallick et al. 2014).

Cadmium (Cd): Cadmium is a soft malleable metal naturally found in soils originating from dust storm, volcanic eruptions, weathering, erosion of rocks and wildfire. Anthropogenic sources of cadmium include coal mining and mineral fertilizers, Ni-Cd batteries, coatings platings, fossil fuel combustion and waste incineration. A survey by the UNEP (United Nations Environment Programme) estimated that the Cd emission of anthropogenic sources is much higher than its natural emission. According to the WHO, 0.003 mg/L of Cd is the allowed limit in drinking water and ingestion through food beyond which cause severe pulmonary, hepatic damage and gastrointestinal problem (Hossain et al. 2012). It has been reported that long-lasting exposure to Cd results in severe impairment to cells and its DNA mainly through the reactive oxygen species production and it may also damage cell membranes and proteins (Hossain and Mukherjee 2012, 2013; Hossain et al. 2014).

Lead (Pb): Lead is a toxic heavy metal abundantly distributed in the environment. Various anthropogenic activities and its non-biodegradable character result in huge accumulation within the environment. Fossil fuels burning, mining, volcanic eruption and different manufacturing procedure such as lead acid batteries and paints increase Pb deposited in the environment. The rate of absorption of ingested Pb by the body is very high and its exposure exerts influence on almost every organ in the body, but predominantly the nervous system in both children and adults (Wani et al. 2015)

Mercury (Hg): Mercury exists in nature as different forms of inorganic and organic compounds. Volcanic eruption, forest fire, weathering of rock, small-scale gold mining, coal burning and some human anthropogenic activities are responsible for the deposition of high-level Hg in the environment. Human toxicity of Hg varies according to the dose and the rate of exposure. The target organ for inhaled or ingested Hg is primarily the nervous system and brain (Rice et al. 2014).

Chromium (Cr): Chromium toxicity and carcinogenicity of plants, animals and humans were already well established. Natural processes like the earth's crust by tectonic and hydrothermal events release low concentration of Cr (VI), however, various anthropogenic activities like metallurgical industries, textile dyes, leather tanning and wood preservation results in discharge of a huge amount of Cr in the environment. Toxicity of Cr varies according to its valence state, where Cr(VI) occurs more toxic and mobile than Cr(III) and various metabolic alterations in human as a result of Cr exposure (Baruthio 1992).

4 Impacts of Wastewater Pollutants on Environmental Health and Human Life

Wastewater contaminants from industries and other sources, especially in developing countries are discharged into the environment and water bodies are mainly affected. The quality and quantity of wastewater effluents are mainly responsible for the pollution of water ecosystem (Fig. 2).

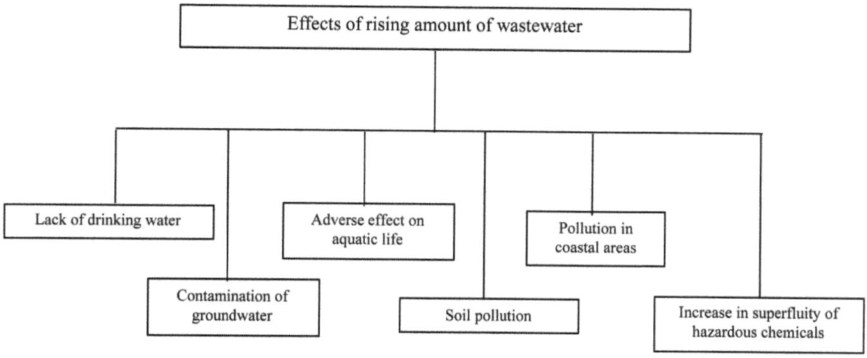

Fig. 2 Effects of wastewater on the environment

4.1 *Impact on Agriculture*

Wastewater has a prime effect on the fertility and productivity of the irrigation soil. The agricultural crop soil has a balanced percentage of various soil ingredients, organic matter, minerals and water content, which together, play a crucial role in providing physical support to the crops. Irrigation done on the contaminated soil result in poor quality and quantity of yield. Additionally, wastewater-contaminated soil exhibit alters in pH that extensively affects the growth of plant growth promoting rhizobacteria as well as plant growth. Gradually, the contaminated soil becomes a wasteland and inhabitable for any crop yield irreversibly.

4.2 *Impact on Human and Animal Health*

Soil pollution, crops grown in polluted land and water pollution directly affect people residing near contaminated wastewater, resulting in serious health and economic problems. Irritation of eyes, skin disease, fever, vomiting sensation, diarrhoea, etc., are such health complications in humans and animals that appear by using wastewater directly or indirectly. In wastewater, amount of dissolved impurities and toxic contaminants such as carbonates, bicarbonates, sulphates and other metal salts along with other colloidal contaminants such as organic waste, silica, clay, etc., determines the pollution level of the particular water. Most of these contaminants present in wastewater have diverse mutagenic, genotoxic, carcinogenic and cytotoxic effects on cells. The wastewater flows into the freshwater systems such as streams or rivers and seeps through the ground into the groundwater as well, thus affecting aquatic organisms such as fish and other aquatic habitats which directly exposed to the toxicants of wastewater. The toxicants ingested aquatic organisms in turn are consumed by higher mammals and reptiles, subsequently, the toxic materials enter into the food chain directly or indirectly, resulting in adversely affecting wildlife and human beings. The concentration of the toxicants intensifies and accumulates as a result of biomagnification within the different tissues of tolerant organisms, which becomes detrimental to the organism (Arnot et al. 2010).

4.3 *Impact on the Ecosystem*

Soil and water are the two most indispensable elements in maintaining the balance of the ecosystem. Contamination of soil and water results in the change of texture, pH, nutritional quality and other essential parameters associated with its quality. Polluted environment comprises less number of beneficial microorganisms in comparison to pathogenic microbes that affect and impair the natural features of the environment. This change in circumstance fluctuates the microbial biomass, which directly contributes to the growth of plants. Thus, the bioavailability of various essential nutrients changes and restricts the growth of the natural flora and fauna. Therefore, pollution of soil and water leads to changes in environmental health that eventually interfere with the balance of the ecosystem.

5 Physiochemical Approaches for Wastewater Management

The rapid escalation of industrialization and urbanization results in the discharge of wastewater at an alarming rate. The direct consequence of which is the decrease in freshwater systems which has a vast impact on the amount of water used for drinking, livestock and irrigation. Nature can neutralize the noxious wastes to a minute degree and, therefore, large quantity of wastewater has to be purified artificially.

Since the last few decades, researchers are investigating divergent wastewater treatment procedures for the removal of contaminants to escalate the availability of clean fresh water. Various physicochemical processes are conventional dependable techniques for wastewater management. However, huge energy consumption and high cost of system establishment are the major disadvantages of such approaches.

5.1 *Ozonation*

Ozone is a highly reactive, unstable allotrope of oxygen with a distinctly pungent smell. It has high oxidizing potential which is exploited for decontamination purposes for the production of potable water for reuse (Fig. 3). Ozonation is a chemical treatment process comprised of the infusion of ozone in water for the clearance of micro-contaminants from wastewater. Ozonation reaction can eliminate microcontaminants by the direct reaction with organic contaminants or it can also react with various substances in alkaline water to form hydroxyl radicals (Margot et al. 2013).

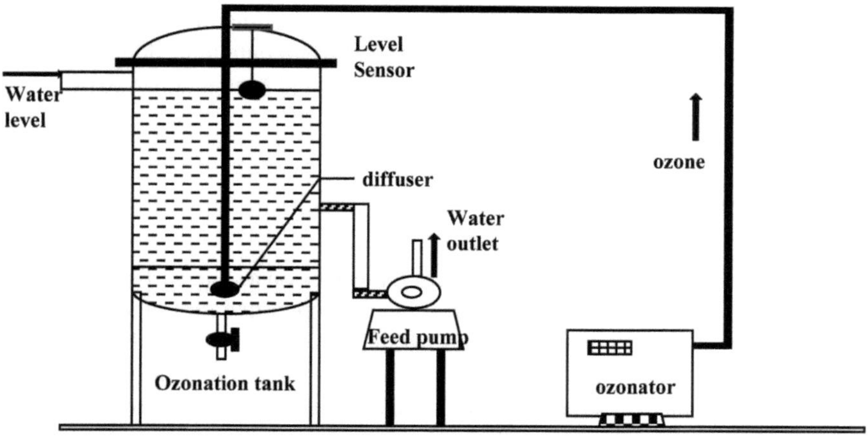

Fig. 3 Ozonation of wastewater

5.2 *Advanced Oxidation Processes*

The advanced oxidation process (AOP) is an innovative procedure, which utilizes ozone, hydrogen peroxide and UV radiation, for decontamination of wastewater. AOPs are very effective in removal of recalcitrant impurities in wastewater by generating potent oxidizing agents such as hydroxyl radicals. The process involves the production of hydroxyl radicals that aids in the conversion of chromogenic groups to CO_2, H_2O and inorganic salts. The AOPs efficiently eliminates the microcontaminants by oxidative reaction targeting selective groups (Al-Kdasi et al. 2004). However, these potential technologies need to be explored at a commercial scale and optimized for the establishment of large-scale management.

5.3 *Fenton Process*

The Fenton process comprises the reactions between the iron ions and hydrogen peroxide (H_2O_2) to generate reactive oxygen species, that can easily oxidize the organic or inorganic compounds present in wastewater as a contaminant. Oxidation of ferrous iron to ferric iron along with the transformation of H_2O_2 to hydroxyl radicals and hydroxide ions are the leader reactions. Most of the hazardous compounds are precipitated along with ferric ions, which could be then efficiently removed from the wastewater. The Fenton process is often used as a pretreatment operation for wastewater purification before biological treatment. High efficiency and operational simplicity for the oxidation of organic and inorganic pollutants are some of the advantages of using the technique. However, the strict acidic pH range of wastewater, high H_2O_2 consumption and accumulation of iron-containing sludge effect the productivity of the Fenton process (Babuponnusami and Muthukumar 2014).

5.4 Adsorption

Adsorption is the technique of adherence of molecules on a solid surface. This technique is often utilized for wastewater treatment, where a fine film of solid material is applied to trap gas or liquid molecules passing through it. Several different kinds of adsorbents are widely available in the market such as activated carbon, silica gel, alumina and polymeric compounds. Activated carbon produced by pyrolysis process is the most cost-effective among other types of adsorbents for wastewater treatment and is accessible in divergent pore dimensions and textures.

5.5 Membrane Filtration

Membrane filtration is a technique used to separate dissolved molecules from solution. Effluents discharged from industries, agricultural fields and municipal sectors are often decontaminated using membrane filtration techniques and the treated water is again collected and reused. Minimum energy exhaustion, simple mechanism and environmental affability are the main advantages of membrane filtration technology. Mainly four different types of membrane filtration processes are very common, such as microfiltration (MF), ultrafiltration (UF), reverse osmosis (RO) and nanofiltration (NF). MF is the conventional filtration operation of wastewater treatment and has a restricted application. Pore sizes of MF membranes have been ranging from 0.1 to 10 μm, and separation operation is effectual at low pressure. Using a dry bath MF can remove different particles and colloidal dyes from suspension. However, soluble organic contaminants may not be removed using MF membranes (Zazouli and Kalankesh 2017). UF technique is applied to remove macromolecules and colloid particles from wastewater, where wastewater is forced through the membrane. UF membranes are porous in character, but have an asymmetric construction with a much tiny pore size and smaller surface porosity that constructs higher hydrodynamic resistance. Usually, UF membrane thickness ranges from 1.0 to 3.0 μm. RO is a technique in which the solvents are pumped through a membrane with excessive solute concentration to a chamber with small solute concentration working under definite pressures. RO technique is applied to remove bacteria, viruses and organic contaminants as well. NF membrane techniques applied for wastewater decontamination using a pressure-driven membrane separation system. NF technique is applied to remove various microcontaminants from effluent and the NF filter can be reused after proper sterilization (Zazouli and Kalankesh 2017).

5.6 Coagulation-Flocculation

The coagulation process weakens the charge of colloidal particles by neutralizing the forces that keep them apart, although, in flocculation the added coagulant forms connections between the particles and bind together to form an aggregate molecule. Coagulants such as alum aluminum sulphate, ferric chloride, ferric sulphate and calcium salts are used in wastewater treatment process depending on the quantity of effluent to be decontaminated. The efficiency of this method depends on other parameters such as effluent pH, coagulant amount and time of treatment.

5.7 Electroflotation

Electroflotation is a combinatorial method comprising of the advantages of electrocoagulation, and flotation for the treatment of wastewater. Electroflotation is a comparatively modern approach, which can produce discrete tiny bubbles that increase the contact area between flocks and bubble, which causes faster clarification than conventional flotation method. Electroflotation can easily be employed with reduced sludge generation, which requires lowest maintenance cost (El-Hosiny et al. 2017).

5.8 Electrocoagulation

Electrocoagulation is an electrochemical procedure used to treat wastewater utilizing electricity rather than dangerous chemicals. It is used to treat soluble or colloidal pollutants separated from various types of industries like food, textile, tanneries, etc. In comparison to conventional coagulation technique, electrocoagulation can efficiently separate very small colloidal particles and has high potential to destabilize and coagulate charged particles. Electrocoagulation technique requires shorter operating time and produces lesser amount of secondary wastes. In this method electrode consisting of metals are converted into polymetallic hydroxides at higher pH due to electrolysis and initiate coagulation. The organic pollutants and colloidal particles are removed by these techniques resulting in poly metal hydroxides which form larger flocks due to aggregation (Butler et al. 2011).

6 Biological Approaches for Wastewater Management

During the last few decades, biological materials are emerging as techniques of decontaminating wastewater exploiting different types of microorganisms like fungi, bacteria and algae, that can effectively absorb and degrade toxic contaminants from wastewater. The efficiency of biological treatment methods in waste removal varies greatly depending on the existing contaminant in the effluents such as organic matter and heavy metals, that is interconnected with the particular nutrient consumption preference of the microorganisms. In the biodegradation process, both aerobic and anaerobic respiration activities are involved during waste removal process. Biological decontamination of waste material from industrial effluents may offer various advantages, such as the conversion of contaminants to carbon dioxide and water and subsequently the formation of less harmful sludge compared to other physicochemical techniques used for wastewater treatment. Biochemical oxygen demand (BOD) is a variable that is used to estimate the relative quantity of oxygen needed by the microorganisms to decompose organic materials present in wastewater and is also used to estimate the waste amount in the treatment plants. Therefore, wastewater treatment by biodegradation method ensures that organic matter consumption by the microbial cells, which in turn reduces the BOD in the aquatic system. Biological treatment structure typically consists of a biological reactor attached to a settling tank to eliminate the generated sludge. Often different pre-treatment and post-treatment operations of the effluents are usually enriched in the wastewater management system using biological methods (Rai et al. 2005; Dutta et al. 2021).

6.1 Aerobic Treatment

The presence and absence of oxygen in biological treatment operations are usually termed as aerobic and anaerobic processes, respectively. In aerobic conditions, microorganisms that are degrading organic impurities of wastewater in the presence of oxygen are termed as aerobes. Oxygen is assimilated into the wastewater by mechanical aeration and the aerobic microorganisms convert organic pollutants into carbon dioxide, water and biomass by using molecular oxygen. Several different aerobic treatment technologies are used for wastewater treatment. In the conventional activated sludge method, the organic materials are degraded by aerobic microorganisms in the reaction tank and the sludge along with biological flocs are then eliminated from the treated water in a separate sedimentation tank. In the moving bed biofilm reactor method, the exploited microorganisms are grown as a biofilm on suspended plastic carriers in the aeration tank. In Membrane bioreactor method, the activated sludge, produced by the micrograms, is passed through the combined membrane filtration system (Kuyukina et al. 2020).

6.2 Anaerobic Digestion

Anaerobic biological treatment is a technique that uses anaerobic microorganisms to degrade organic pollutants in anaerobic environment. It was initially utilized for sludge treatment process, but in the last decades, it is also gradually been employed in the wastewater treatment method. It is the most appropriate method in correspondence to using aerobic treatment method for the transformation and treatment of high carbon content pollutants. Generally, anaerobic wastewater treatment procedures comprise of two phases such as an acidification step, where the anaerobic microorganisms are degraded complex organic materials into low molecular weight evaporative organic acids, and followed by a methane-making step or methanogenesis, where anaerobes produced acetate or formate along with hydrogen and carbon dioxide by the degradation of various organic acids. Thus, by anaerobic digestion microorganisms produce biogas containing approximately 60% methane and 40% carbon dioxide and the activated sludge produced by this process can be further used for green manure because of its high nutrient content. Thus, the most typical class of anaerobic digesters are solid-state anaerobic digester (SSAD), continuous stirred-tank reactor (CSTR) and upflow anaerobic sludge blanket digestion (UASB) (Adekunle and Okolie 2015).

7 Microbial Fuel Cell

The present wastewater treatment technologies face major challenges because of its huge energy exhaustion and high operational costs. Abortive treatment of the gases and soluble molecules like ammonia and phosphates are additional failures of current traditional wastewater treatment procedures. However, it has been recently established that invaluable materials, such as hydrogen, methane, hydrogen peroxide, etc. can be recovered from wastewater using Microbial Fuel Cells (MFCs). Furthermore, MFCs have also become an auspicious resolution for bio-electricity production because of its potency, recyclability and production of less toxic end products.

The main advantage of using MFCs in wastewater treatment process is the displacement and recovery of the contaminants, that contribute to high Chemical Oxygen Demand, such as heavy metals and ammonia, produced by biological degradation of various organic matter, which are in turn used to generate electricity (Trapero et al. 2017). MFCs can comfortably oxidize metal carbonates to carbon dioxide and eventually, through some biochemical reduction it can act as an electron transporter (Rahimnejad et al. 2012). In MFCs, breakdown of the substrates in the anode generates electrons and protons, which are then transported to the cathode via a resistor, where it reacts with oxygen to produce electricity (Sharma and Li 2010). Modern studies on MFCs are focused on the model to scale up from laboratory to big scale of its implementation, feasibility and to overcome operational impediment. Thus, in terms of large-scale application of MFCs, improvement of cathode and anode materials is required for its substantial efficiency. The electrochemical reactions, easy proton mass transfer, good circuit resistance and acceptors are required in MFCs to accomplish cost-effective and energy-saving production processes.

Microbial Fuel Cell is an arrangement that converts natural organic waste materials into electricity utility microorganisms as the biocatalyst (Fig. 4). The upgradation of the biological procedure that employs microbes to generate power is a unique

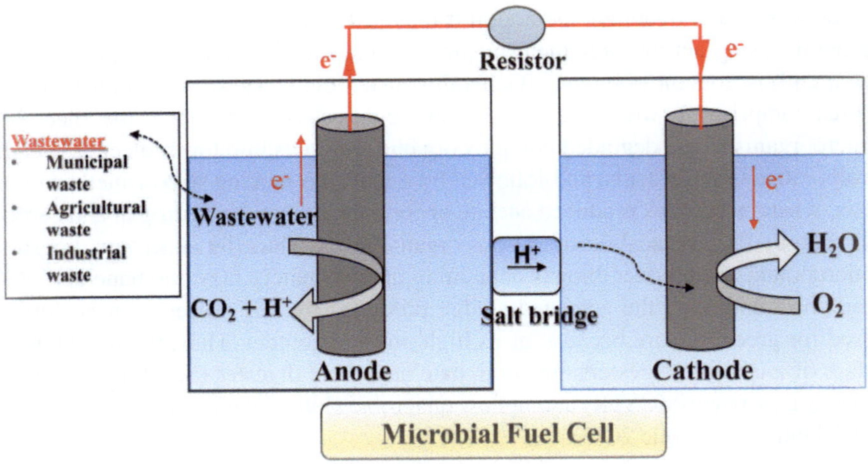

Fig. 4 Microbial fuel cell

approach for bioenergy production and as the microorganisms are self-replicating, causing diverse oxidative environmental conditions for wastewater contaminants. In general, MFCs comprise of two terminals present within single or double chambers that work as a unified reactor and the anode is positioned within the power device or anode chamber. Within the power device, the electrons are discharged from fuel materials due to oxidation and are transferred onto the anode. The free electron is then passed into the cathode room by an electrical connotation and the oxidant present within the cathode chamber picks up the electrons. The positively charged particles formed in the anode chamber transfer into the cathode chamber through a permeable film present in between the two chambers, that completes the electrical circuit. The potential to change wastewater pollutants into electricity is an exciting phenomenon, although basic knowledge of microbiology and challenging refinement are required for innovation. With consistent improvements in microbial energy constituents, it might be possible to change the microbial growth rate, lower generation time and reduce operational expenses.

7.1 MFC Design

Configuration of MFCs is a significant feature for the generation of power in order to obtain better efficiency and reduced running costs in industrial applications. Electrons produced by microorganisms, in the anode chamber are then transported to the cathode chamber, where O_2 is reduced. Therefore, the typical MFC layout can be altered into different blueprints, such as double-chambered MFC, single-chambered MFC or multiple chamber MFC.

7.1.1 Double Chamber MFC

For wastewater treatment the most suitable design is the dual chamber MFCs, however, it has not been applied in situ at a large scale. Dual chamber MFCs is generally working in batch mode and comprises of two electrodes separated by a membrane through which electrons and protons are transported between two chambers. Permeable ceramics or salt forms the electrode separating membrane which allows to transport electrons and protons between the chambers but restricts the contaminants substrates and microorganisms. The electrodes are made up of conductive, non-corrosive materials because it always remains immersed in the water. The anode is immersed in wastewater and the cathode is immersed in electrolyte solutions to aid in electron transportation. However, large-scale application of double-chambered MFC is limited because of high cost of aeration. Double-chambered MFC can be further classified according to their shape and design.

Cylindrical MFC: Cylindrical MFC is generally utilized for wastewater decontamination rather than electricity generation. It uses both series and continual modes to increase electricity production. It has a high loading rate and can also significantly reduce COD. The cylindrical MFCs are comparatively easy to handle since it can be operated in continuous mode. The major advantage of this up-flow MFC is that suspended solids can be easily removed, thus clogging can be avoided (He et al. 2005).

Rectangular MFC: Rectangular MFC generally has a dual-chambered configuration where the anode and cathode sides are designed as rectangular vessels. The two electrodes are separated by a membrane and can be operated in continuous flow. This type of MFC is most commonly used for wastewater treatment because of its easy construction and is generally made up of glass or polymethylmethacrylate which is malleable in nature. These MFCs require constant agitation for the microbes that are in contact with the substrate. High voltage and power generation are the main advantages of using these rectangular MFC, though it has very low Coulombic efficiency and loading rate (Ullah and Zeshan 2020).

Miniature MFC: Miniature-shaped MFC has the ability to produce power from milliliter to microliter scale so that it can increase power generation. In miniature MFC the electrodes take up small volume but high surface area and the fuel liquid fill smaller volume which results in greater energy production. Such configuration can be applied in long run, however, it has a low loading rate and is not satisfactory for wastewater treatment (Fan et al. 2021).

Flat Plate MFC: Flat Plate shaped MFC consists of a single electrode containing both cathode and anode. It has a compact configuration where the cathode is placed into the membrane electrode assembly. The major difference between the flat plate MFC and other MFCs is that there is a minimum separation between the anode and cathode, which reduces the internal resistance and thus enhance electron transfer. The major disadvantage of using flat plate is that the membrane is permeable to O_2 and gets easily distorted with regular use (Song et al. 2016).

Membrane MFC (MBR-MFC): MBR-MFC is a combinatorial technique that uses both active biological treatment and physical filtration methods. The principal objective of the MBR-MFC is to generate high quality reusable water from wastewater by reducing turbidity of effluents along with bacterial contamination. Low sludge-making and high contaminant removal efficiency are the basic advantages of using this MFC. However, membrane fouling, high membrane cost and low volume capacity are some of the disadvantages related to it (Wang et al. 2012).

7.1.2 Single Chamber MFCs

Single chamber MFCs have a simple design because the cathode is not immersed in any solution and is rather exposed to the air and thus, such a configuration decreases the overall cost of assembly. Air acts as the electron acceptor, while electrons are transferred to the highly permeable cathode. Single MFCs have been used to treat diverse constituents of wastewater in terms of COD and organic matter and thus, simultaneously generating electricity. However, for high power density single MFCs are not achievable because the anode and cathode are not padlocked, and therefore, smaller working volume can only be used. Although, large-scale implementation requires huge chamber volume to achieve high power generation for the wastewater treatment (Santoro et al. 2013). The use of diverse types of cathode material can improve power generation. Most commonly, anodes of a single chamber MFC are constructed with graphite fibers, that can efficiently treat wastewater. The distance between the anode and cathode of the MFC has been decreased to minimize the internal resistance, which in turn enhances the efficiency.

7.1.3 Multiple Chambers MFCs

MFCs having multiple chambers for pollutant removal from wastewater and energy production have also been reported. Multiple sections can be linked in series or parallel arrays, which can increase power generation. The overall power generation depends on the efficiency of each microcell in the arrangement. Several designs for multiple chambers MFCs are possible according to its application, nevertheless, most of the configurations are efficient in treating wastewater (Jiang et al. 2011).

7.2 Mode of Operation

Depending on the effluent, electrode material and microbial cell type MFCs can operate in either batch or continuous operation mode.

7.2.1 Batch Operation

In batch operation microorganisms are accumulated within the vessel where they produce soluble redox mediators. The major benefit of batch culture mode over continuous operation is that contaminants, like N_2 or COD materials are removed with higher efficiency because of the extended period of retention times in each cycle. When the anode chamber of MFC is employed in batch mode, only then it is possible to remove microorganisms continuously in succession to have a steady growth and consequently are able to manage more quantity of wastewater. Moreover, different wastewater substrate combinations, like olive mill wastewater mixed with domestic wastewater can achieve greater COD removal efficiency of up to 60% compared with that swine wastewater, is only 22%. (Pepé Sciarria et al. 2013).

7.2.2 Continuous Operation

The microbial biofilm arrangement is enhanced in continuous culture operation, which is used for the electrode transportation using moving shuttling materials, thereby increasing the power generation. However, the rise in flow rate lowers the MFCs efficiency, like removal of COD and Coulombic efficiency, because of the contaminant degradation by the microorganisms takes place for a shorter time period. A decline in COD removal from 79% to 32% was reported when loading rate was altered and subsequently, Coulombic efficiency was decreased from 8% to 0.8% (Jayashree et al. 2015). Nowadays, continuous feed supply technology is also adopted by Up flow MFCs to increase the efficiency of wastewater treatment.

8 Oxidation-Reduction Reactions in MFCs

The microorganism chosen for the MFCs depends on its competency to transfer electrons from substrates to the cathode via anode. Usually, wastewater contains different chemical pollutants with mixed cultures of microorganisms, that are being widely studied for utilization in the MFC and wastewater treatment. The metabolic reactions of particular microorganisms in aerobic or anaerobic conditions are playing an important role in electron transportation. Aerobic microorganisms use O_2 as the sole source of electron acceptor whereas, different electron acceptors like nitrate, sulphate, etc., are used by anaerobic bacteria which grow in the absence of O_2. Microbes can directly transfer electrons to anode surface in the absence of electron acceptor. However, some surfaces cannot transfer electrons and thus use Mn^{4+}, Fe^{3+}, etc., as a mediator. Usually, these mediators are integrated in the anode chamber to enable electron transportation and electrons generated in the anodic chamber can be easily transported to the anode by electron mediators.

Saccharomyces cerevisiae and *Escherichia coli* are the most commonly used microorganisms in MFCs, have a non-conductive outer layer comprising of peptidoglycans and lipopolysaccharides that cause obstruction in the direct transport of electrons to the anode. The electron mediators must penetrate through the microbial outer membrane and react with the internal reductive species in order to become reduced during the microbial metabolism. Methylene blue, riboflavin, neutral red, anthraquinone-2-sulphonate, humic acid, etc., are the typical electron mediators used in various MFCs (Du et al. 2007). The oxidative potential of the reductive metabolite and the electron mediator play a significant role in the efficiency of MFCs and the mediators should not interfere with other microbial metabolites. The reduced mediator then diffuses out of the microbial cell and transported to the anode, where the mediator might be oxidized. The oxidation process of the reduced mediator on electrode surface must be a rapid and mediator in oxidized or reduced form, must not be adsorbed on neither microbial cell surface nor on electrode surface rather necessitated to be chemically inert in the electrolyte liquid. Electrogenesis bacteria have the potential to transfer electrons from substrate material and carry it to the anode employing microbial internal mediators. In the iron-reducing bacterial outer membrane contains an electrochemically active cytochromes and hence, these bacteria can be used in MFCs without any mediator (Kim et al. 2002). Some bacteria like *Shewanella*, photosynthetic *cyanobacteria, Geobacter* and thermophilic fermentative bacteria can able to produce highly conductive nanowire-like biofilm around anode surface, which act as a mediator in MFCs (Gorby et al. 2006). Therefore, it was established that in MFCs, bacterial natural mediating component for electrogenesis is more effective than synthetic redox mediators integrate with the microbes as shown in Fig. 5 (Ieropoulos et al. 2005).

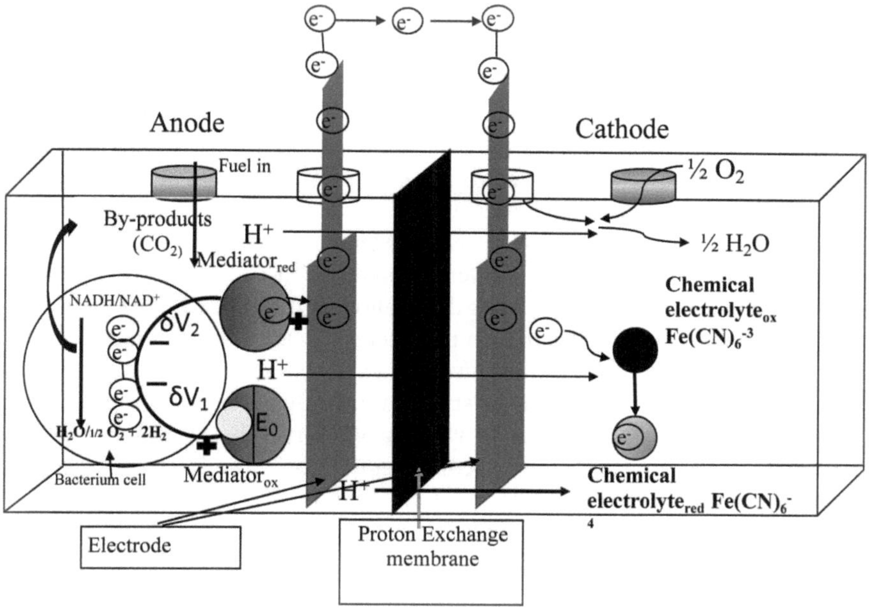

Fig. 5 Redox reactions occurring in microbial fuel cell

9 Advantages and Disadvantages of MFCs

Microbial fuel cells are apparently a new bioelectrochemical technique that focuses to generate electricity by utilizing the electrons gain from various biochemical processes catalyzed by the microbial enzymatic system. The power generated by MFCs is presumed to supply adequate energy to partly need the energy demand in wastewater decontamination. However, the main drawback of MFCs methods is the enormous principal installation charge. The electrodes used as cathode or anodes and the catalysts are also expensive, which in turn increases the continuation price as well and, thereby, making the technique less practicable for industrial wastewater decontamination. The electricity generated is not adequate in comparison with the energy input, however, the energy output of MFCs can decrease the overall running cost and therefore, massive amounts of sludge are not produced. Other disadvantages of MCF techniques are the scale-up, which is attributed to allowing systematic electron movement in electrodes. The electron transferring to cathode is inhibited by microbial biofilms, which are utilized for electron movement in the anode. Although, bacteria like *Geobacter* and *Shewanella* which have a higher transport rate can settle the issue. A mixed group of microorganisms isolated from wastewater can also enhance electron transfer efficiency because they can easily oxidize the organic compounds due to fast electrochemical reactions. The distance between the electrodes can enlarge internal resistance and affect power production, thereby resulting in low MFC efficiency. Hence, if the distance between the electrodes is shortened, the difficulty can be overcome without further investment in MFC design adjustment.

Catalysts like platinum used for electrochemical reactions can further increase the cost, hence to compensate for the high valuation, carbon-based materials made biocathodes can be used. Therefore, to enhance power production in a cost-effective manner carbon-based electrode materials, having great affinity and resistance could be an exciting field of further MFCs research. Membrane separator contamination in MFCs negatively affects the cation movement and as a consequence power generation is affected. This can be avoided by using divergent catalysts like carbon nanotubes like materials. The fouling precipitation on the electrode separator membrane of MBR-MFC is a prime trouble and these negatively charged pollutants can be avoided by the use of electric fields. Wastewater consists of several different final electron acceptors like nitrate, soluble iron and sulfates, which may obstruct the electron transfer rate to the cathode and thus power production can decrease. The different techniques in interrogating with MFCs can enhance COD removal and escalate the creation of high value materials and energy yield. MFCs are also advantageous due to its low CO_2 emission, hence it does not have much negative impact on the environment. Therefore, MFCs are an impeding process for the production of sustainable power energy and wastewater decontamination.

10 Wastewater Detoxification by MFCs

Microbial fuel cells can be used to treat wastewater discharged from all sources, and has proven to minimize the quantity of harmful substances present in wastewater. Pre-treatment of wastewater to separate the harmful substances and non-biodegradable materials is a necessary step for wastewater detoxification. Different arrangements of MFCs having wastewater as a substrate are developing as an alternative renewable energy source with high operational sustainability. MFCs operating under various environmental conditions affect the removal of COD and power generation capacity. The different parameters include effluent type, external resistance, pH, temperature, time and aeration. The electricity production and COD removal efficacy of MFCs can be improved by utilizing the multiple devices in a series or by integrating other techniques along with traditional MFCs.

During the last decades, different MFCs combined hybrid machinery with diverse efforts have been examined. When photocatalytic oxidation is integrated with MFCs, it shows higher efficiency. When combining MFCs with photocatalytic oxidation, the propensity of photo-induced electron and holes rearrangement was restricted to some extent, that could increase the photolytic effectiveness, and correspondingly, the energy created by MFCs could grant the breakdown of the materials (Yuan et al. 2010). A study of distillery wastewater decontamination utilizing fungal pretreatment method pursued by an MFC module has attained majestic removal productivity. In addition, a recent study on distillery wastewater treatment using fungal decontamination as a pretreatment step followed by MFCs has achieved an impressive removal efficiency in both COD removal and energy production (Ghosh Ray and Ghangrekar 2015). With the composite of individual excellence and also the supplement to independent structure, it will be better to choose hybrid methods by preference comparing than respective one to detoxify wastewater under the proposition of sustainability.

In hybrid technology, the mutual benefit of independent methods can increase the convenience of applicability of the methods. Though still a comparatively new method, the capability of MFCs to extract electrical energy from wastewater treatment is encouraging and has need for ample research into modifying the technology. The expansion of energy outputs concomitantly with high grade effluent is the primary aim in these composite methods and the efficiency is purely controlled by reactor volume. Therefore, substantial research is needed for intrinsic improvement of electricity production as well as water purification. Specifically, some important steps need to improve such as accelerating the movement of electrons liberated from microorganisms to the anode which can remarkably lessen activation losses within the arrangement. New procedures like the investigation of low cost anode materials or improvements of anode materials can be taken as an important goal. To acquire desirable performance improvement of proton transfer from anode to cathode as an essential benchmark. Accordingly, the evolution of effectual and cost-effective ion exchange separators or membranes should also be considered. Uplifting the oxygen reduction rate on the cathode by the process of surface alterations can be taken as a critical measure. The potential energy in average-size wastewater is three times more powerful than the energy ingestion in orthodox wastewater treatment methods. That implies that a fundamental self-sustained or even a net energy extraction from wastewater system is the prospective save or recover power. Large scale use of these hybrid arrangements is also a barrier and remains to be resolved because most of the reported documents have been performed at lab scale. Though it is clearly decisive to recognize the intricate system in detail for scaling up possible active applications. The firmness and extended accomplishment of hybrid systems are another critical concern. As previously stated that MFCs, it is powerful in terms of the variety of materials and its broad span of concentration, which built MFC hybrid structures competently functioning under a variable habitat. With continued collaboration through different interdisciplinary specialists such as microbiologists, engineers, material scientist, electrochemists, etc., MFCs-based wastewater treatment technology provides a conceivably sustainable resolution of the wastewater problem with the production of sustainable power energy.

11 Future Perspective of MFCs in Wastewater Management

MFC is a technology that comprises of wastewater decontamination and power generation by eco-friendly processes. In MFC technology, the power generated is frequently recirculated use in the method and thus reduces the expenses related to energy exhaustion and skirt the utilization of fossil fuels for the procedure. Microorganisms associated with the reaction in anode electrode and in contaminant degradation, interplay a pivotal function in the electron transport. Satisfactory results in wastewater treatment are often accomplished with mixed microflora since it can proficiently oxidize the complex organic materials in wastewater compared to that of single microorganism. Nevertheless, upcoming research in microbial genetic engineering to escalate the electrogenic effectiveness, would intensify the electron movement, and thus the reduction of organic contaminants is improved and a greater MFC functioning is attained. Expeditiously developing technologies have shown that, by combining MFCs into existing methods, wastewater treatment effectiveness can do better with the subsequent production of high amount of electricity in an eco-friendly way. Although, it features a highly motivated research collaboration through the different interdisciplinary specialists such as microbiologists, engineers, material scientists, electrochemists, etc., are needed for intrinsic improvement of electricity production as well as water purification.

References

Adekunle K, Okolie J (2015) A review of biochemical process of anaerobic digestion. Adv Biosci Biotechnol 6(3):205–212

Al-Kdasi A, Idris A, Saed K, Guan CT (2004) Treatment of textile wastewater by advanced oxidation processes—a review. Glob Nest J 6(3):222–230

Arnot JA, Mackay D, Parkerton TF, Zaleski RT, Warren CS (2010) Multimedia modeling of human exposure to chemical substances: the roles of food web biomagnification and biotransformation. Environ Toxicol Chem 29(1):45–55

Babuponnusami A, Muthukumar K (2014) A review on Fenton and improvements to the Fenton process for wastewater treatment. J Environ Chem Eng 2(1):557–572

Baruthio F (1992) Toxic effects of chromium and its compounds. Biol Trace Elem Res 32:145–153

Basu S, Dutta A, Mukherjee SK, Hossain ST (2021) Exploration of green technology for arsenic removal from groundwater by oxidation and adsorption using arsenic-oxidizing bacteria and metal nanoparticles. In: Shah M, Rodriguez-Couto S, Kumar V (eds) New trends in removal of heavy metals from industrial wastewater. Elsevier, Amsterdam, pp 177–211. ISBN 9780128229651

Butler E, Hung Y-T, Yeh RY-L, Al Ahmad MS (2011) Electrocoagulation in wastewater treatment. Water 3(2):495–525

Chojnacka K (2010) Biosorption and bioaccumulation—the prospects for practical applications. Environ Int 36(3):299–307

Du Z, Li H, Gu T (2007) A state of the art review on microbial fuel cells: a promising technology for wastewater treatment and bioenergy. Biotechnol Adv 25(5):464–482

Dutta A, Basu S, Mukherjee SK, Hossain ST (2021) Wastewater treatment by microbial biofilm: a distinct possibility. In: Shah M, Rodriguez-Couto S (eds) Microbial ecology of wastewater treatment plants. Elsevier, Amsterdam, pp 435–468. ISBN 97801282250352021

El-Hosiny F, AbdelKhalek MA, Selim K, Osama I (2017) A designed electro-flotation cell for dye removal from wastewater. J Appl Res Indus Eng 4(2):133–147

Fan Y, Qian F, Huang Y, Sifat I, Zhang C, Depasquale A, Wang L, Li B (2021) Miniature microbial fuel cells integrated with triggered power management systems to power wastewater sensors in an uninterrupted mode. Appl Energy 302:117556

Ghosh Ray S, Ghangrekar MM (2015) Enhancing organic matter removal, biopolymer recovery and electricity generation from distillery wastewater by combining fungal fermentation and microbial fuel cell. Bioresour Technol 176:8–14

Gorby YA, Yanina S, McLean JS, Rosso KM, Moyles D, Dohnalkova A, Beveridge TJ, Chang IN, Kim BH, Kim KS, Culley DE, Reed SB, Romine MF, Saffarini DA, Hill EA, Shi L, Elias DA, Kennedy DW, Pinchuk G, Watanabe K, Ishii S, Logan B, Nealson KH, Fredrickson JK (2006) Electrically conductive bacterial nanowires produced by Shewanella oneidensis strain MR-1 and other microorganisms. Proc Natl Acad Sci USA 103(30):11358–11363

He Z, Minteer SD, Angenent LT (2005) Electricity generation from artificial wastewater using an upflow microbial fuel cell. Environ Sci Technol 39(14):5262–5267

Hossain ST, Mukherjee SK (2012) CdO nanoparticle toxicity on growth, morphology, and cell division in *Escherichia coli*. Langmuir 28(48):16614–16622

Hossain ST, Mukherjee SK (2013) Toxicity of cadmium sulfide (CdS) nanoparticles against *Escherichia coli* and HeLa cells. J Hazard Mater 260:1073–1082

Hossain ST, Das P, Mukherjee SK (2014) Toxicity of cadmium nanoparticles to *Bacillus subtilis*. Toxicol Environ Chem 260:1073–1082

Hossain ST, Mallick I, Mukherjee SK (2012) Cadmium toxicity in *Escherichia coli*: cell morphology, Z-ring formation and intracellular oxidative balance. Ecotoxicol Environ Saf 86:54–59

Ieropoulos I, Melhuish C, Greenman J, Horsfield I (2005) Artificial symbiosis: towards a robot-microbe partnership. In: Proceedings of Towards Autonomous Robotic Systems (TAROS '05) Conference, pp 89–93

Jayashree C, Sweta S, Arulazhagan P, Yeom IT, Iqbal MII, Banu JR (2015) Electricity generation from retting wastewater consisting of recalcitrant compounds using continuous upflow microbial fuel cell. Biotechnol Bioprocess Eng 20:753–759

Jiang D, Curtis M, Troop E, Scheible K, McGrath J, Hu B, Suib S, Raymond D, Li B (2011) A pilot-scale study on utilizing multi-anode/cathode microbial fuel cells (MAC MFCs) to enhance the power production in wastewater treatment. Int J Hydrog Energy 36(1):876–884

Kansal A, Siddiqui NA, Gautam A (2013) Assessment of heavy metals and their interrelationships with some physicochemical parameters in eco-efficient rivers of Himalayan region. Environ Monit Assess 85(3):2553–2563

Kim HJ, Park HS, Hyun MS, Chang IS, Kim M, Kim BH (2002) A mediator-less microbial fuel cell using a metal reducing bacterium, *Shewanella putrefaciens*. Enzyme Microb Technol 30(2):145–152

Kuyukina MS, Krivoruchko AV, Ivshina IB (2020) Advanced bioreactor treatments of hydrocarbon-containing wastewater. Appl Sci 10(3):831

Logan BE, Hamelers B, Rozendal R, Schröder U, Keller J, Freguia S, Aelterman P, Verstraete W, Rabaey K (2006) Microbial fuel cells: methodology and technology. Environ Sci Technol 40(17):5181–5192

Lokhande RS, Singare PU, Pimple DS (2011) Toxicity study of heavy metals pollutants in waste water effluent samples collected from Taloja Industrial Estate of Mumbai, India. Resour Environ 1(1):13–19

Mallick I, Hossain ST, Sinha S, Mukherjee SK (2014) *Brevibacillus* sp. KUMAs2, a bacterial isolate for possible bioremediation of arsenic in rhizosphere. Ecotoxicol Environ Saf 107:236–244

Margot J, Kienle C, Magnet A, Weil M, Rossi L, De Alencastro LF, Abegglen C, Thonney D, Chevre N, Schärer M, Barry DA (2013) Treatment of micropollutants in municipal wastewater: ozone or powdered activated carbon? Sci Total Environ 461–462:480–498

Pepé Sciarria T, Tenca A, D'Epifanio A, Mecheri B, Merlino G, Barbato M, Borin S, Licoccia S, Garavaglia V, Adani F (2013) Using olive mill wastewater to improve performance in producing electricity from domestic wastewater by using single-chamber microbial fuel cell. Bioresour Technol 147:246–253

Rahimnejad M, Ghoreyshi AA, Najafpour GD, Younesi H, Shakeri M (2012) A novel microbial fuel cell stack for continuous production of clean energy. Int J Hydrogen Energy 37(7):5992–6000

Rai HS, Bhattacharyya MS, Singh J, Bansal TK, Vats P, Banerjee UC (2005) Removal of dyes from the effluent of textile and dyestuff manufacturing industry: a review of emerging techniques with reference to biological treatment. Crit Rev Environ Sci Technol 35(3):219–238

Rice KM, Walker EM Jr, Wu M, Gillette C, Blough ER (2014) Environmental mercury and its toxic effects. J Prev Med Public Health 47(2):74–83

Santoro C, Ieropoulos I, Greenman J, Cristiani P, Vadas T, Mackay A, Li B (2013) Current generation in membraneless single chamber microbial fuel cells (MFCs) treating urine. J Power Sources 238:190–196

Shah MP (2020) Microbial bioremediation & biodegradation. Springer, Singapore

Shah MP (2021) Removal of refractory pollutants from wastewater treatment plants. CRC Press, Boca Raton

Sharma Y, Li B (2010) Optimizing energy harvest in wastewater treatment by combining anaerobic hydrogen producing biofermentor (HPB) and microbial fuel cell (MFC). Int J Hydrogen Energy 35(8):3789–3797

Song YE, Boghani HC, Kim HS, Kim BG, Lee T, Jeon BH, Premier GC, Kim JR (2016) Maximum power point tracking to increase the power production and treatment efficiency of a continuously operated flat-plate microbial fuel cell. Energy Technol 4(11):1427–1434

Trapero JR, Horcajada L, Linares JJ, Lobato J (2017) Is microbial fuel cell technology ready? An economic answer towards industrial commercialization. Appl Energy 185(1):698–707

Ullah Z, Zeshan S (2020) Effect of substrate type and concentration on the performance of a double chamber microbial fuel cell. Water Sci Technol 81(7):1336–1344

Upadhyayula VK, Deng S, Mitchell MC, Smith GB (2009) Application of carbon nanotube technology for removal of contaminants in drinking water: a review. Sci Total Environ 408(1):1–13

Wang Y-P, Liu X-W, Li W-W, Li F, Wang Y-K, Sheng G-P, Zeng RJ, Yu H-Q (2012) A microbial fuel cell-membrane bioreactor integrated system for cost-effective wastewater treatment. Appl Energy 98:230–235

Wani AL, Ara A, Usmani JA (2015) Lead toxicity: a review. Interdiscip Toxicol 8(2):55–64

Yuan SJ, Sheng GP, Li WW, Lin ZQ, Zeng RJ, Tong ZH, Yu HQ (2010) Degradation of organic pollutants in a photoelectrocatalytic system enhanced by a microbial fuel cell. Environ Sci Technol 44(14):5575–5580

Zazouli MA, Kalankesh LR (2017) Removal of precursors and disinfection by-products (DBPs) by membrane filtration from water; a review. J Environ Health Sci Engineer 15:25

Bioremediation of Arsenic: Microbial Biotransformation, Molecular Mechanisms, and Multi-omics Approach

Juan Gerardo Flores-Iga, Lizbeth Alejandra Ibarra-Muñoz, Aldo Almeida-Robles, Miriam P. Luévanos-Escareño, and Nagamani Balagurusamy

1 Introduction

Arsenic (As) is a metalloid and is the 20th most abundant element found on earth, mainly in sediments, soil and groundwater. Its concentration in different zones depends on the crust depth and mineral composition of seismic tectonic plates and as well as due to anthropogenic activities such as mining, pesticides and groundwater exploitation, leading to natural and man-made contamination of soil and water (Lièvremont et al. 2009; Rae 2020; Podgorski and Berg 2020), exceeding the limits prescribed by WHO. As contamination is widely reported in many countries from South Asia, Latin America, and other countries including Australia. La Comarca Lagunera, located in the North-East of Mexico is one of the regions, where concentration of arsenic in groundwater ranges from 0.02 mg As/L to 0.65 mg As/L (Sariñana-Ruiz et al. 2017; Dorjderem et al. 2020). There is concern around the world due to globalized exportation of agricultural and other products from these contaminated regions (Rae 2020; Irshad et al. 2021).

Arsenic exists in different states and its speciation relies on physical and chemical characteristics and as well as microbial activity in these environments (Zhu et al. 2017; Rae 2020). The four oxidation states of As namely are arsenide (As(-III)), elemental arsenic (As(0)), arsenite (As(III)), and arsenate (As(V)) and the last two being the most abundant arsenic oxidation states, mostly found in anoxic and

J. G. Flores-Iga · L. A. Ibarra-Muñoz · A. Almeida-Robles · M. P. Luévanos-Escareño · N. Balagurusamy (✉)
Laboratorio de Biorremediación, Facultad de Ciencias Biológicas, Ciudad Universitaria de La Universidad Autónoma de Coahuila, 27000 Torreón, Coahuila, Mexico
e-mail: bnagamani@uadec.edu.mx

A. Almeida-Robles
Nørregade 10, 1165 København, Denmark

M. P. Shah (ed.), *Microbial Technologies in Industrial Wastewater Treatment*,
https://doi.org/10.1007/978-981-99-2435-6_6

oxygenic environments, respectively (Hayat et al. 2017; Irshad et al. 2021). Inorganic species of arsenic are commonly linked covalently with oxygen as arsenite (AsO_3^{3-}) and arsenate (AsO_4^{3-}) (Rae 2020). Arsenic toxicity on living organisms varies according to speciation and bioavailability. The exposition of plants, animals, and humans to As through daily water intake causes several deteriorations to their wellness and health. Arsenic causes many immediate symptoms such as arsenicosis, long-term exposure to arsenic leads to leukemia, skin and kidney cancer (Podgorski and Berg 2020).

Trivalent form of arsenic binds with sulfur-containing amino acids, structural proteins and important metabolites, and subsequently disrupts metabolic pathways and vital cell functions. On the other hand, pentavalent arsenic species do not commonly bind to thiol groups but they resemble with form and electron configuration of phosphate, and is harmful to energy pathways such as glycolysis and oxidative phosphorylation (Rae 2020). Furthermore, several organic arsenic species, e.g., monomethylated As III (MMA^{III}) apart from retaining the toxicity properties of As(III), forms an even tougher binding complex with essential cell molecules (Irshad et al. 2021). Hence, As remediation and removing them from water is an essential process to eliminate its toxic properties.

In general, physico-chemical methods are commonly used for the removal of arsenic from wastewaters; e.g., ion-exchange resins and inverse osmosis, etc. But, these processes involve high costs, difficult to operate, and generate toxic residual wastes, which would require an expensive downstream process for removal. A nonexpensive alternative technology that is being applied is the use of suitable microorganisms, known as bioremediation (Alka et al. 2021). In general, bioremediation involves biotransformation of toxic or persistent species into less toxic species and easy to removal forms. Omics-sciences is an emerging approach where essential molecules for life such as genes, transcripts, proteins and metabolites are assessed to understand the dynamic and mechanistic interactions between single cells and entire microbiomes for bioremediation of As and other metal-contaminated sites (Malla et al. 2018; Shah 2020). The evaluation at molecular level aid in the identification of potential functional genes, which can either be cloned or modified to enhance their bioremediation potential by employing metabolic engineering techniques (Yamamura and Amachi 2014; Zhu et al. 2017; Plewniak et al. 2018).

2 Arsenic Pollution in Water and Microbial Relevance

As mentioned previously, natural resources like soil and water can be contaminated with arsenic and other heavy metals by natural origins and as well as by anthropogenic activities (Garbinski et al. 2019; Podgorski and Berg 2020; Alka et al. 2021). Solubilization and speciation of arsenic in water depends on the physicochemical and microbiological conditions of the environment. Arsenic can be found in different oxidation states being arsine (As(-III)), elemental arsenic (As(0)), arsenite (As(III)), and arsenate (As(V)). As(III) and As(V) are the most abundant species that are commonly linked covalently with oxygen as arsenite (AsO_3^{3-}) and arsenate (AsO_4^{3-}). As(III) arsenic species are more toxic than As(V) derived species and trivalent methylated arsenic forms are the most toxic (Garbinski et al. 2019). In general, As(V) is present at higher concentrations in oxygenic environments while As(III) at anoxic conditions (Andres and Bertin 2016; Yan et al. 2019; Rae 2020). Microorganisms play a key role in the biogeochemical cycles of As through oxidation, reduction, and methylation processes (Zhu et al. 2017; Sher and Rehman 2019). For instance, microorganisms can oxidize As(III) to a less harmful species As(V). In the same manner, few microorganisms as well reduce As(V) to As(III), which is 100 times more toxic than As(V). However, it can be pumped out easily by an As(III) specific membrane transporter or can be used as an electron donor for energetic purposes by few bacterial species (Mujawar et al. 2019).

Furthermore, methylation of As is considered as an important defense mechanism for prokaryotic and eukaryotic microorganisms due its transformation to volatile compounds that are released passively out from the cell (Qin et al. 2006). In addition, methane biosynthetic pathway is linked to As volatilization under anaerobic conditions since coenzyme M is capable of methylating As(III) and produce arsine-methylated gases (Mohapatra et al. 2008; Sodhi et al. 2019). The application of the biotransformation mechanisms found in microorganisms, i.e., bacteria, yeasts, fungi, and microalgae can be employed in the bioremediation of As polluted aquatic environments and wastewaters (Sher and Rehman 2019; Sodhi et al. 2019).

3 Microbial Uptake of As

The uptake of As(III) and As(V) by microorganisms are generally described through aquaglyceroporin and phosphate transporters (Pi), respectively, since the chemical structure of As(III) and As(V) resemble the substrates of these transporters (Kruger et al. 2013; Garbinski et al. 2019). In bacteria, there are two phosphate transporters involved for As(V), Pit, and Pst (Tsai et al. 2009; Jia et al. 2019). An example is *Escherichia coli*, where arsenic is taken up through the nonspecific phosphate transporter Pit. Other bacteria may express a specific phosphate transporter, Pst, which is less efficient in arsenate transport (Kruger et al. 2013). On the other hand, As(III) is taken up by a member of the glycerol channels of the major intrinsic protein (MIP) family, called the glycerol transporter GlpF (Yan et al. 2019). In addition, the transport of As(V) is considered as an active transport in contrast with As(III) where no energy is necessary to pass through the membrane (Kruger et al. 2013). Fungi and yeast possess homology mechanisms of As uptake from those identified in bacterial cells. *Saccharomyces cerevisiae* employs the phosphate transporters; Pho84p and Pho87p to mediate As(V) transport (Shah et al. 2010), although, they possess low affinity As(V) proteins, Pho90p and Pho91p (Maciaszczyk-Dziubinska et al. 2012). In the case of As(III), two different transport systems are used. One is aquaglyceroporin Fps1p, which is a glycerol transporter encoded by the FPS1 gene. When prolonged exposure of As(III) occurs, it is released from the same transporter aquaglyceroporin Fps1p, as it is a bidirectional channel for arsenite efflux, which can transport substrates found in high concentrations on cells to its environment release (Maciaszczyk-Dziubinska et al. 2012). Another way for As(III) uptake described in eukaryotes is via hexose permeases. However, it has been described that the As(III) uptake through this channels occurs exclusively on the absence of hexoses, i.e., glucose as it has highest affinity. Hence, As(III) transport through hexose transporters is inhibited by its primary substrates (Liu et al. 2004; Maciaszczyk-Dziubinska et al. 2012).

Arsenic uptake in microalgae, i.e., *Chlorella sp.* species is guided by protein transporters that are incrusted in the algal plasma membrane. Similar to bacteria and fungi, As(V) enters algae cells through phosphate transporters; however, it has been proved that algae can take up As independently of phosphate transporters. Further, it has been recorded that As(V) is a competitive inhibitor of phosphate uptake but nitrate presence affects entrance to the cell due to the anionic competitive effect; however, this competition is not completely clarified (Hussain et al. 2021). On the other hand, As(III) enters the plasma membrane through aquaglyceroporins and hexose permeases such as other eukaryotic cells described above (Xie et al. 2018; Hussain et al. 2021; Leong and Chang 2020).

4 Biotransformation of Arsenic: Microbial Machines to Cope with Arsenic

Biotransformation of arsenic involves chemical changes carried out by microorganisms in which arsenic is transformed from oxidized to reduced states and vice versa by addition or removal of electrons and functional groups, catalyzed by As-related enzymes and also by nonenzymatic antioxidants (Sher and Rehman 2019). Microbial systems developed these mechanisms either as defense mechanism against arsenic toxicity or to generate energy (Rahman and Hassler 2014). Arsenic biotransformation is considered as As detoxification pathway in microbes that can be utilized to bioremediate the high toxicity As species into less toxic species with better ion removal performance or to convert As to low boiling point As methylated compounds for its dissipation through the cell membrane (Hu et al. 2021). There are many gram-positive and gram-negative bacteria that are resistant to As(III). Several operons containing As-specific genes (Ars) have been reported in plasmids and chromosomal DNA of bacteria, such as *Pseudomonas sp., Bacillus subtilis, Mycobacterium sp,* etc. (Hamood Altowayti et al. 2020). Ars operons are primarily responsible for the detoxification of inorganic arsenicals (Lièvremont et al. 2009; Yan et al. 2019; Zhou et al. 2020). In general, *Ars* operon contains three genes, it is called Ars RBC. Ars R is a gene that encodes a trans-acting repressor protein that has also been associated to As accumulation in *E. coli* (Kostal et al. 2004; Hamood Altowayti et al. 2020). The Ars B gene encodes an As(III)-specific transmembrane transporter that codifies an As(III) efflux pump. When As possess its reduced form, As(III) is expelled by this encoded transporter (Hamood Altowayti et al. 2020). Lastly, Ars C encodes an arsenate reductase enzyme that is responsible for the reduction of As(V) to As(III). As(V) binds to a recognition domain composed of Arg Residues, where a disulfide bond is formed between the cysteine residues of Ars C and the reducing equivalents. The reduction of the disulfide bond through electron transfer results in the reduction of As(V) to As(III) using glutathione (GSH) as electron donor (Zhou et al. 2020). Most bacteria reduce As(V) to As(III) when the Ars operon acts and then discharges As(III) through its efflux pump, and this mechanism is reported as an *Ars* operon-dependent detoxification pathway. Additionally, there is a five-gene Ars operon called Ars RDABC that contains two more genes in its composition. This Ars RDABC operon has been found in plasmids of gram-negative bacteria. Ars A gene encodes an intracellular soluble ATPase subunit that binds to ArsB and converts the As(III) transporter proteins into an efflux pump that is driven by ATP. Most bacteria employ the Ars A/B system, while some bacteria rely solely on ArsB (Tsai et al. 2009; Ben Fekih et al. 2018). In addition to the Ars AB complex formed by the addition of Ars A gene, this operon contains two trans-acting regulatory elements, ArsR and Ars D (Ben Fekih et al. 2018). Arsenic biotransformation in eukaryotic cells is described as the reduction of As(V) by an homologous arsenate reductase (ARC) that allows the production of volatile As species and sequestration by nonenzymatic antioxidants. Moreover, As organic species can be produced by the sequential methylation of arsM gene found in prokaryotic and eukaryotic cells. In the case of bacteria, this gene is found in operons under the regulation of an arsR gene (Ben Fekih et al. 2018; Ghosh et al. 2015). The mechanism of As detoxification found in bacteria containing arsRBC and arsM operon is illustrated in Fig. 1.

5 Oxidation and Reduction of As in Microbial Cells

In the bacterial cell, arsenic can work as an electron donor or acceptor in the electron transport chain on the basis of its oxidation state (Mazumder et al. 2020). The respiratory reduction of As(V) to As(III) is carried out by the arsenate reduction system (*arr*), composed of the *arr*A genes. As(V) reduction is catalyzed by the enzyme arsenate reductase (Arr) under anoxic conditions and is a periplasmic dimethylsulfoxide (DMSO) reductase that is composed of two heterologous subunits, one large (ArrA) of 87 kDa and one small (ArrB) of 29 kDa (Yan et al. 2019; Zhai et al. 2020; Singh et al. 2021). The catalytic activity of ArrA is dependent on a molybdopterin and a [4Fe-4S] cluster, whereas ArrB can be composed of three (Jia et al. 2019; Yan et al. 2019; Zhou et al. 2020) or four [4Fe-4S] clusters (Kumari and Jagadevan 2016). Arsenate reducing microorganisms (ARMs) reduce As(V) to As(III) as a final electron acceptor to survive the high arsenic concentrations in the various environments in which they thrive. However, these microorganisms do not obtain energy during the process (Kumari and Jagadevan 2016). The first eukaryotic arsenate reductase enzyme, Acr2p, was identified in yeast (Mukhopadhyay and Rosen 1998). It is known that As(V) reduction is an As tolerance mechanism utilized by fungi and

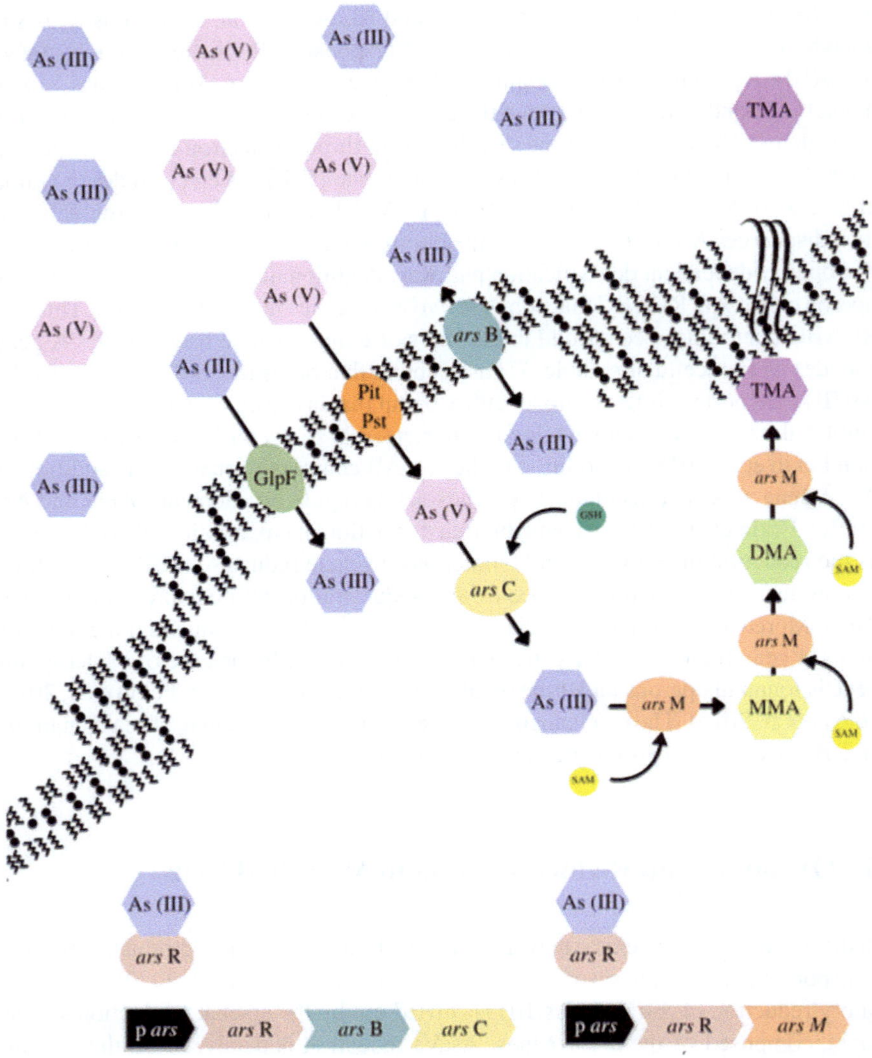

Fig. 1 Uptake and detoxification mechanism found in bacteria. ArsRBC and ArsM operons are identified to encode enzymes related to As tolerance. The arsRBC contains three encoded proteins, Ars R is a regulator protein, Ars B is a membranal As efflux pump, and ArsC encodes an arsenite reductase enzyme that utilizes GSH as electron donor. On the other hand, Ars M operon is related to biovolatilization. It contains genes for a regulator protein and an As methyl transferase enzyme that utilizes S-Adenosine-Methionine (SAM) as a methyl donor

microalgae since it is necessary for further methylation and complexation with thiol-rich molecules such as GSH (Garbinski et al. 2019). Two ArsC homologous genes associated with arsenic reduction in *C. reinhardtii* microalgae species were identified by Yin et al. (2011).

On the other hand, the respiratory oxidation of As(III) requires the use of an As(III) oxidase enzyme encoded by aioA and aioB genes under aio operon regulation in bacteria. Anderson et al. (1992) purified this enzyme and discovered its two-subunit composition. The large 85 kDa subunit of arsenite oxidase is codified by aio A gene while the small 14 kDa subunit is encoded by aioB, these dimeric parts are composed of a molybdenum center and a [3Fe-4S] cluster with similarities to members of DMSO reductase and a [2Fe-2S] protein domains, respectively (Yamamura and Amachi 2014; Guo et al. 2019). Even though there is no evidence of an As(III) oxidase enzyme in eukaryotic cells, it has been demonstrated the extracellular enzymatic oxidation of As(III) by *chlorella sp.* is carried out by carbonic anhydrase and phosphatase (Qin et al. 2009).

6 Biovolatilization of Arsenic

Arsenic biovolatilization depends on the methylation of inorganic As species; i.e., As(V) and As(III) forming low boiling point metabolites that dissipate passively out from the cell. It is recognized as a detoxification pathway but not all of the products from these processes are less toxic to initial species. The volatilization of these products is affected by environmental factors, i.e., pH, temperature, organic matter, and moisture content (Cullen 2014). Addition of Methyl groups to As usually occurs after the reduction of As(V) to As(III) by an arsenite adenosylmethionine methyltransferase (As(III)-SAM) enzyme encoded by a microbial ArsM homologous gene that utilizes S-adenosylmethionine as the methyl donor and proceeds in three subsequent additions that changes the chemical properties of As being first to methylarsenite (MMA (III)), then dimethylarsenite (DMA(III)), and finally trimethylarsenite (TMA(III)) (Yan et al. 2019). The arsM gene has been found in operons in bacteria and independently in eukaryotic species (Cullen 2014; Ben Fekih et al. 2018). Under aerobic conditions, As(III) intermediates can be oxidized to less toxic As(V) methylated arsenicals such as methylarsenite (MAs(V)), dimethylarsenate (DMAs(V)), and trimethylarsenite oxide (TMAs(V)) since trivalent arsenicals are considerably more toxic than pentavalent arsenicals due its higher affinity to sulfur amino acids of proteins (Garbinski et al. 2019; Yan et al. 2019). The As-related SAM enzymes have four cysteine residues segments found along the amino acid sequence which are conserved among prokaryotic and eukaryotic methyltransferases that are responsible for the As(III) binding due to the interaction with thiol groups of the protein (Yang et al. 2020). ArsM-encoded enzymes have been characterized from arsenite methylating bacteria and eukaryotes such as *Rhodopseudomonas palustris, Streptomyces sp. strain GSRB54, Methanosarcina acetivorans C2A, and W. aurantiaca* (Zhang et al. 2015; Verma et al. 2016; Uppal et al. 2017). Interestingly, this process under anaerobic conditions has been linked to methanogens as a mechanism of As detoxification (Mohapatra et al. 2008; Webster et al. 2016; Yang et al. 2020). Methanogens can produce a series of arsine and methylarsine gases from arsenic species in As polluted environments utilizing the methyl donors methylcobalamin and coenzyme M present in the final steps of methane pathways. They have been reported to produce mono-, di- and tri-methylarsine gases from arsenite, when there are limited final electron acceptors (Chen et al. 2019; Webster et al. 2016).

7 Sequestration of Arsenic

Arsenic resistance also includes the process of extracellular and intracellular sequestration due its interaction with polyanionic components and thiol-rich molecules, respectively. The extracellular sequestration can be seen as biofiltration where biofilms of microorganisms assembled on polymeric substances, i.e., exopolysaccharides (ESP) provide the ability to passively immobilize the toxic As species and can therefore be used to detoxify arsenic-contaminated water (Saba et al. 2019). In addition, the coprecipitation of As ions in the presence of ferric ions shows a high affinity toward the As ion, and leads to As precipitation in the EPS of the biofilm, since As is soluble over a wide pH range (Lièvremont et al. 2009; Hayat et al. 2017; Ojuederie and Babalola 2017). Microalgal species possess functional groups in the cell walls such as carboxyl, sulfhydryl, and amino groups that adsorb As in outer cell membrane and prevent As to reach the intracellular components of the cell (Wang et al. 2015). Moreover, It is known that As can be accumulated intracellularly via microorganisms, this is an intracellular sequestration fate for As species. After the uptake of As, it tends to bind to cytoplasmic chemical compounds that result in physical sequestration in cellular compartments, such as the vacuole (Garbinski et al. 2019). Arsenic can bind to chelating peptides containing thiol ligands that are predominantly nonenzymatic antioxidants, i.e., glutathione (GSH), metallothionines (MTs), and phytochelatin (PCs) and form inactive complexes that disable As species to react with other cellular components (Wang et al. 2015; Garbinski et al. 2019). GSH is known for functioning as a reducing agent of As(V) to further transport to vacuoles as As(III), MTs have been characterized as related to As binding in algae, as is the case in *Fucus vesiculosus* and PCs are enzymatically synthesized cysteine-rich peptides that bind As(III) mainly in eukaryotic photosynthetic organisms as is the case of several microalgae strains. It's been reported that there are other As chelators, mainly low molecular weight proteins with high content of sulfur residues (Tsai et al. 2009; Wang et al. 2015; Garbinski et al. 2019).

8 Application of Microbes for Treatment of Arsenic Polluted Wastewater

As is conventionally removed through physico-chemical methods that require the pre-oxidation of As species and efficient downstream removal to ensure its removal, which often have low energy efficiency and hence is an expensive process. Hence, the application of As-tolerant microorganisms to biotransform toxic species into a less toxic, and as well as limiting the bioavailability of arsenic highlighted their potential use in As remediation and use of microbial technology have been reported to be cost-effective and efficient (Hayat et al. 2017; Irshad et al. 2021). It has been proposed that volatilization of As is one of the principal mechanisms to detoxify waters since methylation into low boiling point compounds produces volatile species that go out from water to the atmosphere. However, there is a concern since breathing volatile As species is a risk for human health (Hayat et al. 2017). Reduction of As(V) leads to a laborious removal of As(III), since in pH range of 4–10, the neutral charge makes difficult the whole removal of As.

Thus, oxidation of As(III) and As species sequestration are the preferred mechanisms of bioremediation due its ionization state and the easy removal (Lièvremont et al. 2009; de Alencar et al. 2017; Hayat et al. 2017; Irshad et al. 2021). Microbial bioremediation of As in wastewater has been studied mainly under laboratory and pilot scale conditions with promising results. For instance, Sher et al. (2020) studied a metal-resistant *Micrococcus luteus* strain AS2 isolated from industrial wastewater that showed a 99% of biomass removal efficiency after 10 h of incubation with 1000 mM As(III). However, there are very few field trials using microbial technology until now (Table 1).

9 Genetically Engineered Microorganisms in Arsenic Bioremediation

The modification of microorganisms either by addition, deletion, or modification of expression of different genes taking advantage of molecular biology and gene editing tools are well known and are known as genetically modified organisms (GMOs). Application of GMOs, especially microorganisms such as bacteria and fungi to improve As biotransformation in terms of efficiency and selected speciation has been studied by various authors (Kostal et al. 2004; Yuan et al. 2008; Shah et al. 2010; Liu et al. 2011). Enhancement of volatilization and sequestration of As species through overexpression and gene cloning have been utilized for arsenic bioremediation since both lead to easier removal and reduce the bioavailability of As, respectively. ArsR and Fps1p & Hxt7p genes have been associated to As accumulation in bacteria and yeast. Kostal et al. (2004) overexpressed the ArsR gene in *E. coli*, which is a regulatory protein from the Ars operon that contains a high-specificity

Table 1 Potential arsenic-resistant microorganisms for As bioremediation in wastewater

Microorganisms	As resistance genes	Activity	Bioremediation efficiency	Reference
Bacillus licheniformis	*n.d*	• As(III) oxidation	• Oxidize 86% at 48 h and 98% at 96 h of 100 μg/ml As(III)	Sher et al.(2021)
***Micrococcus luteus strain* AS2**	*aioB, arsC1, arsC2, ACR3 and arsR*	• As(III) removal with bacterial biomass	• Remove 99% of 1000 mM As(III) at 10 h	Sher et al. (2020)
***Staphylococcus sp. strain* AS6**	*arsC1, arsC2, arsB, arsR, arsA*, and *arsD*	• As(III) oxidation • As removal with bacterial biomass	• Oxidize 91% of 250 mM As(III) at 96 h • Removed 93% of 1000 mM at 10 h	Sher et al. (2020)
***Klebsiella pneumoniae strain* SSSW7**	*aioA*	• As(III) oxidation	• Oxidize 10 mM of As(III) at 24 h	Mujawar et al. (2019)
Bacillus cereus	*aox*	• As(III) oxidation	• Oxidize 96% of As(III) at 6 days	Naureen and Rehman (2016)
Acinetobacter junii	*aox*	• As(III) oxidation	• Oxidize 88% of As(III) at 6 days	Naureen and Rehman (2016)
***Anoxybacillus flavithermus strain* TCC9-4**	*Low homology aio*	• As(III) oxidation	• Oxidize 90% of 100 mg/L of As(III) at 36 h	Jiang et al. (2015)
Cladophora algae	*n.d*	• As species biosorption	• Biosorpts 99.8% of 6 mg/L As species at 10 days	Jasrotia et al. (2014)
***Thiomonas Arsenivorans* Strain b6**	*n.d*	• As(III) oxidation	• Oxidize 98.9% of 100 mg/L of As(III) at 72 h	Dastidar and Wang (2009)
Bacillus sp. strain DJ-1	Absence of arsB and arsC	• As bioaccumulation	• Bioaccumulates 6.14 mg/g of As(III) and 9.8 mg/g of As(V) with respect to cell dry weight at 24 h	Joshi et al. (2009)

binding site for As(III) and As(V). The mutant strain overexpressing ArsR demonstrated higher accumulation capability in growing and resting cells resulting in a 100% accumulation of 1 μm of arsenite contained in the polluted water in contrast to the relatively low levels of accumulation by the wild type. Moreover, overexpression of As transporters Fps1p and Hxt7p in *S. cerevisiae* at lower and high concentrations; imitating groundwater amounts of arsenite, accumulated 2–fourfold of arsenic than control (200–300 μg As/g cells) (Yuan et al. 2008). In both cases, the potential of these mutant microorganisms as arsenic ligand microorganisms for arsenic biotreatment in contaminated wastewater was assessed. In aqueous solution, ArsM gene known to be involved in the biovolatilization of As in *Rhodopseudomonas palustris* was expressed in the strains *S.desiccabilis* and *B.idriensis* conferring resistance to higher concentration of As to these bacteria and generated more than tenfold higher concentrations of gaseous arsenic in comparison to the wildtype strains. Although, volatilization of As(III) and As(V) methylated forms is being considered less toxic than inorganic species, the toxicity of gaseous As species is still debated since the identification of its cancerogenic effects (Liu et al. 2011; Hayat et al. 2017).

10 Assessment and Screening of As Biotransformation Mechanisms: A Multi-omics Approach

The bioremediation mechanisms utilized by single species and as well by synergistically acting consortia on removal of As can be assessed through omics-science data (Fig. 2). Omics sciences provide data from biological systems such as DNA, RNA, proteins, and metabolites that can be processed by bioinformatic tools to screen for innovative strategies and better-performance microorganism under environmental conditions (Malla et al. 2018). Arsenic tolerance and biotransformation have been studied in this approach on natural polluted environments to reveal microbial interactions, participating genes, their regulation, and biotransformation pathways for elucidation purposes (Andres and Bertin 2016). High-throughput omics data from As stressed microorganisms can be processed in analysis of a given biomolecule, i.e., genomics, transcriptomics, proteomics, metabolomics, etc., or through a multi-omics approach where a dataset of molecules is studied to relate the biological As decontaminant machineries in a better way. Genomics examines the genetic material of living systems making possible the characterization of genes in novel organisms, recreation of biochemical putative pathways, and defense systems to cope with As stress (Li et al. 2014; Rahman et al. 2016; Plewniak et al. 2018; Castro-Severyn et al. 2020; Sher et al. 2020). There have been several isolated organisms found in As-polluted water that possess potential genes for As tolerance (Yan et al. 2019; Sher et al. 2020). For instance, Sher et al. (2020) isolated a *Staphylococcus sp.* strain AS6 from As polluted industrial wastewater. Genomic sequencing data identified genes related to As reduction (arsC, arsB, arsR, arsA, and arsD) and further cell accumulation. Metagenomics, a branch of genomics, evaluates the DNA content from an entire consortium of nonculturable organisms from a given environment that enables the construction of metagenomic assembled genomes and inference of those As related mechanisms, assess prospective microbes and genes to develop biotechnological bioremediation strategies. It also enables the understanding of the microbial taxonomic interaction in a microbiome such as natural and anthropogenic As polluted waterbodies (Danczak et al. 2019; Jiang et al. 2019; Shi et al. 2019; Wang et al. 2020). Biomobilization of As and other heavy metals in groundwater was evaluated by Danczak et al. (2019) through metagenomics, which revealed the widespread functionality of arsB and ArsC, arsenic resistance genes, in distinct taxonomies. However, a potential gene for arsenic aerobic oxidation, aioA, was found in several members of methylocystaceae bacterial family. Those results shed light on a potential family of microorganisms for bioremediation since As oxidation is often a preferred mechanism for arsenic bioremediation. Genomics and metagenomics reveal the identification of microbial capability genomes and genes for As bioremediation; however, there is a lack of gene activity comprehension from those sciences. Thus, transcriptomics analyses are required to explain the expression of protein and nonprotein coding genes as a response environment under different biotic and abiotic conditions including As pollution, hence, enabling the spatial and temporal comprehension of As related genes activation and repression through metatranscriptomics is necessary (Malla et al. 2018). Similarly, proteomics evaluates the whole set of proteins to stress conditions, permitting their qualitative and quantitative characterization to elucidate their functional and stability properties such as kinetics, optimum pH, and thermal tolerance. Even though proteomics and transcriptomics can provide a complete understanding of genomics and metagenomics data under As stress, there are not many studies in this regard in natural environment. However, Sher and Rehman (2021) identified 7 proteins with clear differential expression under arsenite stress in the arsenic resistance strain *Micrococcus luteus* isolated from wastewaters. Four of these seven proteins upregulated are involved in DNA protection against ROS and protein folding and three proteins were found to be downregulated and are related to fatty acid biosynthesis and enzymatic antioxidant activity. On the other hand, gene expression analysis identified no significative differential expression of aioB gene which indicates the naturality capacity of the arsenic oxidation by this bacterium in the presence of As. Metabolomics and its derived science fluxomic have not been employed for As bioremediation even though their importance to describe metabolic products released from microorganisms and their metabolic rates. However, it is necessary to have a complete point of view on As transformation by microbes through multi-omics approach to fully explain microbial interaction and their dynamics to understand microbial community synergy, elucidation of genes and its expression and enzymatic parameters in natural environments in order to develop innovative bioremediation strategies (Zhu et al. 2017; Plewniak et al. 2018).

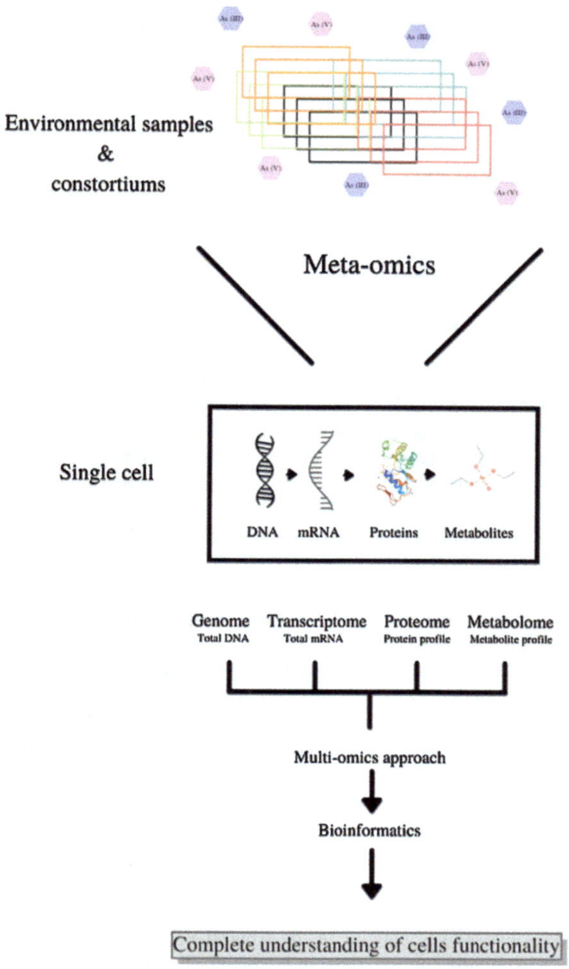

Fig. 2 Multi-omics approach for screening of potential mechanisms found in single cell and consortia for bioremediation of arsenic

11 Conclusion and Future Perspectives

Arsenic pollution at toxic concentrations is widely found on earth causing direct and nondirect global health problems. As presence in wastewater is mainly due to industries that utilize As in their processes and discharge in wastewaters and remain as they are not efficiently removed. Conventional methods of As removal are described as of low efficiency and high cost. Bioremediation technologies seem to be a promising technology as they employ microbial detoxification pathways to either enhance the As removal by changing the speciation of arsenic or biosequestrate the metalloid inside cells. However, this technology has not been employed at industrial scales. It has been mentioned that a combination of psycho-chemical and bioremediation technologies can result in efficient means of arsenic removal (Hubadillah et al. 2020). In addition, multi-omics approach offers the possibility of developing/selecting potential synergistically acting consortia for As bioremediation.

References

Alka S, Shahir S, Ibrahim N, Ndejiko MJ, Vo D-VN, Manan FA (2021) Arsenic removal technologies and future trends: a mini review. J Clean Prod 278:123805. https://doi.org/10.1016/j.jclepro.2020.123805

Anderson GL, Williams J, Hille R (1992) The purification and characterization of arsenite oxidase from Alcaligenes faecalis, a molybdenum-containing hydroxylase. J Biol Chem 267(33):23674–23682. https://doi.org/10.1016/S0021-9258(18)35891-5

Andres J, Bertin PN (2016) The microbial genomics of arsenic. FEMS Microbiol Rev 40(2):299–322. https://doi.org/10.1093/femsre/fuv050

Ben Fekih I, Zhang C, Li YP, Zhao Y, Alwathnani HA, Saquib Q, Rensing C, Cervantes C (2018) Distribution of arsenic resistance genes in prokaryotes. Front Microbiol 9:2473. https://doi.org/10.3389/fmicb.2018.02473

Castro-Severyn J, Pardo-Esté C, Mendez KN, Morales N, Marquez SL, Molina F, Remonsellez F, Castro-Nallar E, Saavedra CP (2020) Genomic variation and arsenic tolerance emerged as niche specific adaptations by different Exiguobacterium strains isolated from the extreme Salar de Huasco environment in Chilean—Altiplano. Front Microbiol 11:1632. https://doi.org/10.3389/fmicb.2020.01632

Chen C, Li L, Huang K, Zhang J, Xie W-Y, Lu Y, Dong X, Zhao F-J (2019) Sulfate-reducing bacteria and methanogens are involved in arsenic methylation and demethylation in paddy soils. ISME J 13(10):2523–2535. https://doi.org/10.1038/s41396-019-0451-7

Cullen WR (2014) Chemical mechanism of arsenic biomethylation. Chem Res Toxicol 27(4):457–461. https://doi.org/10.1021/tx400441h

Danczak RE, Johnston MD, Kenah C, Slattery M, Wilkins MJ (2019) Capability for arsenic mobilization in groundwater is distributed across broad phylogenetic lineages. PLoS ONE 14(9):e0221694. https://doi.org/10.1371/journal.pone.0221694

Dastidar A, Wang YT (2009) Arsenite oxidation by batch cultures of Thiomonas arsenivorans strain b6. J Environ Engg 135(8):708–715

de Alencar FLS, Navoni JA, de Amaral VS (2017) The use of bacterial bioremediation of metals in aquatic environments in the twenty-first century: a systematic review. Environ Sci Pollut Res 24(20):16545–16559. https://doi.org/10.1007/s11356-017-9129-8

Dorjderem B, Torres-Martínez JA, Mahlknecht J (2020) Intensive long-term pumping in the Principal-Lagunera Region aquifer (Mexico) causing heavy impact on groundwater quality. Energy Rep 6:862–867. https://doi.org/10.1016/j.egyr.2019.11.020

Garbinski LD, Rosen BP, Chen J (2019) Pathways of arsenic uptake and efflux. Environ Int 126:585–597. https://doi.org/10.1016/j.envint.2019.02.058

Ghosh P, Rathinasabapathi B, Teplitski M, Ma LQ (2015) Bacterial ability in AsIII oxidation and AsV reduction: relation to arsenic tolerance, P uptake, and siderophore production. Chemosphere 138:995–1000. https://doi.org/10.1016/j.chemosphere.2014.12.046

Guo T, Li L, Zhai W, Xu B, Yin X, He Y, Xu J, Zhang T, Tang X (2019) Distribution of arsenic and its biotransformation genes in sediments from the East China Sea. Environ Pollut 253:949–958. https://doi.org/10.1016/j.envpol.2019.07.091

Hamood Altowayti WA, Almoalemi H, Shahir S, Othman N (2020) Comparison of culture-independent and dependent approaches for identification of native arsenic-resistant bacteria and their potential use for arsenic bioremediation. Ecotoxicol Environ Saf 205:111267. https://doi.org/10.1016/j.ecoenv.2020.111267

Hayat K, Menhas S, Bundschuh J, Chaudhary HJ (2017) Microbial biotechnology as an emerging industrial wastewater treatment process for arsenic mitigation: a critical review. J Clean Prod 151:427–438. https://doi.org/10.1016/j.jclepro.2017.03.084
Hu L, Nie Z, Wang W, Zhang D, Long Y, Fang C (2021) Arsenic transformation behavior mediated by arsenic functional genes in landfills. J Hazard Mater 403:123687. https://doi.org/10.1016/j.jhazmat.2020.123687
Hubadillah SK, Othman MHD, Gani P, Sunar NM, Tai ZS, Koo KN, ... & Zahari SSNS (2020) Integrated green membrane distillation-microalgae bioremediation for arsenic removal from Pengorak River Kuantan, Malaysia. Chem Engg Proces-Process Intensification 153:107996.
Hussain MM, Wang J, Bibi I, Shahid M, Niazi NK, Iqbal J, Mian IA, Shaheen SM, Bashir S, Shah NS, Hina K, Rinklebe J (2021) Arsenic speciation and biotransformation pathways in the aquatic ecosystem: the significance of algae. J Hazard Mater 403:124027. https://doi.org/10.1016/j.jhazmat.2020.124027
Irshad S, Xie Z, Mehmood S, Nawaz A, Ditta A, Mahmood Q (2021) Insights into conventional and recent technologies for arsenic bioremediation: a systematic review. Environ Sci Pollut Res 28(15):18870–18892. https://doi.org/10.1007/s11356-021-12487-8
Jasrotia S, Kansal A, Kishore V. V. N. (2014). Arsenic phyco-remediation by Cladophora algae and measurement of arsenic speciation and location of active absorption site using electron microscopy. Microchem J 114:197–202
Jia M-R, Tang N, Cao Y, Chen Y, Han Y-H, Ma LQ (2019) Efficient arsenate reduction by As-resistant bacterium Bacillus sp. strain PVR-YHB1-1: characterization and genome analysis. Chemosphere 218:1061–1070. https://doi.org/10.1016/j.chemosphere.2018.11.145
Jiang D, Li P, Jiang Z, Dai X, Zhang R, Wang Y, ... Wang Y (2015) Chemolithoautotrophic arsenite oxidation by a thermophilic Anoxybacillus flavithermus strain TCC9-4 from a hot spring in Tengchong of Yunnan, China. Front Microbiol 6:360
Jiang Z, Li P, Wang Y, Liu H, Wei D, Yuan C, Wang H (2019) Arsenic mobilization in a high arsenic groundwater revealed by metagenomic and Geochip analyses. Sci Rep 9(1):12972. https://doi.org/10.1038/s41598-019-49365-w
Joshi DN, Flora SJS, Kalia K (2009) Bacillus sp. strain DJ-1, potent arsenic hypertolerant bacterium isolated from the industrial effluent of India. J Hazardous Mater 166(2–3):1500–1505
Kostal J, Yang R, Wu CH, Mulchandani A, Chen W (2004) Enhanced arsenic accumulation in engineered bacterial cells expressing ArsR. Appl Environ Microbiol 70(8):4582–4587. https://doi.org/10.1128/AEM.70.8.4582-4587.2004
Kruger MC, Bertin PN, Heipieper HJ, Arsène-Ploetze F (2013) Bacterial metabolism of environmental arsenic—mechanisms and biotechnological applications. Appl Microbiol Biotechnol 97(9):3827–3841. https://doi.org/10.1007/s00253-013-4838-5
Kumari N, Jagadevan S (2016) Genetic identification of arsenate reductase and arsenite oxidase in redox transformations carried out by arsenic metabolising prokaryotes—a comprehensive review. Chemosphere 163:400–412. https://doi.org/10.1016/j.chemosphere.2016.08.044
Leong YK, Chang J-S (2020) Bioremediation of heavy metals using microalgae: recent advances and mechanisms. Biores Technol 303:122886. https://doi.org/10.1016/j.biortech.2020.122886
Li X, Zhang L, Wang G (2014) Genomic evidence reveals the extreme diversity and wide distribution of the arsenic-related genes in Burkholderiales. PLoS ONE 9(3):e92236. https://doi.org/10.1371/journal.pone.0092236
Lièvremont D, Bertin PN, Lett M-C (2009) Arsenic in contaminated waters: biogeochemical cycle, microbial metabolism and biotreatment processes. Biochimie 91(10):1229–1237. https://doi.org/10.1016/j.biochi.2009.06.016
Liu S, Zhang F, Chen J, Sun G (2011) Arsenic removal from contaminated soil via biovolatilization by genetically engineered bacteria under laboratory conditions. J Environ Sci 23(9):1544–1550. https://doi.org/10.1016/S1001-0742(10)60570-0
Liu Z, Boles E, Rosen BP (2004) Arsenic trioxide uptake by hexose permeases in Saccharomyces cerevisiae. J Biol Chem 279(17):17312–17318

Maciaszczyk-Dziubinska E, Wawrzycka D, Wysocki R (2012) Arsenic and antimony transporters in eukaryotes. Int J Mol Sci 13(3):3527–3548. https://doi.org/10.3390/ijms13033527

Malla MA, Dubey A, Yadav S, Kumar A, Hashem A, Abd Allah EF (2018) Understanding and designing the strategies for the microbe-mediated remediation of environmental contaminants using omics approaches. Front Microbiol 9:1132. https://doi.org/10.3389/fmicb.2018.01132

Mazumder P, Sharma SK, Taki K, Kalamdhad AS, Kumar M (2020) Microbes involved in arsenic mobilization and respiration: a review on isolation, identification, isolates and implications. Environ Geochem Health 42(10):3443–3469. https://doi.org/10.1007/s10653-020-00549-8

Mohapatra D, Mishra D, Chaudhury GR, Das RP (2008) Removal of arsenic from arsenic rich sludge by volatilization using anaerobic microorganisms treated with cow dung. Soil Sediment Contam: Int J 17(3):301–311. https://doi.org/10.1080/15320380802007069

Mujawar SY, Shamim K, Vaigankar DC, Dubey SK (2019) Arsenite biotransformation and bioaccumulation by Klebsiella pneumoniae strain SSSW7 possessing arsenite oxidase (aioA) gene. Biometals 32(1):65–76. https://doi.org/10.1007/s10534-018-0158-7

Mukhopadhyay R, Rosen BP (1998) *Saccharomyces cerevisiae ACR2* gene encodes an arsenate reductase. FEMS Microbiol Lett 168(1):127–136. https://doi.org/10.1111/j.1574-6968.1998.tb13265.x

Naureen A, Rehman A (2016) Arsenite oxidizing multiple metal resistant bacteria isolated from industrial effluent: their potential use in wastewater treatment. World J Microbio Biotechnol 32:1–9

Ojuederie O, Babalola O (2017) Microbial and plant-assisted bioremediation of heavy metal polluted environments: a review. Int J Environ Res Public Health 14(12):1504. https://doi.org/10.3390/ijerph14121504

Plewniak F, Crognale S, Rossetti S, Bertin PN (2018) A genomic outlook on bioremediation: the case of arsenic removal. Front Microbiol 9:820. https://doi.org/10.3389/fmicb.2018.00820

Podgorski J, Berg M (2020) Global threat of arsenic in groundwater. Science 368(6493):845–850. https://doi.org/10.1126/science.aba1510

Qin J, Rosen BP, Zhang Y, Wang G, Franke S, Rensing C (2006) Arsenic detoxification and evolution of trimethylarsine gas by a microbial arsenite *S*-adenosylmethionine methyltransferase. Proc Natl Acad Sci 103(7):2075–2080. https://doi.org/10.1073/pnas.0506836103

Qin J, Lehr CR, Yuan C, Le XC, McDermott TR, Rosen BP (2009) Biotransformation of arsenic by a Yellowstone thermoacidophilic eukaryotic alga. Proc Natl Acad Sci 106(13):5213–5217. https://doi.org/10.1073/pnas.0900238106

Rae ID (2020) Arsenic: its chemistry, its occurrence in the earth and its release into industry and the environment. ChemTexts 6(4):25. https://doi.org/10.1007/s40828-020-00118-7

Rahman MA, Hassler C (2014) Is arsenic biotransformation a detoxification mechanism for microorganisms? Aquat Toxicol 146:212–219. https://doi.org/10.1016/j.aquatox.2013.11.009

Rahman A, Nahar N, Jass J, Olsson B, Mandal A (2016) Complete genome sequence of Lysinibacillus sphaericus B1-CDA, a bacterium that accumulates arsenic. Genome Announc 4(1). https://doi.org/10.1128/genomeA.00999-15

Saba RY, Ahmed M, Sabri AN (2019) Potential role of bacterial extracellular polymeric substances as biosorbent material for arsenic bioremediation. Bioremediat J 23(2):72–81. https://doi.org/10.1080/10889868.2019.1602107

Sariñana-Ruiz YA, Vazquez-Arenas J, Sosa-Rodríguez FS, Labastida I, Armienta MaA, Aragón-Piña A, Escobedo-Bretado MA, González-Valdez LS, Ponce-Peña P, Ramírez-Aldaba H, Lara RH (2017) Assessment of arsenic and fluorine in surface soil to determine environmental and health risk factors in the Comarca Lagunera, Mexico. Chemosphere 178:391–401. https://doi.org/10.1016/j.chemosphere.2017.03.032

Shah MP (2020) Microbial bioremediation & biodegradation. Springer, Singapore

Shah D, Shen MWY, Chen W, Da Silva NA (2010) Enhanced arsenic accumulation in Saccharomyces cerevisiae overexpressing transporters Fps1p or Hxt7p. J Biotechnol 150(1):101–107. https://doi.org/10.1016/j.jbiotec.2010.07.012

Sher S, Hussain SZ, Rehman A (2020) Phenotypic and genomic analysis of multiple heavy metal–resistant Micrococcus luteus strain AS2 isolated from industrial waste water and its potential use in arsenic bioremediation. Appl Microbiol Biotechnol 104(5):2243–2254. https://doi.org/10.1007/s00253-020-10351-2

Sher S, Rehman A (2019) Use of heavy metals resistant bacteria—a strategy for arsenic bioremediation. Appl Microbiol Biotechnol 103(15):6007–6021. https://doi.org/10.1007/s00253-019-09933-6

Sher S, Rehman A (2021) Proteomics and transcriptomic analysis of Micrococcus luteus strain AS2 under arsenite stress and its potential role in arsenic removal. Curr Res Microb Sci 2:100020. https://doi.org/10.1016/j.crmicr.2021.100020

Sher S, Sultan S, Rehman A (2021) Characterization of multiple metal resistant Bacillus licheniformis and its potential use in arsenic contaminated industrial wastewater. Appl Water Sci 11:1–7

Shi L-D, Chen Y-S, Du J-J, Hu Y-Q, Shapleigh JP, Zhao H-P (2019) Metagenomic evidence for a methylocystis species capable of bioremediation of diverse heavy metals. Front Microbiol 9:3297. https://doi.org/10.3389/fmicb.2018.03297

Singh N, Ghosh PK, Chakraborty S, Majumdar S (2021) Decoding the pathways of arsenic biotransformation in bacteria. Environ Sustain 4(1):63–85. https://doi.org/10.1007/s42398-021-00162-0

Sodhi KK, Kumar M, Agrawal PK, Singh DK (2019) Perspectives on arsenic toxicity, carcinogenicity and its systemic remediation strategies. Environ Technol Innov 16:100462. https://doi.org/10.1016/j.eti.2019.100462

Tsai S-L, Singh S, Chen W (2009) Arsenic metabolism by microbes in nature and the impact on arsenic remediation. Curr Opin Biotechnol 20(6):659–667. https://doi.org/10.1016/j.copbio.2009.09.013

Uppal JS, Shuai Q, Li Z, Le XC (2017) Arsenic biotransformation and an arsenite S-adenosylmethionine methyltransferase in plankton. J Environ Sci 61:118–121. https://doi.org/10.1016/j.jes.2017.11.010

Verma S, Verma PK, Meher AK, Dwivedi S, Bansiwal AK, Pande V, Srivastava PK, Verma PC, Tripathi RD, Chakrabarty D (2016) A novel arsenic methyltransferase gene of Westerdykella aurantiaca isolated from arsenic contaminated soil: phylogenetic, physiological, and biochemical studies and its role in arsenic bioremediation. Metallomics 8(3):344–353. https://doi.org/10.1039/C5MT00277J

Wang L, Yin Z, Jing C (2020) Metagenomic insights into microbial arsenic metabolism in shallow groundwater of Datong basin, China. Chemosphere 245:125603. https://doi.org/10.1016/j.chemosphere.2019.125603

Wang Y, Wang S, Xu P, Liu C, Liu M, Wang Y, Wang C, Zhang C, Ge Y (2015) Review of arsenic speciation, toxicity and metabolism in microalgae. Rev Environ Sci Bio/Technol 14(3):427–451. https://doi.org/10.1007/s11157-015-9371-9

Webster TM, Reddy RR, Tan JY, Van Nostrand JD, Zhou J, Hayes KF, Raskin L (2016) Anaerobic disposal of arsenic-bearing wastes results in low microbially mediated arsenic volatilization. Environ Sci Technol 50(20):10951–10959. https://doi.org/10.1021/acs.est.6b02286

Xie S, Liu J, Yang F, Feng H, Wei C, Wu F (2018) Arsenic uptake, transformation, and release by three freshwater algae under conditions with and without growth stress. Environ Sci Pollut Res 25(20):19413–19422. https://doi.org/10.1007/s11356-018-2152-6

Yamamura S, Amachi S (2014) Microbiology of inorganic arsenic: from metabolism to bioremediation. J Biosci Bioeng 118(1):1–9. https://doi.org/10.1016/j.jbiosc.2013.12.011

Yan G, Chen X, Du S, Deng Z, Wang L, Chen S (2019) Genetic mechanisms of arsenic detoxification and metabolism in bacteria. Curr Genet 65(2):329–338. https://doi.org/10.1007/s00294-018-0894-9

Yang P, Ke C, Zhao C, kuang Q, Liu B, Xue X, Rensing C, Yang S (2020) ArsM-mediated arsenite volatilization is limited by efflux catalyzed by As efflux transporters. Chemosphere 239:124822. https://doi.org/10.1016/j.chemosphere.2019.124822

Yin X, Wang L, Duan G, Sun G (2011) Characterization of arsenate transformation and identification of arsenate reductase in a green alga Chlamydomonas reinhardtii. J Environ Sci 23(7):1186–1193. https://doi.org/10.1016/S1001-0742(10)60492-5

Yuan C, Lu X, Qin J, Rosen BP, Le XC (2008) Volatile arsenic species released from *Escherichia coli* expressing the AsIII S-adenosylmethionine methyltransferase gene. Environ Sci Technol 42(9):3201–3206. https://doi.org/10.1021/es702910g

Zhai W, Qin T, Li L, Guo T, Yin X, Khan MI, Hashmi MZ, Liu X, Tang X, Xu J (2020) Abundance and diversity of microbial arsenic biotransformation genes in the sludge of full-scale anaerobic digesters from a municipal wastewater treatment plant. Environ Int 138:105535. https://doi.org/10.1016/j.envint.2020.105535

Zhang J, Cao T, Tang Z, Shen Q, Rosen BP, Zhao F-J (2015) Arsenic methylation and volatilization by arsenite *S*-adenosylmethionine methyltransferase in Pseudomonas alcaligenes NBRC14159. Appl Environ Microbiol 81(8):2852–2860. https://doi.org/10.1128/AEM.03804-14

Zhou X, Kang F, Qu X, Fu H, Alvarez PJJ, Tao S, Zhu D (2020) Role of extracellular polymeric substances in microbial reduction of arsenate to arsenite by *Escherichia coli* and *Bacillus subtilis*. Environ Sci Technol 54(10):6185–6193. https://doi.org/10.1021/acs.est.0c01186

Zhu Y-G, Xue X-M, Kappler A, Rosen BP, Meharg AA (2017) Linking genes to microbial biogeochemical cycling: lessons from arsenic. Environ Sci Technol 51(13):7326–7339. https://doi.org/10.1021/acs.est.7b00689

Microbial Biofilms in Wastewater Remediation

Ayushi Sharma and Sahil Dhiman

1 Introduction

The growth in industrial activity has resulted in significant pollution of the aquatic environment. Wastewaters contain high quantities of organic pollutants, detergents, pesticides, heavy metals, humic substances, and phenols. These compounds are toxic to the living population and their accumulation in the food chain results in hazardous consequences (Asri et al. 2019). Wastewater treatment is necessary since insufficiently treated wastewaters cause microbial contamination of natural environments, toxicity, and eutrophication (Turki et al. 2017). According to the World Health Organization, decontamination of the drinking water and the resultant diarrhoeal infections is estimated to cause 485,000 deaths each year (WHO 2019). The WHO regards polluted water as the cause of 30% of the total diseases and subsequent 40% deaths worldwide (Sehar and Naz 2016). The government is therefore imposing various regulations for discharge of industrial wastewaters.

Undesirable elements of the wastewater are removed using processes such as membrane treatment, adsorption, and flocculation (Asri et al. 2019). Increasing quantities of wastewater, the strict quality control regulations, and limited land spaces have increased the demand of novel wastewater treatment strategies relying on biofilm-based techniques (Andersson et al. 2008). Remediation using biofilms is a cost-effective and environment-friendly approach for cleaning environmental pollutants (Mitra and Mukhopadhyay 2016). Immobilization of bacteria on to solid surfaces

A. Sharma (✉)
Department of Biotechnology and Bioinformatics, Jaypee University of Information Technology, Waknaghat, Solan, Himachal Pradesh 173234, India
e-mail: ayushisharma3194@gmail.com

S. Dhiman
Department of Civil Engineering, National Institute of Technology, Hamirpur, Himachal Pradesh 177005, India

M. P. Shah (ed.), *Microbial Technologies in Industrial Wastewater Treatment*,
https://doi.org/10.1007/978-981-99-2435-6_7

either by attachment or entrapment leads to the development of biofilms. The fixed-film reactors are generally more efficient in wastewater treatment in comparison to the freely suspended bacteria. An increased rate of biofilm formation is evident in fixed-film reactors when compared to activated sludges (Schneider and Topalova 2013). Biofilms of wastewater treatment facilities show increased stability to toxic compounds, pollutants, and environmental changes. High volumetric wastewater loadings are efficiently treated by the coexisting aerobic and anoxic microbes within a single processing unit (Sfaelou et al. 2015; Shah Maulin 2020). Biofilms control the reaction rate and mechanisms of the resident population in a more efficient manner (Wang et al. 2019). Moreover, the use of smaller volumes of the microbial biomass leads to a cost-efficient treatment process (Schneider and Topalova 2013; Sfaelou et al. 2015).

The chapter summarizes the pre-existent knowledge of the use of biofilms in wastewater treatment. It discusses the beneficial role of microbial biofilms in bioremediation. It also presents an overview of the conventional and engineered strategies for screening wastewater biofilms, along with a special focus on the growth sectors in biofilm-related wastewater treatment research.

2 Microbial Biofilms

Microbes form biofilms by attaching to substrates. Biofilm-forming microbes are enclosed within a self-generated extracellular polymeric matrix. The phenomenon was firstly observed by Antoine von Leeuwenhoek on his own teeth (Jamal et al. 2018; Shah Maulin 2021a, b). Biofilm development encompasses five stages: initial adhesion of planktonic microbes to surfaces submerged in aqueous medium, irreversible attachment and the production of extracellular matrix, monolayer microcolony formation by the replicating early colonizers, biofilm maturation by attachment of debris and new planktonic cells, and finally biofilm dispersion (Fig. 1). The electrostatic forces of attraction, interactions between acids and bases, and Brownian motion forces are the major forces that facilitate microbial adhesion to substrates (Lasa and Penadés 2006).

Despite the negative facets of biofilms, bacterial biofilms are beneficial in the industrial and agricultural settings. The beneficial biofilm biomass is being used as biological control agents to enhance the production of crops. In addition, biofilms have been exploited for protecting the marine ecosystems, bioremediation of hazardous pollutants, treatment of wastewater, and prevention of membrane corrosion (Muhammad et al. 2020). Therefore, the formation of beneficial biofilms is being encouraged by manipulating quorum sensing signals, adhesion surfaces, and environmental conditions.

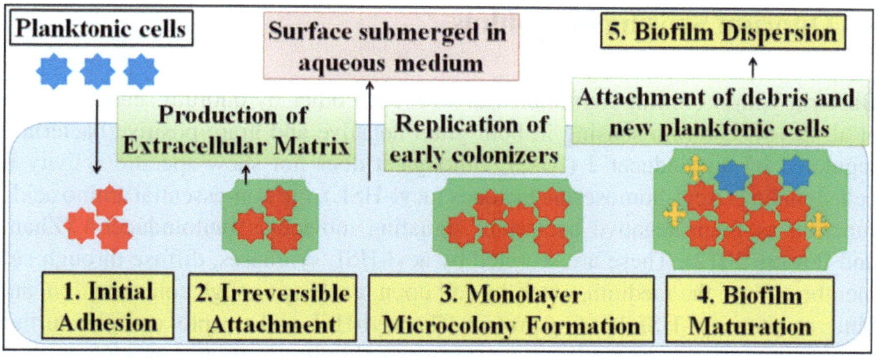

Fig. 1 Biofilm development stages

3 Biofilm Extracellular Polymeric Substances

50–90% of the total organic carbon in biofilms is embedded in extracellular polymeric substances (EPSs) (Donlan 2002) that provide a matrix for the formation of aggregates. The EPS provides mechanical stability to biofilms. Usually, the secreted matrix is 0.2–1.0 μm thick for a bacterial biofilm of size ranging between 10–30 nm (Jamal et al. 2018). Abundant carbon and limited availability of phosphate, potassium, and nitrogen aid in thickening the sessile structures. EPSs are usually composed of about 97% water (Jamal et al. 2015). Water channels enclosed within heterogeneous communities facilitate transportation of oxygen, water, and nutrients to the resident cells together with collection of harmful metabolic products (Richards and Ojha 2014). Most of the accumulated products are both hydrophobic and hydrophilic polymers of high molecular weight formed from sugar residues that are secreted into surroundings by bacteria. Presence of channels also provides a platform for establishing cell communication within cell aggregates (Donlan 2002).

Biofilm matrix protects the resident cells from the action of environmental stresses, pollutants, solvents, and toxic chemicals (Mitra and Mukhopadhyay 2016). The matrix generates an area with low oxygen concentration at the center of the biofilm. This area attracts anaerobes and aerobes, nitrifiers and heterotrophs, and both the sulfate reducers and oxidizers, correspondingly promoting degradation of pollutants in the natural and engineered systems (Field et al. 1995). Heavy metals are removed from the aqueous phase by the matrix secreted by cyanobacteria which acts as a biosorbent (De Philippis et al. 2011). Besides, the extracellular enzymes of the biofilm matrix are capable of decontaminating organic compounds (Flemming and Wingender 2010). Phosphorous is removed and recovered from wastewater by the extracellular matrix of phosphorous-accumulating microorganisms (Zhang et al. 2013).

4 Quorum Sensing in Biofilms

Bacteria within biofilm communicates with each other by quorum sensing (Jamal et al. 2018). Quorum sensing in both gram-negative and gram-positive bacteria is regulated by autoinducer-2 (AI-2), although it does not show specific activity in cell signaling. Acyl-homoserine lactones (acyl-HSL), the non-essential amino acids, function as gram-negative bacterium signaling molecules (autoinducers) (Zhang and Powers 2012). These are secreted by acyl-HSL synthases, diffuse through cell membrane into the medium, enter the cell upon reaching enough concentration, and bind receptor acyl-HSL binding protein. The acyl-HSL and receptor complex further activate transcription by binding target virulent or biofilm-forming genes (Gupta et al. 2016).

Oligopeptides on the contrary are utilized as signaling molecules by gram-positive bacterium. ABC protein complex transports oligopeptides from cell interior to the extracellular space between different cells, from where these bind to protein kinases and receptor complexes upon reaching an essentially high concentration in the medium. Oligopeptides bind, phosphorylate, and activate protein kinases; simultaneously turning on regulatory proteins which bind particular target genes and start up transcription. This system of double regulation detects changes in peptides and patterns of gene expression (Singh et al. 2017).

Quinolones are signaling molecules that belong to LuxR family of proteins. An increase of 95% was reported in the biofilm mass of *Halanaerobium praevalence* on addition of 100 nM quinolone-type signaling molecules (Monzon et al. 2016). This simultaneously led to a 30% increase in the power density. Valle et al. (2004) showed that phenol-degrading activated sludge contained seven proteobacterial strains exhibiting acyl-HSL activity. The reports suggest that the bacterial community communicates via cell–cell interactions mediated by quorum sensing.

5 Factors Affecting Biofilm Diversity

Several factors play a role in mediating the process by which the microbes form biofilms. The following section describes the major factors influencing microbial biofilm formation.

i. *Nutrient composition*

Abundance of nutrients promotes bacterial biofilm formation, whereas a decline in nutrients causes dispersal of biofilm cells (Sehar and Naz 2016).

ii. *Concentration of H^+ ions*

Changes in pH also affect biofilm development (Sharma et al. 2021b). Bacteria are efficient in adjusting to internal or external pH fluctuations. However, cellular processes such as the production of extracellular polysaccharides are negatively affected by the changes in pH.

iii. *Growth temperature*

Variation in the optimum growth temperature impacts microbial activities. Slightest change in the temperature affects enzyme reaction rates and may reduce the efficiency of bacterial growth to a great extent. Optimum temperature for bacteria thriving in cooling water systems ranges around 40 °C (Ells and Hansen 2006).

iv. *Surface topography*

Microbial attachment is further influenced by surface topography. Surface roughness generally enhances bacterial adhesion to substrates by reducing the shear forces acting on bacterial cells (Donlan 2002). Furthermore, porous surfaces favor cell adhesion when compared to regular surfaces (Hoh et al. 2016).

v. *Substrate charge, elasticity, and cell surface hydrophobicity*

Bacterial attachment varies with substrate charge, elasticity, and hydrophobicity (Sharma et al. 2022a, 2022b; Prakash et al. 2003). Cell surface hydrophobicity tends to increase bacterial interactions with the surface (Donlan 2002). Hydrophobicity reduces the forces of repulsion between the colonizing substratum and the bacterial surface (Muhammad et al. 2020). Kesaano and Sims (2014) reported that microbes show increased adhesion to hydrophobic surfaces such as Perspex, titanium, and stainless steel. Also, since bacterial cells are negatively charged due to the presence of phosphate, carboxyl, and amino groups, positively charged surfaces promote bacterial attachment (Muhammad et al. 2020).

vi. *Hydrodynamics*

The flow velocity of water, turbulence, and water hydrodynamics also influence the rates of biofilm formation and development (Simoes et al. 2007).

vii. *Extracellular DNA*

Extracellular DNA, an important component secreted by bacterial extracellular matrix, serves as a nutrient source for the growth of biofilms (Das et al. 2013).

viii. *Divalent cations*

Divalent cations like Ca^{2+} modify bacterial cell surface and impart antimicrobial and detergent resistance to biofilms (Mulcahy et al. 2008). These cations facilitate the attachment of microbial cell clusters to anaerobic sludge granules and activated sludge flocs by bridging the negative charges on extracellular polymers (De Kerchove and Elimelech 2008).

ix. *Ionic strength of the aqueous medium*

Ionic strength of the aqueous medium can also be exploited for increasing the density and mechanical strength of biofilms, making the biofilms durable to the flow of wastewater (Donlan 2002).

6 Biofilms in Wastewater Treatment

The use of biofilm for removal of pollutants is serving as an appealing, cost-effective, and environment-friendly strategy (Asri et al. 2019). Differential gene expression within bacterial biofilms leads to metabolic shifts that aid degradation of pollutants (Mitra and Mukhopadhyay 2016). Biofilm-based bioreactors are used on a commercial scale for sorption and biochemical conversion of hydrocarbons and heavy metals from municipal and industrial wastewaters. The reactors are advantageous as they offer several advantages over the conventional treatment processes. The reactors retain biofilm biomass for longer durations, exhibit increased tolerance to toxic compounds, have large mass transfer areas, favor the coexistence of aerobic and anoxic microbes, and show an enhanced volumetric biodegradation capacity (Mitra and Mukhopadhyay 2016).

The attached growth system of wastewater treatment utilizes microbial biofilms attached to support medium. Rocks, sand, gravels, stones, and soil act as the natural support medium, whereas plastic and rubber act as artificial support systems. The media, interactions between cells, and surface polymers influence biomass attachment to surfaces. Moreover, biomass agglomerates (granules) itself act as the support for wastewater treatment using microbes. Biofilms survive in wastewater by feeding off the nutrients and organic matter flowing over them (Sehar and Naz 2016). Trickling filters, constructed wetlands, rotating biological contactors, and membrane bioreactors are the commonly used wastewater treatment systems that rely on attached biomass growth (Dhiman and Sharma 2022) (Fig. 2). Microbes in the aerobic reactors utilize dissolved oxygen for converting organic compounds to carbon dioxide and biomass, whereas the anaerobes convert organic compounds to water and carbon dioxide in the absence of oxygen. Generally, aerobic reactors are preferred for treating wastewater containing chemical oxygen demand (COD) concentrations less than 1000 mg/L (Cakir and Stenstrom 2005). On the contrary, anaerobic reactors are used for treating wastewater with over 4000 mg/L COD (Chan et al. 2009). However, anaerobic reactors work on less energy and lead to less nutrient recovery when compared to aerobic reactors (Asri et al. 2019).

The ideal support systems for wastewater treatment must be non-toxic to the microbes, have stable chemical properties, good mass transfer characteristics, and an appreciable structural strength (Nie and Gu 2010). Carbon fibers (Matsumoto et al. 2012), polypropylene, polystyrene, tire-derived rubber (Khatoon et al. 2014), and pebbles (Khan et al. 2015) have been exploited as support mediums in fixed-film bioreactors. Sfaelou et al. (2015) show that the reactors made up of either hydrophobic polyethylene or polar polyvinyl alcohol materials are equally efficient in studying biofilm formation. The results show no significant variations between the two reactors in terms of removal efficiencies of organic matter, phosphorous, or nitrogen. However, biofilms formed using supports having polar surface groups accumulated more microbial biomass for wastewater treatment. In addition, the elemental composition of the filter medium can be quantified by spectroscopic techniques such as

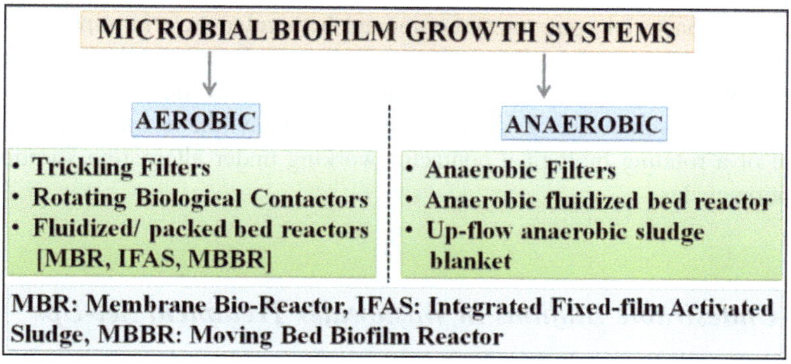

Fig. 2 Microbial biofilm growth systems in wastewater treatment

energy-dispersive X-ray spectroscopy (EDS/ EDX/ XEDS), or X-ray photoelectron spectroscopy (XPS) (Crist 2000; Hafner 2006).

Microbial communities attached to inert solid surfaces are serving as a promising technology for treating wastewater (Sfaelou et al. 2015). The technology has been applied to remove nitrogen (Seo et al. 2001), phthalic acid esters (Wang et al. 2003), several organic constituents (Inamori et al. 1989), and toxic compounds such as 2,4-dichlorophenol (Quan et al. 2003), Cr(VI) (Papadimitriou et al. 2010), and phenol (Wang et al. 1995; Papadimitriou et al. 2006) from wastewater. Water contaminated with heavy metals like nickel, zinc, copper, cobalt, and cadmium has been treated using biofilm-forming sulphate-reducing bacteria. The metals are converted into their respective sulfides with the aid of bacteria involved in biofilm formation (Muyzer and Stams 2008). Gebara (1999) utilized an activated sludge biofilm wastewater treatment system for boosting the biochemical oxygen demand (BOD) removal and the settling efficiencies. The system can be beneficial for treating wastewater in areas with land constraints. Andersson et al. (2008) studied the role of single and mixed bacterial strains in the treatment of wastewater. They studied the ability of 13 bacterial strains, *Acinetobacter calcoaceticus*, *Aeromonas hydrophila* L6, *Bacillus cereus* SJV, *Brachymonas denitrificans* B79, *B. denitrificans* SJV, *Comamonas denitrificans* 110, *C. denitrificans* 123, *Delftia* sp. SJV, *Escherichia coli* AF1000, *E. coli* K-12, *Klebsiella pneumoniae* SJV, *Pseudomonas aeruginosa*, and *Zoogloea ramigera*, for wastewater treatment. Also, competition between the mixed microbial populations was evaluated for supporting the development of biofilm-based wastewater treatment process. Overall, the study aimed at providing information useful for the design of biofilm-based wastewater treatment set-up and facilitating the choice of inoculums for the same. The synergistic and antagonistic associations were suggested to be exploited for forming tailored systems for clearing municipal and industrial wastewater.

A bench scale biofilm reactor showed excellent TN, NH_4–N, and COD_{Cr} removal efficiencies of 70.6%, 98.5%, and 92.8%, respectively, from an influent having 220 mg/L COD_{Cr}, a salinity of 3%, and containing 32 mg/L NH_4–N (Tian et al. 2017). The system exhibited stability and showed potential for treating domestic saline sewage. The bacterial community comprised betaproteobacteria, alphaproteobacteria, gammaproteobacteria, flavobacteria, and anaerolineae, while nitrososphaera showed up as the dominant nitrifier. In another study, Costley and Wallis (2001) showed an 84% removal of Zn^{2+}, Cu^{2+}, and Cd^{2+} in synthetic wastewater through the use of a rotating biological contactor working under alternating sorption and desorption cycles.

6.1 Undesirable Biofilms in Wastewater Treatment Set-Ups

Biofilms can largely cause material corrosion and degradation besides being useful for wastewater treatment (Hamadi et al. 2005). The major biofilm-related problems can be observed in corrosion of sewer pipelines and the filtration membranes in drinking and wastewater treatment systems (Lewandowski and Boltz 2011). Microbial biofilms can be problematic in dairy plants owing to generation of foul odor and blockading of pipes. Biofilms in the dairy industry can somehow reduce treatment capacity, and add on to the cleaning and maintenance costs. Excessive build-up of the biofilms therefore needs to be monitored. Dixon et al. (2018) studied the bacterial nutritional requirements and the effects of ions in the growth and development of bacteria forming biofilms in dairy wastewater. The study showed that addition of calcium up to 0.1 M stimulated bacterial biofilm formation, whereas addition of calcium over 0.5 M decreased biofilm formation ability of the bacteria.

7 Strategies for Screening Wastewater Biofilms

7.1 Conventional Strategies

i. *Determination of biomass weight and optical density*

A digital weighing balance can be used to determine the biofilm dry and wet weight under aseptic conditions. However, weight of the natural support systems can be evaluated after drying in oven at 60 °C. Biofilms of the wastewater treatment systems can also be recovered using sonication. The quantity of biofilms is then estimated using optical density method (Sehar and Naz 2016).

ii. *Potentiometric mass titration and diffuse reflectance spectroscopy*

Polymers do not absorb strongly in the visible and ultraviolet regions of the spectrum. Diffuse reflectance (DR) spectroscopy has been specifically used to observe biofilms formed on polymers owing to its ability to absorb light in the stated spectrums. Also, Potentiometric mass titration (PMT) technique helps decipher more information on characteristics such as the nature of biofilms and the content of hydrogen ions consumed by biofilms; more the hydrogen ion consumption, more the quantity of biofilms formed. Sfaelou et al. (2015) monitored biofilm formation on high-density polyethylene and porous polyvinyl alcohol gel biocarriers using PMTs and DR spectroscopy. The biofilms were evaluated for their efficiency to treat wastewater using these easy and quick methods.

iii. *Microscopic investigation*

The non-invasive techniques such as light and electron microscopy are more accurate in biofilm visualization as these accurately estimate the amount of biofilm formed on support media. Scanning electron microscopy efficiently examines the morphology and topography of biofilms (Sharma et al. 2021b). Advanced techniques such as laser scanning microscopy (LSM), magnetic resonance imaging (MRI), scanning transmission X-ray microscopy (STXM), and confocal laser scanning microscopy (CLSM) are also being used for analyzing the composition, structure, and dynamics of microbial communities (Palmer and Sternberg 1999).

iv. *Determination of metabolic activity*

Biofilm metabolic activity can be determined by examining the rate of conversion of specific substrates. Naz et al. (2016) evaluated the activity of *Nitrosomonas* spp. by measuring nitrite strength from known concentration of $(NH_4)_2SO_4$.

7.2 Engineered Molecular Strategies

i. *Cloning and sequencing*

Biofilm communities can be studied by cloning techniques. Nucleic acid is extracted from the biofilms and the 16S rRNA is amplified using either bacterial or archaeal universal primers. The polymerase chain reaction (PCR) products are then cloned into a plasmid and transformed into cells. Isolation of plasmid DNA from the clones further helps in creating clone libraries. Finally, phylogenetic softwares and tools like PHYLIP, SeqLab, ARB, and PAUP are used for identifying the respective clones. Cloning is combined with other advanced techniques for exploring biofilm communities of the wastewater treatment systems (Sehar and Naz 2016).

ii. *Denaturing gradient gel electrophoresis*

Microbial fingerprinting methods are highly efficient in revealing the profile of biofilm communities. Denaturing gradient gel electrophoresis (DGGE) is a technique that generates genetic fingerprints of complex microbial populations. A direct comparison between different microbial communities has been done previously using DGGE. The technique utilized 16S rDNA for analyzing the structural diversity (Muyzer et al. 1993). Turki et al. (2017) also assessed the structure and the diversity of bacterial biofilm community in wastewater treatment using a rotating bio-disk system, DGGE, and 16S rRNA method. Their results revealed that the rotating bio-disk was more efficient in wastewater purification during the summer season, when compared to the purification efficiency during the winter season. The bacterial communities present in the summers were more complex, variable, and abundant.

iii. *Fluorescence in situ hybridization*

Fluorescence in situ hybridization (FISH) targets 16S, 18S, 23S rRNA, and mRNA. Specific fluorescent probes are used to differentiate individual species from a mixed microbial population (Amann et al. 1990). It is an easy and fast approach for visualizing and quantifying microorganisms (Sanz and Köchling 2007). FISH is used with CLSM for obtaining three-dimensional images of thick samples containing high load of biofilms and sludge flocs. The technique works by fixing the specimen either by cross-linking agents like aldehydes, or precipitation agents such as methanol or ethanol. Surfaces of the specimen are subjected to pre-treatment by coating with either poly-L-lysine or gelatin. Further, mixture of salts, detergents, fluorescent probes, and formamide are mixed inside a dark humid chamber at a temperature ranging between 37 and 50 °C and used for sample hybridization. The unbound probe is rinsed, dried, and mounted for visualizing samples using a conventional epifluorescence microscope. The apparatus is coupled to a charged coupled device and a camera for measuring the activity of individual cells within biofilms. FISH-based methods are increasingly used to investigate the morphology and composition of microbial biofilms prevalent in wastewater treatment units (Moter and Göbel 2000).

iv. *Terminal restriction fragment length polymorphism*

Terminal restriction fragment length polymorphism (T-RFLP) is a method that relies on nucleic acids for identifying microbial populations (Malik et al. 2008). For RFLP, DNA or RNA is extracted from a microbial community. The extracted nucleic acids are amplified using a PCR. The PCR products are digested enzymatically with restriction enzymes. Finally, the amplified fragments are separated electrophoretically by using an agarose gel. The sequences can be compared to sequence databases for confirmation of the fragments. T-RFLP is more sensitive than DGGE as it is capable of detecting the sequences low in number (Von Mering et al. 2007).

v. *Next-generation sequencing*

Next-generation sequencing (NGS) methods are increasingly being used for determining the dynamics and functions of microbes in the environment. The techniques generate large amounts of high-throughput data with maximum precision and a comparatively low cost. Roche 454 is a well-known platform for analyzing microbial communities. It is based on pyrophosphate release principle. Illumina is another common NGS approach. Other NGS technologies are SOLiD, DNA nanoball sequencing, ion torrent, and Qiagen gene reader. The third-generation sequencing encompasses the Heliscope, a sequencer based on single molecule real time approach, and the Oxford nanopore sequencing technologies. Several bioinformatics tools and softwares such as MOTHUR, MOCAT, DADA2, MEGAN, and GBDP are accordingly used to manage the extensive data generated from NGS approaches (Sharma et al. 2021a).

8 Growth Sectors in Biofilm-Related Wastewater Treatment Research

8.1 *Biofilms in the Dairy Industry*

The total nitrate reductase and dehydrogenase activities, the enzyme indicators of biofilm biomass have been exploited for dairy wastewater treatment using a sequencing batch biofilm reactor under anaerobic conditions (Schneider and Topalova 2013). Anaerobic treatment was done for 282 days. Denitrifying bacteria, anaerobic spore-forming microbes, and anaerobic heterotrophs were exploited for investigating the microbial structure of biofilms. Synergistic associations of the fixed and suspended biomass in the system were efficient in removing up to 67% nitrates and up to 90% mineralized organic pollutants from the dairy wastewater.

8.2 *Biofilms in the Wine Industry*

Wineries generate wastewaters high in organic load, having high carbon–nitrogen ratios, and low pH. The discharge pollutes surface and ground water and contaminates the soil. Rotating biological contactors are cheap alternatives to endure the fluctuating organic loads of wine industries. A rotating biological contactor efficiently reduced the influent COD of the effluent discharged from a wine industry by 23%, exhibiting a decline from 3828 mg/L to 2910 mg/L. Simultaneously, the pH increased from 5.77 to 6.13 (0.95 unit increase) at 1 h of the average retention time (Coetzee et al. 2004).

8.3 *Biofilms in the Tannery Industry*

Untreated wastewater from tanneries contains heavy metals that present a major environmental challenge to the society. The inorganic metals such as lead, cadmium, mercury, and chromium need to be remediated before the tannery effluent is discharged into the environment. Bacterial biofilms show biosorption capacity for synergistically removing the toxic heavy metals from wastewaters (Igiri et al. 2018). Bioremediation using biofilms is an environment-friendly approach in nature revitalization using cost-effective techniques (Hrynkiewicz and Baum 2014). However, the approach suffers drawbacks such as the inefficient degradation of heavy metals, and the production of toxic metabolites by the degrading microbes. Grujić et al. (2017) showed that the metal removal efficiency of *Rhodotorula mucilaginosa* biofilm cells (91.71 - 95.39%) was much higher than that of the corresponding planktonic cells (4.79–10.25%).

8.4 Microalgal Biofilms

Microalgal biofilms are complex consortia of photosynthetic and heterotrophic inhabitants that are both unicellular and multicellular. The inhabitants comprise diatoms, cyanobacteria, fungi, and bacteria (Miranda et al. 2017). These can be established on surfaces receiving sufficient moisture and light (Osorio et al. 2021). Microalgae efficiently remove the primary nutrients such as phosphorous, nitrogen, carbon, micropollutants, and heavy metals from municipal, mining, and animal wastewaters (Miranda et al. 2017). They are advantageous in synthesizing high-value biomass. Algal biomass can also be utilized for animal feedstock, biofertilizer, and biofuels. Moreover, remediation using microalgae generates oxygen and carbon dioxide instead of the greenhouse gases methane and carbon dioxide generated by the usual aerobic digesters and anaerobic fermenters used for wastewater treatment (Hu et al. 2021). Microalgae are used in the secondary and tertiary wastewater treatment units for removing coliform bacteria, reducing both the BOD and COD, capturing carbon dioxide, and accumulating and degrading polycyclic aromatic hydrocarbons (Osorio et al. 2021). Miranda et al. (2017) showed that the microalgal biofilms can be used for the treatment of commercial wastewaters by following high-efficiency and low-energy strategies. These biofilms were also shown as serving as novel sustainable feedstocks for producing bio-hydrogen and biodiesel as renewable bioenergy. However, microalgal biofilms are at the development stage for industrial implementation (Hu et al. 2021). Nevertheless, some researchers highlight the adverse outcomes of the use of microalgal biofilms for wastewater treatment in view of clogging issues (Kesaano and Sims 2014).

8.5 Bioenergy Production

Microalgae can also be exploited for biofuel production. However, production of biodiesel using microalgae incurs heavy expenses, accounting for up to 30–40% of the total cost. Algal biofilms offer an advantage for cost reduction by concentrating microalgae to make harvesting and dewatering a lot cheaper and easier (Miranda et al. 2017). They are advantageous since these do not require arable lands and therefore do not compete with crops. The use of algal biomass for biogas production should however be a coupled process for overcoming the high cost of fuel production. Algae can be cultivated in the digestate. It can then serve as nutrient input for generating carbon dioxide which can be utilized as biomethane and biogas by the conversion technologies (Zabed et al. 2020; Osorio et al. 2021).

8.6 *Microbial Fuel Cells*

Biofilms can be engineered for human benefits such as fuel and electricity production, and pollution remediation (Sehar and Naz 2016). Biofilms have also been exploited to control energy crisis using microbial fuel cells (MFCs). The process is run using electrogenic microbes that either accept or donate electrons to electrodes, using the chemical energy of the organic and inorganic compounds (Jamal et al. 2018). The microbes are known to form biofilms called electroactive biofilms on the surface of electrodes. The recalcitrant heavy metals can be efficiently remediated from wastewaters using MFCs (Igiri et al. 2018). MFCs generally utilize carbon-based support materials such as carbon felt, carbon paper, and carbon cloth (Asri et al. 2019). MFCs are run using algal and bacterial species; however, a plethora of microbes can be exploited for green energy production (Saba et al. 2017).

A typical MFC works on biochemical and electrochemical reactions. It consists of anodic and cathodic chambers enmeshed in a conducting material. Industries such as the refinery (Zhang et al. 2014) and food-processing units (Blanchet et al. 2015) allow their wastewater to be used as inoculums in the anode chamber of the MFCs. Moreover, heavy metal-rich wastewater is increasingly being used as alternative electron acceptors. In addition, wastewaters with COD concentrations over 8000 mg/L are attractively used for running MFCs. MFCs have hence been claimed as a promising approach for both the removal of pollutants and the generation of electricity. The approach is energy efficient, leads to lowered biomass production, and aids COD removal without additional oxygenation (Asri et al. 2019).

Irrespective of the use of MFCs for wastewater treatment and energy generation, MFC-based biosensors are highly sensitive and stable. The biosensors are capable of working in remote locations deprived of electricity supplies. Moreover, they are utilized for testing microbial load, the BOD, detect the presence of cytotoxic elements and corrosive biofilms, and monitor microbial activity (Angelaalincy et al. 2018).

9 Control and Prevention of Corrosion in Water Systems

Microbial activity can both induce the formation of, and inhibit corrosion of substrates (Zuo 2007). The beneficial biofilms are gaining interest for preventing corrosion and serving as an efficient, cost-effective, and environment-friendly approach to combat corrosion. The biofilms work by removing the corrosive substances such as oxygen by exploiting the respiration mechanism of aerobic bacteria. The sulfate-reducing bacteria, generally known for inducing corrosion are then inhibited by the antimicrobial compounds secreted by biofilms. Further, biofilms produce the γ-polyglutamate protective coats. Lastly, the biofilms serve as a barrier to diffusion for hindering metal dissolution (Guo et al. 2018). The use of beneficial biofilms for preventing corrosion has been reported for aluminum, carbon steel, stainless steel, and copper. The use of bacterial biofilms is an emerging area and requires extensive research. However, the literature provides some studies exploiting beneficial biofilms for corrosion control. Zuo et al. (2004) report curtail in the corrosion of mild steel by biofilms formed by gramicidin-S-producing *Bacillus brevis*. The bacterium suppressed the growth of *Desulfosporosinus orientis*, the sulfate-reducing bacterium, and *Leptothrix discophora* SP-6, the iron-oxidizing bacterium. Another study shows suppression in metal corrosion by inhibition of the sulfate-reducing bacteria, *D. gigas,* and *D. vulgaris*, by genetically engineered *B. subtilis* (Zuo 2007). The bacterium was able to show anti-corrosion activity through the action of antimicrobial compounds such as probactenecin, bactenecin, and indolicidin, secreted by its biofilms.

10 Conclusion

Biofilm-mediated treatment of wastewater is an attractive choice for environmental pollution mitigation. The disadvantages associated with the use of either aerobic or anaerobic reactors can be minimized by using aerobic-anaerobic systems. Combination of treatment systems relying on planktonic bacteria and biofilm reactors will maximize the strength of individual treatment systems. Furthermore, recalcitrant pollutants can be remediated by synergistic association of chemical treatment and phytoremediation. In other cases, bacteria-fungi interactions can be exploited for degrading xenobiotic compounds from wastewaters. Increased use of microalgal biofilms for wastewater treatment will further efficiently remediate wastewater and lead to the production of high-value biomass. Apart from the use of conventional techniques, the use of advanced strategies such as genetic fingerprinting, microbial sequencing, and next-generation sequencing for profiling biofilm communities is increasing stability, robustness, and performance of biofilm reactors. However, continued efforts are needed to develop stable biofilm-based reactors for wastewater remediation. The issues concerning membrane fouling also need to be addressed with the increasing scientific interventions.

References

Amann RI, Binder BJ, Olson RJ, Chisholm SW, Devereux R, Stahl DA (1990) Combination of 16S rRNA-targeted oligonucleotide probes with flow cytometry for analyzing mixed microbial populations. Appl Environ Microbiol 56:1919–1925

Andersson S, Kuttuva Rajarao G, Land CJ, Dalhammar G (2008) Biofilm formation and interactions of bacterial strains found in wastewater treatment systems. FEMS Microbiol Lett 283:83–90

Angelaalincy MJ, Navanietha Krishnaraj R, Shakambari G, Ashokkumar B, Kathiresan S, Varalakshmi P (2018) Biofilm engineering approaches for improving the performance of microbial fuel cells and bioelectrochemical systems. Front Energy Res 6:63

Asri M, Elabed S, Koraichi SI, El Ghachtouli N (2019) Biofilm-based systems for industrial wastewater treatment. In: Handbook of environmental materials management. Springer Nature Switzerland, pp 1767–1787

Blanchet E, Desmond E, Erable B, Bridier A, Bouchez T, Bergel A (2015) Comparison of synthetic medium and wastewater used as dilution medium to design scalable microbial anodes: application to food waste treatment. Biores Technol 185:106–115

Cakir FY, Stenstrom MK (2005) Greenhouse gas production: a comparison between aerobic and anaerobic wastewater treatment technology. Water Res 39:4197–4203

Chan YJ, Chong MF, Law CL, Hassell DG (2009) A review on anaerobic–aerobic treatment of industrial and municipal wastewater. Chem Eng J 155:1–8

Coetzee G, Malandra L, Wolfaardt GM, Viljoen-Bloom M (2004) Dynamics of a microbial biofilm in a rotating biological contactor for the treatment of winery effluent. Water SA 30:407–412

Costley SC, Wallis FM (2001) Bioremediation of heavy metals in a synthetic wastewater using a rotating biological contactor. Water Res 35:3715–3723

Crist BV (2000) XPS handbook: elements and native oxides. Wiley, New York, p 458

Das T, Sehar S, Manefield M (2013) The roles of extracellular DNA in the structural integrity of extracellular polymeric substance and bacterial biofilm development. Environ Microbiol Rep 5:778–786

De Kerchove AJ, Elimelech M (2008) Calcium and magnesium cations enhance the adhesion of motile and nonmotile *Pseudomonas aeruginosa* on alginate films. Langmuir 24:3392–3399

De Philippis R, Colica G, Micheletti E (2011) Exopolysaccharide-producing cyanobacteria in heavy metal removal from water: molecular basis and practical applicability of the biosorption process. Appl Microbiol Biotechnol 92:697–708

Dhiman S, Sharma A (2022) Secondary clarification of wastewater relying on biological treatment processes: advancements and drawbacks. In: Wastewater Treatment Molecular Tools Techniques and Applications. CRC Press Boca Raton, pp 157–168

Dixon MJ, Flint SH, Palmer JS, Love R, Chabas C, Beuger AL (2018) The effect of calcium on biofilm formation in dairy wastewater. Water Pract Technol 13:400–409

Donlan RM (2002) Biofilms: microbial life on surfaces. Emerg Infect Dis 8:881

Ells TC, Hansen LT (2006) Strain and growth temperature influence *Listeria* spp. attachment to intact and cut cabbage. Int J Food Microbiol 111:34–42

Field JA, Stams AJ, Kato M, Schraa G (1995) Enhanced biodegradation of aromatic pollutants in cocultures of anaerobic and aerobic bacterial consortia. Antonie van Leeuwenhoek 67:47–77

Flemming HC, Wingender J (2010) The biofilm matrix. Nat Rev Microbiol 8:623–633

Gebara F (1999) Activated sludge biofilm wastewater treatment system. Water Res 33:230–238

Grujić S, Vasić S, Radojević I, Čomić L, Ostojić A (2017) Comparison of the *Rhodotorula mucilaginosa* biofilm and planktonic culture on heavy metal susceptibility and removal potential. Water Air Soil Pollut 228:73

Guo J, Yuan S, Jiang W, Lv L, Liang B, Pehkonen SO (2018) Polymers for combating biocorrosion. Frontiers in Materials. 5:10

Gupta P, Sarkar S, Das B, Bhattacharjee S, Tribedi P (2016) Biofilm, pathogenesis and prevention—a journey to break the wall: a review. Arch Microbiol 198:1–5

Hafner B (2006) Energy dispersive spectroscopy on the SEM: a primer. Characterization Facility, University of Minnesota, pp 1–26. http://www.charfac.umn.edu/instruments/eds_on_sem_primer.pdf

Hamadi F, Latrache H, Mabrrouki M, Elghmari A, Outzourhit A, Ellouali M, Chtaini A (2005) Effect of pH on distribution and adhesion of Staphylococcus aureus to glass. J Adhes Sci Technol 19:73–85

Hoh D, Watson S, Kan E (2016) Algal biofilm reactors for integrated wastewater treatment and biofuel production: a review. Chem Eng J 287:466–473

Hrynkiewicz K, Baum C (2014) Application of microorganisms in bioremediation of environment from heavy metals. In: Environmental deterioration and human health. Springer, Dordrecht, pp 215–227

Hu Y, Xiao Y, Liao K, Leng Y, Lu Q (2021) Development of microalgal biofilm for wastewater remediation: from mechanism to practical application. J Chem Technol Biotechnol 96:2993–3008

Igiri BE, Okoduwa SI, Idoko GO, Akabuogu EP, Adeyi AO, Ejiogu IK (2018) Toxicity and bioremediation of heavy metals contaminated ecosystem from tannery wastewater: a review. J Toxicol

Inamori Y, Matushige K, Sudo R, Kikuchi H (1989) Effect of organochlorine compounds on existence and growth of soil organisms. Water Sci Technol 21:1887–1890

Jamal M, Ahmad W, Andleeb S, Jalil F, Imran M, Nawaz MA, Hussain T, Ali M, Rafiq M, Kamil MA (2018) Bacterial biofilm and associated infections. J Chin Med Assoc 81:7–11

Jamal M, Tasneem U, Hussain T, Andleeb S (2015) Bacterial biofilm: its composition, formation and role in human infections. Res Rev: J Microbiol Biotechnol 4:1–14

Kesaano M, Sims RC (2014) Algal biofilm based technology for wastewater treatment. Algal Res 5:231–240

Khan ZU, Naz I, Rehman A, Rafiq M, Ali N, Ahmed S (2015) Performance efficiency of an integrated stone media fixed biofilm reactor and sand filter for sewage treatment. Desalin Water Treat 54:2638–2647

Khatoon N, Naz I, Ali MI, Ali N, Jamal A, Hameed A, Ahmed S (2014) Bacterial succession and degradative changes by biofilm on plastic medium for wastewater treatment. J Basic Microbiol 54:739–749

Lasa I, Penadés JR (2006) Bap: a family of surface proteins involved in biofilm formation. Res Microbiol 157:99–107

Lewandowski Z, Boltz JP (2011) Biofilms in water and wastewater treatment. Elsevier, pp 529–570

Malik S, Beer M, Megharaj M, Naidu R (2008) The use of molecular techniques to characterize the microbial communities in contaminated soil and water. Environ Int 34:265–276

Matsumoto S, Ohtaki A, Hori K (2012) Carbon fiber as an excellent support material for wastewater treatment biofilms. Environ Sci Technol 46:10175–10181

Miranda AF, Ramkumar N, Andriotis C, Höltkemeier T, Yasmin A, Rochfort S, Wlodkowic D, Morrison P, Roddick F, Spangenberg G, Lal B (2017) Applications of microalgal biofilms for wastewater treatment and bioenergy production. Biotechnol Biofuels 10:1–23

Mitra A, Mukhopadhyay S (2016) Biofilm mediated decontamination of pollutants from the environment. AIMS Bioengineering 3:44–59

Monzon O, Yang Y, Li Q, Alvarez PJ (2016) Quorum sensing autoinducers enhance biofilm formation and power production in a hypersaline microbial fuel cell. Biochem Eng J 109:222–227

Moter A, Göbel UB (2000) Fluorescence in situ hybridization (FISH) for direct visualization of microorganisms. J Microbiol Methods 41:85–112

Muhammad MH, Idris AL, Fan X, Guo Y, Yu Y, Jin X, Qiu J, Guan X, Huang T (2020) Beyond risk: bacterial biofilms and their regulating approaches. Front Microbiol 11:928

Mulcahy H, Charron ML, Lewenza S (2008) Extracellular DNA chelates cations and induces antimicrobial resistance in *Escherichia coli* biofilms. PLoS Pathogen 4:1213–1218

Muyzer G, De Waal EC, Uitterlinden AG (1993) Profiling of complex microbial populations by denaturing gradient gel electrophoresis analysis of polymerase chain reaction-amplified genes coding for 16S rRNA. Appl Environ Microbiol 59:695–700

Muyzer G, Stams AJ (2008) The ecology and biotechnology of sulphate-reducing bacteria. Nat Rev Microbiol 6:441–454

Naz I, Ullah W, Sehar S, Rehman A, Khan ZU, Ali N, Ahmed S (2016) Performance evaluation of stone-media pro-type pilot-scale trickling biofilter system for municipal wastewater treatment. Desalin Water Treat 57:15792–15805

Nie Q, Gu F (2010) Preparation of composite carrier material immobilized activated sludge and application in treating wastewater. Adv Mater Res 113:301–304

Osorio JHM, Pollio A, Frunzo L, Lens PN, Esposito G (2021) A review of microalgal biofilm technologies: definition, applications, settings and analysing. Front Chem Eng 12:43

Palmer RJ, Sternberg C (1999) Modern microscopy in biofilm research: confocal microscopy and other approaches. Curr Opin Biotechnol 10:263–268
Papadimitriou CA, Dabou X, Samaras P, Sakellaropoulos GP (2006) Coke oven wastewater treatment by two activated sludge systems. Global NEST J 8:16–22
Papadimitriou CA, Karapanagioti HK, Samaras P, Sakellaropoulos GP (2010) Treatment efficiency and sludge characteristics in conventional and suspended PVA gel beads activated sludge treating Cr (VI) containing wastewater. Desalin Water Treat 23:199–205
Prakash B, Veeregowda BM, Krishnappa G (2003) Biofilms: a survival strategy of bacteria. Curr Sci, 1299–1307
Quan X, Shi H, Wang J, Qian Y (2003) Biodegradation of 2, 4-dichlorophenol in sequencing batch reactors augmented with immobilized mixed culture. Chemosphere 50:1069–1074
Richards JP, Ojha AK (2014) Mycobacterial biofilms. Microbiol Spectr 2:2–5
Saba B, Christy AD, Yu Z, Co AC, Islam R, Tuovinen OH (2017) Characterization and performance of anodic mixed culture biofilms in submersed microbial fuel cells. Bioelectrochemistry 113:79–84
Shah Maulin P (2020) Microbial bioremediation & biodegradation. Springer
Sanz JL, Köchling T (2007) Molecular biology techniques used in wastewater treatment: an overview. Process Biochem 42:119–133
Schneider I, Topalova Y (2013) Microbial structure and functions of biofilm during wastewater treatment in the dairy industry. Biotechnol Biotechnol Equip 27:3782–3786
Shah Maulin P (2021a) Removal of emerging contaminants through microbial processes. Springer
Shah Maulin P (2021b) Removal of refractory pollutants from wastewater treatment plants. CRC Press
Sehar S, Naz I (2016) Role of the biofilms in wastewater treatment. In: Microbial biofilms-importance and applications, pp 121–144
Seo JK, Jung IH, Kim MR, Kim BJ, Nam SW, Kim SK (2001) Nitrification performance of nitrifiers immobilized in PVA (polyvinyl alcohol) for a marine recirculating aquarium system. Aquacult Eng 24:181–194
Sfaelou S, Karapanagioti HK, Vakros J (2015) Studying the formation of biofilms on supports with different polarity and their efficiency to treat wastewater. J Chem
Sharma A, Vashistt J, Shrivastava R (2021a) Next-generation omics technologies to explore microbial diversity. In: Microbes in land use change management. Elsevier, pp 541–563
Sharma A, Vashistt J, Shrivastava R (2021b) Response surface modeling integrated microtiter plate assay for *Mycobacterium fortuitum* biofilm quantification. Biofouling 37:830–843
Sharma A, Vashistt J, Shrivastava R (2022a) Knockdown of the Type-II Fatty acid synthase gene hadC in mycobacterium fortuitum does not affect its growth biofilm formation and survival under stress. Int J Mycobacteriology 11(2):159. https://doi.org/10.4103/ijmy.ijmy_46_22
Sharma A, Vashistt J, Shrivastava R (2022b) Mycobacterium fortuitum fabG4 knockdown studies: implication as pellicle and biofilm specific drug target. J Basic Microbiol 62(12):1504–1513. https://doi.org/10.1002/jobm.202200230
Simoes M, Pereira MO, Sillankorva S, Azeredo J, Vieira MJ (2007) The effect of hydrodynamic conditions on the phenotype of *Pseudomonas fluorescens* biofilms. Biofouling 23:249–258
Singh S, Singh SK, Chowdhury I, Singh R (2017) Understanding the mechanism of bacterial biofilms resistance to antimicrobial agents. Open Microbiol J 11:53
Tian H, Liu J, Feng T, Li H, Wu X, Li B (2017) Assessing the performance and microbial structure of biofilms adhering on aerated membranes for domestic saline sewage treatment. RSC Adv 7:27198–27205
Turki Y, Mehri I, Lajnef R, Rejab AB, Khessairi A, Cherif H, Ouzari H, Hassen A (2017) Biofilms in bioremediation and wastewater treatment: characterization of bacterial community structure and diversity during seasons in municipal wastewater treatment process. Environ Sci Pollut Res 24:3519–3530

Valle A, Bailey MJ, Whiteley AS, Manefield M (2004) N-acyl-L-homoserine lactones (AHLs) affect microbial community composition and function in activated sludge. Environ Microbiol 6:424–433

Von Mering C, Hugenholtz P, Raes J, Tringe SG, Doerks T, Jensen LJ, Ward N, Bork P (2007) Quantitative phylogenetic assessment of microbial communities in diverse environments. Science 315:1126–1130

Wang JL, Hou WH, Qian Y (1995) Immobilization of microbial cells using polyvinyl alcohol (PVA)—polyacrylamide gels. Biotechnol Tech 9:203–208

Wang JL, Ye YC, Wu WZ (2003) Comparison of di-n-methyl phthalate biodegradation by free and immobilized microbial cells. Biomed Environ Sci 16:126–132

Wang S, Parajuli S, Sivalingam V, Bakke R (2019) Biofilm in moving bed biofilm process for wastewater treatment. In: Bacterial Biofilms, pp 1–5

World Health Organization (2019) Drinking-water report 2019. Geneva, Switzerland: World Health Organization. https://www.who.int/news-room/fact-sheets/detail/tuberculosis

Zabed HM, Akter S, Yun J, Zhang G, Zhang Y, Qi X (2020) Biogas from microalgae: technologies, challenges and opportunities. Renew Sustain Energy Rev 117:109503

Zhang B, Powers R (2012) Analysis of bacterial biofilms using NMR-based metabolomics. Future Med Chem 4:1273–1306

Zhang F, Ahn Y, Logan BE (2014) Treating refinery wastewaters in microbial fuel cells using separator electrode assembly or spaced electrode configurations. Biores Technol 152:46–52

Zhang HL, Fang W, Wang YP, Sheng GP, Zeng RJ, Li WW, Yu HQ (2013) Phosphorus removal in an enhanced biological phosphorus removal process: roles of extracellular polymeric substances. Environ Sci Technol 47:11482–11489

Zuo R (2007) Biofilms: strategies for metal corrosion inhibition employing microorganisms. Appl Microbiol Biotechnol 76:1245–1253

Zuo R, Örnek D, Syrett BC, Green RM, Hsu CH, Mansfeld FB, Wood TK (2004) Inhibiting mild steel corrosion from sulfate-reducing bacteria using antimicrobial-producing biofilms in Three-Mile-Island process water. Appl Microbiol Biotechnol 64:275–283

Green Nano-Bioremediation Process for Ultimate Water Treatment Purpose

Aishwarya Das, Ranjana Das, and Chiranjib Bhattacharjee

1 Introduction

The increased amount of effluents generated from the industries in recent times is alarming. To mitigate this issue, a sustainable, eco-friendly, feasible yet advanced treatment process is required. The conventional methods, namely coagulation, flocculation, precipitation, adsorption, and activated sludge mechanism practiced in present days for water treatment purposes have several shortcomings in respect of foot print, less efficiency, use of chemicals, cost factors, and environmental compatibility issues. Bioremediation processes associated with water treatment have high potency to remediate the recalcitrant wastes and recycle water in an eco-friendly green route, but may not be efficient for high toxic level wastes that are harmful to microorganisms. Nano materials have emerged as unique material with high potency for treatment of several types of contaminants with ultimate eco-friendly discharge, but with limitation of efficient synthesis protocols. Nanomaterials have vast application in wastewater management practices involving photocatalytic degradation, adsorption, filtration through nanoparticles, and detection of different contaminants and pollutants in waste streams (Hussain and Palit 2020). Extensive studies have also been reported on the application of zero valent metal nanoparticles, carbon nanotubes, metal oxide nanoparticles, and nanocomposites for waste water treatment (Hussain and Palit 2020). Various nanomaterials have been reported in published literature, based on carbon, metals, and their oxide, which have been exploited for treatment of industrial effluents. Integration of the concept of nanomaterial with bioremediation has also become a thrust area of research in water treatment purposes. Biosynthesis of nanomaterials by microorganisms is an exciting approach toward the development of greener methods of waste water treatment. Microbial nanotechnology has

A. Das · R. Das (✉) · C. Bhattacharjee
Chemical Engineering Department, Jadavpur University, Kolkata 700032, India
e-mail: ranjanads78@gmail.com

M. P. Shah (ed.), *Microbial Technologies in Industrial Wastewater Treatment*,
https://doi.org/10.1007/978-981-99-2435-6_8

emerged as a subject of extensive research in recent years, which couples microorganisms in the production of nanoparticles. The microbial-mediated biosynthesis of metallic (also alloys), non-metallic, or metal oxide nanoparticles have been reported for many microbial strains of bacteria, yeast, mold, and microalgae (Hulkoti and Taranath 2014). Biogenic production of nanoparticles is a result of the enzymes secreted by microorganisms and the phytochemicals released by plants, involving oxidation/reduction reaction. The enzymes and phytochemicals have the potency to reduce metal ions to their respective nanoparticles. Nanoparticles exhibit some unique physical and chemical properties that render them as a suitable choice for treating waste water. Nanomaterials synthesized by microbes have been substantially explored and analyzed. They reveal distinct advantages and features like, (i) definite chemical composition, size, and morphology of synthesized nanomaterial, (ii) convenient handling and cultivation of microbes, (iii) scale-up of cell culture is possible, (iv) the biosynthesis could be carried out at mild physico-chemical environment, and (v) in-vivo medication can be performed in synthesized nanomaterials through changing the experimental parameters or by genetic engineering. Integration of nanotechnology and bioremediation together puts forward an innovative strategy to treat pollutants in waste water. This combined approach can include a wider range of potential applications with reduced costs and minimum negative impacts on the environment (Rizwan et al. 2014) for treating pollutants in groundwater and wastewater (Yogalakshmi et al. 2020), sediments polluted with heavy metals and hydrocarbons (De Gisi et al. 2017), and either organic or inorganic compounds in soil (Bharagava et al. 2020). This chapter provides an insight about microbial bioremediation integrated with microbial nanotechnology and its application in waste water treatment with specific emphasis on the potential microbial systems for waste water treatment and their modification as per requirement.

2 Microbes: An Eco-Friendly Alternative of Conventional Waste Water Treatment Process

Microorganisms are ubiquitous in nature. They are found in several extreme environmental niches and possess the ability to tolerate pollutants. They are responsible for remediation of the recalcitrant wastes and balancing the environment. They have developed certain properties to tolerate the unfavorable conditions. Various microorganisms have modified their cell wall structures that protect themselves from the damages caused by the environmental hazards and toxic elements. The microbes have developed few mechanisms like adsorption, absorption, nutrient cycling, and oxidation/reduction to tolerate these stress conditions. The different potential microorganisms used in waste treatment primarily belong to aerobic, anaerobic, and facultative group. The aerobic microorganism uses oxygen within water to breakdown the pollutants in sewage water and utilize it as energy source. Anaerobic species are majorly used to reduce the volume of sludge and produce biogas. Facultative origins can switch their roles between aerobic and anaerobic depending upon conditions and need. The enzyme secreted by microorganisms play a key role in the biological degradation of organic pollutants in waste water.

Bioremediation is a process that involves microorganisms in cleaning up the wastes in an eco-friendly route. The microbes enhance the biodegradation of the toxic pollutants from water to harmless or less harmful end products. The technological innovations and advancements have increased the acceptance of bioremediation as an alternative to conventional method. The microbes used in bioremediation commonly function by redox reactions, in which either oxygen, commonly used as an electron acceptor is added to enhance the oxidation of the reduced contaminants (e.g., hydrocarbons) or, an organic substrate plays the role of an electron donor to reduce oxidized pollutants (e.g., nitrates, oxidized metals, chlorinated solvents etc.). In both approaches of redox reaction, the conditions may be required to be optimized by enriching the growth media of the microbes. Various specialized bioremediation processes have gained importance in modern waste water purification research, namely bioaugmentation where specialized enriched microbial cultures are used to enhance the in situ bioremediation, bioventing, bioleaching, bioslurping, and biostimulation. Figure 1 shows various bioseparation processes practiced in different sectors.

The microorganisms are exploited in various ways to treat the contaminated sites. They are selected in accordance with the kind of contaminated area and are supplied with essential nutrients which make them destroy the pollutants at the particular site. Several environmental conditions play an important role in degrading the pollutants by microorganisms based upon the situation present. Variation in pH, temperature, growth nutrients of the microorganism, and even the presence of percentage of moisture determine how far the process will occur. The various microorganisms use their catabolic pathway for degrading these pollutants. The biodegradation of toxic compounds depends upon the condition of any microorganisms which are sensitive toward various environmental factors (Dzionek et al. 2016). The catabolic pathway is defined as a set of metabolic reactions which occur in the presence of oxygen and releases energy which is used in other reactions like the anabolic reactions (Sharma et al. 2018).

Bacteria and fungi are considered to be the most potential microorganisms used in the biological process of degrading contaminants. When the microbes are exposed to specific metal ions they can reduce it to the respective nanoparticles. Integrating bioremediation with nanoparticles is an effective method for waste water treatment. Various bacteria have the capability to degrade heavy metals by a process called nano-bioremediation. The use of microbes in metal ion removals is considered as an effective method. They have evolved certain resistance mechanisms against metal toxicity. Some of the genes identified for metal degradation include the merA reductase gene which encodes for the mercuric ion reductase which is highly toxic to humans (Brim et al. 2000). Arsenic resistance genes are organized in operons present on the chromosome as well as in plasmids of bacteria (Bruins et al. 2000). Marine bacterial species are important in heavy metal removal from waste water as they

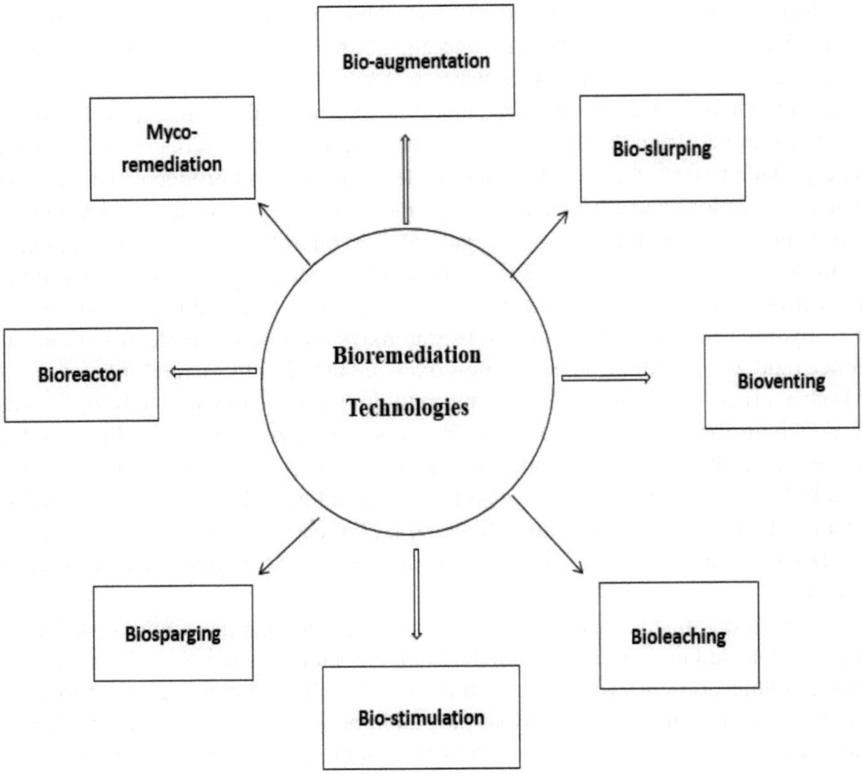

Fig. 1 Schematic representation of types of bioremediation using microbes

thrive in harsh and adverse environment. Sodium and potassium are the major nutrients utilized by the marine bacteria for their growth. Among them, sodium is utilized in the production of indole from tryptophan which helps in transport of substrates into the cell, which indirectly indicates how these marine bacteria if utilized will penetrate deep inside with the help of nanomaterials/nanoparticles and help in cleaning the polluted sites. Recently a team of researchers has developed a technique on attaching bacteria onto the surface of the inexpensive natural nanoparticles known as "Halloysites" which roughens the surface and makes bacteria gets easily attached to the surface and helps in the degradation of oil. The bacterium used in this research is *Alcanivorax borkumensis* which is an alkanotrophic bacterium and is usually found in the marine habitats (Panchal et al. 2018).

Algae are of current interest to researchers apart from bacteria and fungi. It is being used in waste water treatment and is also used to remove heavy metals, organic wastes, and other wastes that are discharged from the industries. Upon the research (Farah and Shakoori 2007) have identified that this algae (*Distigmaproteus*) have found to remove the Cadmium contamination for 48% in 2 days, 68% in 4 days, 80% after 6 days, and 90% after 8 days which shows that this species of algae is a potent source for metal detoxification and can be used in the environmental clean-up (Farah and Shakoori 2007). *Distigmaproteus* belongs to Euglenophyta, isolated from industrial waters. In another research, a group of researchers have discovered that *Trentipholia* (filamentous chlorophyte green algae) which is found growing on the surface of buildings and which is used in the detection of accumulating metals from the surrounding environment which can act as bio-indicators for metal pollution in the environment. These researchers have confirmed the presence of these particular algae by various analyses like SEM, X-ray fluorescence spectroscopy, and SEM–EDS (Enmala et al. 2019).

Cyanobacteria strains have also shown their potential in degrading the hydrocarbon-polluted sites and several strains like *Microcoleus chthonoplastes* and *Phormidium corium* have the ability to degrade then-alkanes. The reason for using the cyanobacterial strains is that they provide oxygen for the efficient degradation of hydrocarbons when available with a group of other organisms. Akoijam et al., have studied the various strains of cyanobacteria species available at the site and they could find that the *Leptolyngbya* and *Planktothrix*, *Tolypothrix* sps., *Anabaena* sps., *Oscillatoriales* sp., *Arthronema* sps., *Tolypothrix*, *Anabena*, *Arthronema*, and *Oscillatorieles* were present and which were responsible for degradation of hydrocarbons with various bacterial strains present at the hydrocarbon contaminated site (Akoijam et al. 2015).

3 Nanobiotechnology and Its Applications

This section involves a theoretical basis of nanobiotechnology sector and the major applications of the nanobiotechnology approach in diverse research and technological developments. In recent research advancements, new technologies have always gained importance. These have largely evolved due to high economic and social value and also keeping in concern the environmental issues. Emergence of nanotechnology has already opened new dimensions in research and integrating it with biotechnological approach has mitigated various issues. Nanobiotechnology is a combination of nanotechnology and biotechnology. Nanotechnology designs and produces various nanostructured materials by controlling the shape, size at nanometer scale. Biotechnology deals with the techniques of biology to alter and manipulate mechanism in cellular, molecular, and genetic level. The hybridization of these two disciplines helps to explore the structure of nanomaterials and biomolecules for novel functional application in medical, environmental, and agricultural fields. Integration of organic molecules with inorganic nanomaterials has been exploited to produce interesting observation in building nano-devices. Organic molecules are used as surface coatings, to inhibit unwanted aggregation of nanoparticles. They have embarked higher functional capabilities into inorganic nanostructures. Recently, utilization of microorganism to produce nanoparticles and its functional application in various fields is of great interest. Thus, these processes have become attractive in current green nano biotechnological approach. Nano-bioscience involves studies in mechanical, electrical, optical, thermal, and biological properties. It is a combination study of structural and mechanistic analysis of biological process at nanoscale. Nanobiotechnology uses most of the fundamentals based on existing nanotechnology. It is thus, a multidisciplinary stream involving activities associated with materials such as biosensors, nanomedicine, nanodevice, nanomachines, nanobioremediation, etc. This new aged multidisciplinary technology exhibits lot of promising methods that will exploit nanobiology in a different way in future. The biological systems are indigenously in nanoscale, thus nanoscience is needed to merge with biology and biotechnology to provide an extended pathway in nanotechnology research.

3.1 *Role of Nanoparticles in Biotechnological Applications*

The materials having at least one dimension sized between 1 and 100 nm are described as nanoparticles. Nanomaterials are produced from different bulk materials and they function according to their unique physico-chemical properties. They are explored extensively in recent times due to their unique properties, like optical, structural, mechanical, magnetic, etc., and these properties have made nanoparticles significantly different and more acceptable than conventional methods. Nanoparticles exhibit a variety of characteristics namely reactivity, catalysis, and adsorption. The properties are unique in relative to bulk material because of the smaller size which enables nanoparticle to enclose electrons and thus producing quantum effects.

In the past decades, nanomaterials have been under active research and development and have been successfully applied in many fields, such as catalysis (Parmon 2008), medicine (Liang et al. 2012), sensing (Kusior et al. 2013), and biology (Bujoli et al. 2006). Biogenic synthesis of NPs is currently being preferred over conventional methods because they are safe, economical, and clean and can be scaled up with ease. The utility of nanomaterials in various fields is thus very prominent. The mobility of nanomaterials in solution is high (Khin et al. 2012). Heavy metals (Tang et al. 2014), organic pollutants (Yan et al. 2015), inorganic anions (Liu et al. 2014), and bacteria (Kalhapure et al. 2015) have been reported to be successfully removed by various kinds of nanomaterials (Lu et al. 2016). This cutting-edge technology has great application in biological field also and its application is increasing at high pace due to its unique combination of features like biocompatibility, anti-inflammatory, antimicrobial properties, effective drug delivery mechanism, bioactivity, bioavailability, and biosorption.

Bioremediation provides a flexible and productive recovery strategy for various pollutants, may it be done by biostimulation, bioaugmentaion, or bioventing. The process has become less effective while dealing with very high concentration of recalcitrant materials. It slows down the treatment efficiency and makes the recovery time unsustainable. To proceed with this process beyond its limitations, integration of nanotechnology with bioremediation is an innovative pathway. The combined approach of nanoparticles and biotic sources like plants, and microorganisms such as bacteria, fungi, and algae are termed as nano-bioremediation. This combined approach can include a wider range of potential applications with reduced costs and minimum negative impacts on the environment (Rizwan et al. 2014) for treating pollutants in groundwater and wastewater (Yogalakshmi et al. 2020), sediments polluted with heavy metals and hydrocarbons (De Gisi et al. 2017), and either organic or inorganic compounds in soil (Bharagava et al. 2020). This emerging technology, with recent advancement serves as an eco-friendly and feasible option to treat waste water. The process helps in converting the hazardous contaminants into safer components assisted by microbes in combination with specific nanomaterial. Various nanoparticles have been synthesized by microbes and have been applied in remediation of recalcitrant. The iron and zinc nanoparticles are widely used to remediate waste water. Microbial-assisted synthesis of iron nanoparticles can be carried out by magnetic bacteria and is currently used in cleaning contaminated sites, removal of heavy metals, etc. The zinc nanoparticles are used as photo catalyst which has the capability to degrade dyes, and pharmaceutical wastes that are left in the sewage system. When compared with various metal nanoparticles the iron nanoparticles were found to possess more applications in degrading pollutants like pesticides dyes hydrocarbons, etc. (Davis et al. 2017).

A new and propitious method for treating waste water is photo-catalysis. In the presence of light and a catalyst, the pollutants are degraded by oxidizing them into low molecular weight intermediates before being converted into compounds like H_2O, CO_2, NO_3^-, PO_4^{3-}, and Cl^-. The most common type of photo-catalyst is a member of the metal oxide class or a sulfide semiconductor. TiO_2 is listed as the material that has been investigated the most in the existing literature. TiO_2 is the most outstanding photo-catalyst to date because of its strong features such as photo-catalytic activity, affordable price, photo-stability, and chemical and biological stability (Guesh et al. 2016; Imamura et al. 2013; Rawal et al. 2013). TiO_2 NPs have low selectivity and can degrade a wide range of recalcitrants namely, chlorinated organic compounds (Ohsaka et al. 2008), polycyclic aromatic hydrocarbons (Guo et al. 2015), dyes (Lee et al. 2008), phenols (Nguyen et al. 2016), pesticides (Alalm et al. 2015), arsenic (Moon et al. 2014), cyanide (Kim et al. 2016), and heavy metals (Chen et al. 2016) as reported in previous literature. TiO_2 NPs can impair the functionality and structure of different cells by producing hydroxyl radicals under UV irradiation (400 nm) (Mills and Hunte 1997). TiO_2 NPs have photo-catalytic capabilities that enable them to destroy a broad range of microbes, including viruses, fungi, algae, and both gram-positive and gram-negative bacteria (Foster et al. 2011).

Because of their various distinct qualities, ZnO NPs have emerged as another effective participant in the field of photo catalysis, in addition to TiO_2 NPs, for the treatment of water and waste water (Janotti and deWalle 2009; Reynolds et al. 1999; Chen et al. 1998). ZnO NPs are suitable for the treatment of water and wastewater because they are environmentally safe and consistent with organisms (Schmidt and MacManus 2007). As a result of their nearly identical band gap energies, ZnO NPs and TiO_2 NPs' photocatalytic abilities are comparable. In contrast to TiO_2 NPs, ZnO NPs provide the benefit of being less expensive (Daneshvar et al. 2004). In comparison to other semiconducting metal oxides, ZnO NPs may absorb a greater variety of solar spectra and more light quanta (Behnajady et al. 2006). In order to eliminate the usage of chemicals, different methods are employed to create Zn nanoparticles. Zn nanoparticles are produced using algae, fungus, and *Aspergillus terreus*, as well as *Sargassum muticum.* It has been suggested that the bacterium *Bacillus licheniform* is used in the synthesis of biologically active nanoparticles (Zare et al. 2013). There have been reports of the yeast *Pichia kudravzevii* being used in the synthesis of new Zn nanostructures (Zare et al. 2013).

With regard to applications in the remediation of water pollution, iron (Fe) has a number of distinct benefits over zinc (Zn), including superior retention capabilities, the ability to precipitate and oxidize (when dissolved oxygen is present), and inexpensive (Lu et al. 2016). Consequently, zerovalent iron nanoparticles have received the most research attention of all zerovalent metal nanoparticles. nZVI has good adsorption characteristics and a potent reducing capacity because of its exceptionally small size and high specific surface area (Matheson and Tratnyek 1994). Iron nanoparticles hold great promise for removing heavy metals from sewage. Iron nanoparticle synthesis using a more environmentally friendly process, such as iron oxide or zerovalent iron, aids in re-establishing ecological sustainability. Bimetallic Fe/Pd nanoparticle stabilization was accomplished by He and Zhao (He and Zhao 2005) using water soluble starch. In this work, it was discovered that starch has a substantial role in the dispersion and stabilization of iron nanoparticles. For the first time, Jegan et al. (2011) created a well-dispersed magnetite (Fe_3O_4) agar nanocomposite by co-precipitating Fe (III) and Fe (II) ions. *Actinobacter* sp. was used by Bharde et al. (2005) to create spherical iron oxide nanoparticles in an aerobic environment. Scientists are currently looking at *Fusarium oxysporum* and *Verticillium* sp. for the synthesis of magnetic nanoparticles. For the synthesis of iron nanoparticles, Kaul et al. (2012) investigated five different species of fungi, including *P. chlamydosporium, A. fumigates, A. wentii, C. lunata, and C. globosum,* as well as two bacteria, *A. faecalis and B. coagulans* (Pavani and Kumar 2013). Another significant natural source for the creation of nanoparticles has been algae. Iron oxide nanoparticles (Fe_3O_4 NPs) were biosynthesized by Mahdavi et al. (2013) by reducing ferric chloride solution with the brown seaweed (*Sargassum muticum*) macroalgae extract. Using soil microalgae of the *Chlorococcum* sp. species and an iron chloride precursor, Subramaniyam et al. (Subramaniyam et al. 2015) created spherical nano-iron with a size range of 20–50 nm. The eradication of germs from water using silver nanoparticles has received extensive research. Because they are extremely poisonous to microorganisms, silver nanoparticles (Ag NPs) exert potent antibacterial effects on a variety of microbes, such as viruses (Borrego et al. 2016), bacteria (Kalhapure et al. 2015), and fungi (Krishnaraj et al. 2012). According to the research, silver nanoparticles can be produced by *Bacillus cereus, Bacillus subtilis, Escherichia coli, Klebsiella pneumoniae, and Shewanella* sp.

Carbon Nanostructures are appealing for both basic research and cutting-edge, practical activities. They are particularly well recognized for being effective adsorbents. Due to their distinctive qualities, they are a part of the fascinating group of materials. Due to their superior ability to adsorb a wide variety of pollutants, quick kinetics, large specific surface area, and selectivity toward aromatics, they are advantageous for the remediation of wastewater. Carbon nanotubes (CNTs), carbon beads, carbon fibers, and nano-porous carbon are just a few examples of the multiple kinds that CNMs can take. The most emphasis has been paid to CNTs, which have advanced quickly in recent years. The mechanical, optical, electrical, and adsorption properties of carbon nanotubes are frequently enhanced by the addition of additional metals or supports (Ray and Shipley 2015). The CNTs' specific surface area is improved by the functionalization, which also increases the quantity of oxygen, nitrogen, or other groups on their surface and improves their dispensability (Adeleye et al. 2016; Li et al. 2002; Adeleye and Keller 2014). For instance, Gupta et al. reported on a work that used CNTs as a framework for magnetic iron oxide (Gupta et al. 2011). Divalent heavy metals like Cu^{2+}, Zn^{2+}, Pb^{2+}, Cd^{2+}, and Co^{2+} have been employed as adsorbents using CNT sheets (Tofighy and Mohammadi 2011). Nanocomposites are also becoming more prevalent, have become crucial in the treatment of waste water, and have been the focus of recent research and material fabrication. By chemically depositing nZVI on CNTs, Azari et al. created a unique nanocomposite in 2014. According to the outcomes of their experiment, the nanocomposite functions well as an adsorbent and can effectively remove nitrate from water (Azari et al. 2014). Additionally, the adsorbent may be effortlessly removed from the solution by the magnet due to its distinct magnetic property (Azari et al. 2014). There have been reports of filamentous fungus producing nanocomposites like selenium nanoparticles. According to published reports, *Aspergillus terreus* and *A. oryzae* are employed in the bioproduction of polymer nanocomposites of se NPs (Zare et al. 2013).

3.2 Microbe-Assisted Fabrication of Metal and Metal Oxide Nanoparticles

Nanotechnology has revolutionized in recent era. There are several developments in the recent time regarding the process development of nanomaterial (Fig. 2). The physical and chemical synthesis of NPs offer high production, controlled and specific size, but the methods are not eco-friendly, requires high cost, and produce hazardous elements which are toxic to environment. Moreover, a previous study demonstrated that the chemical synthesis of NPs is toxic and less biocompatible. In need of safer, biocompatible, and economically feasible method, green synthesis has gained attention as an alternative to synthesize nanoparticles.

The biological synthesis of metal and metal oxide NPs involves unicellular and multicellular biological entities including bacteria (Kundu et al. 2014), yeast (Moghaddam et al. 2017), fungi (Shamsuzzaman et al. 2013), virus (Nam et al. 2006), and algae (Azizi et al. 2014). Microbes possess the ability to survive in heavy metal rich adverse environment and can accumulate the metal ion and reduce it to nanoparticles, thus serving as important nano factories.

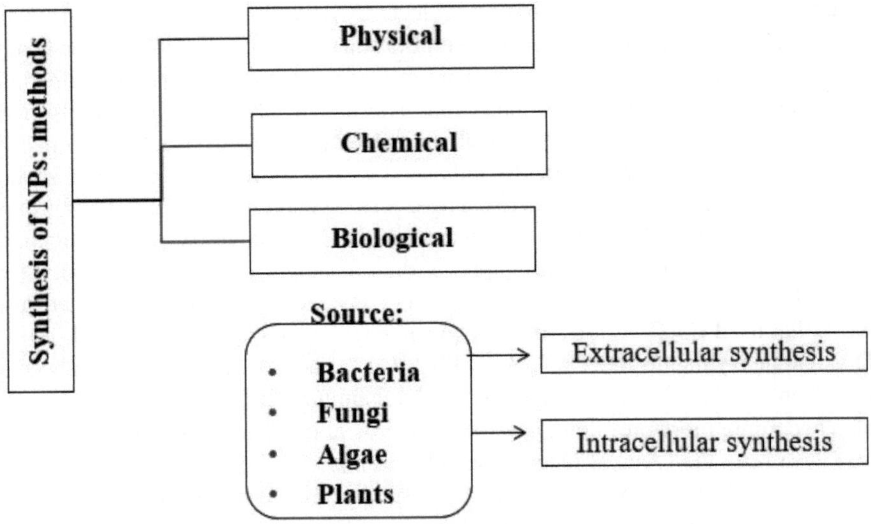

Fig. 2 Flowchart of methods of synthesis of NPs

Nanoparticle synthesis is categorized into two processes, namely "top-down approach" and "bottom-up approach" as represented in Fig. 3.

Top-down approach—a bulk material is broken down into nano-sized structures. This technique is extension to those that produce materials in micron size. Top-down fabrication method involves simpler technique to fabricate nanomaterial of desired size and properties.

The main drawback of the process is imperfection in structure. Physical techniques are primarily used like machining, coating, atomization, lithography, and etching.

Bottom-up approach— atoms and molecules are assembled to molecular structure in nanometer range. The process is more economical and produces less waste. Chemical and biological process uses this fabrication method.

Inorganic materials produced by microbes are either intracellular or extracellular in nature and often in nanoscale range. To survive in metal rich, adverse condition microbes use chemical detoxification pathway, as well as ion efflux by membrane proteins in energy independent way. The membrane proteins function as ATPase, chemiosmotic cation, or proton anti-transporters. Thus microbial system has this special ability to detoxify the metal ions through reduction or precipitating the soluble toxic inorganic ions to insoluble non-toxic metal nanoclusters. Microbes can adapt to higher concentrations of metals and have the potential to reduce inorganic materials into NPs through their extracellular or intracellular routes (Salem and Fouda 2021; Fariq et al. 2017). Microbes absorb metal ions from their surrounding environment/media and convert these metallic ions into elemental form via an enzymatic reduction (Li et al. 2011). Extracellular and intracellular mechanisms of NPs synthesis are well documented in previous literature by Patil and Chandrasekaran (2020).

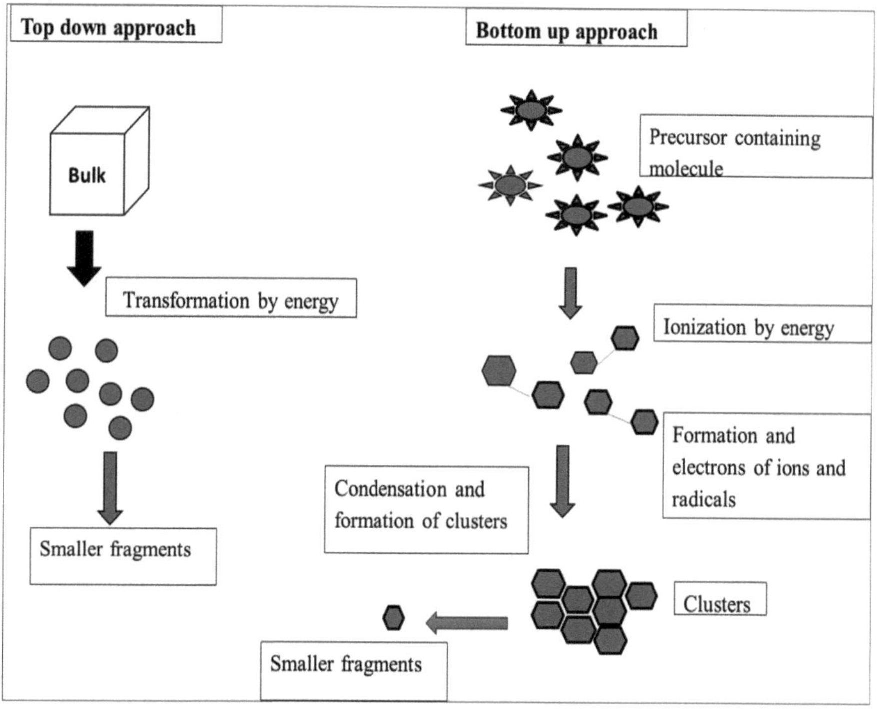

Fig. 3 Schematic diagram of top-down and bottom-up approach

3.3 *Microbial Mechanism of NPs Synthesis*

Biosynthesis route involves, synthesis of nano-dimension materials within the microorganisms by binding with targeted toxic metal ions and converting these toxic metal ions into the corresponding element metal through cellular enzymes. This section involves a detailed illustration of the various mechanism involved in the synthesis of nanomaterials by microbial route and the effect on various targeted geogenic contaminants. The extracellular and intracellular pathways of nanomaterial synthesis have been detailed in this section.

Extracellular mechanism used by microorganisms to synthesize NPs utilizes enzymes and proteins, cell wall components of bacteria and fungi to reduce the metal ions. The extracellular mode involves trapping the metal ions on the cell surface and reducing ions in the presence of enzymes (Das et al. 2014). The microbes are cultured in suitable growth media, centrifuged, and the supernatant is collected. The supernatant consists of reductive enzymes used for NPs synthesis (Yadav et al. 2015; Al-Dhabi et al. 2018). The enzymes present in the supernatant are allowed to react with metal ions in a separate vessel. The bio-reduction of metal ions in cell-free supernatant results in the formation of NPs (Marooufpour et al. 2019). The reaction conditions like temperature, pH, pressure, culture medium, and added nutrients are optimized according to the microbes used. Nitrate reductase enzyme, an extracellular enzyme from *Fusarium oxysporium* is responsible for the formation of silver NPs. Anil Kumar et al. (2007) in his studies revealed that another NADH-dependent nitrate reductase was used in the production of AgNPs. The NADH-dependent oxidoreductase produced by fungi is also used.

The intracellular process involves initial electrostatic attraction of metal ions by carboxyl groups of the microbial cell wall, resulting in passage of metal ions through the cells and reduction by intracellular proteins and cofactors to produce NPs (Siddiqi et al. 2018). Microbial cellular mechanism is involved in intracellular fabrication of NPs. The microbial cultures are maintained in appropriate liquid media and the microbial biomass is washed with sterile distilled water followed by centrifugation to obtain the biomass pellet (Shah et al. 2015; Castro et al. 2014; Fernández-Llamosas et al. 2017). The pellet obtained is then allowed to interact with aqueous solution of metal ions. The solution of metal ions and cell biomass is cultivated under optimized reaction conditions and awaited for a chromatic change. A specific chromatic alteration denotes the formation of NPs. When a whitish yellow to yellow color appears, it suggests the synthesis of zinc and manganese NPs. While the appearance of a pale yellow to pinkish color suggests the synthesis of gold NPs, and a pale yellow to brownish color suggests the synthesis of AgNPs (Waghmare et al. 2015). Microbial cell walls are negatively charged, and through this process the positively charged metal ions are trapped or absorbed conveniently within the cell wall. The microbial enzyme-assisted action is responsible for electrostatically bioreducing the metal ions, forming nanoclusters. The microbial cells also produce some adhesive components that may be responsible for the stabilization of the NPs. The NPs further diffuse to the solution from the cell wall of the microbes, thus involves in a mechanism of nutrient exchange and substance diffusion as stated in literature (Marooufpour et al. 2019). Both intracellular and extracellular mechanism is represented schematically in Fig. 4.

Bacteria-mediated production of NPs has become an important area of research contributing to waste water treatment. Bacterial cells have served as nanofactories for fabrication of various metal nanoparticles such as Ag, Au, Fe, Cu, and metal oxides NPs like Ag_2O, ZnO, TiO_2, MnO_2, MgO, and Fe_2O_3 (Grasso et al. 2020). The potentiality of bacteria to tolerate the abiotic stress and to bioreduce metal ions present them as effective bioagent for NPs production. The special metal binding abilities of the bacterial cells and S-layers make them useful for technical applications in bioremediation and nanotechnology. *Lactobacillus*, *Bacillus* sp., *Acinetobacter*, *Klebsiella pneumoniae*, and *Corynebacterium* sp. are commonly explored species for the biosynthesis of NPs. In a study conducted by Priyabrata et al., states that in the formation of silver NPs, most of the particles remain affixed to the cytoplasmic membrane (Priyabrata et al. 2001). Studies have revealed that the fabrication method may be depended upon various parameters like the addition of buffers, the growth phase, and incubation time of the microbe. Holmes et al. (1995) cultured the bacteria *Klebsiella aerogens* under various buffer conditions (Tris, Bistris propane, Tricine, Bes, ortho-phosphate compounds, TES (a solution made up of Tris, EDTA and NaCl), or 4- (2-hydroxyethyl)-1-piperazineethanesulfonic acid) and utilized the bacteria to

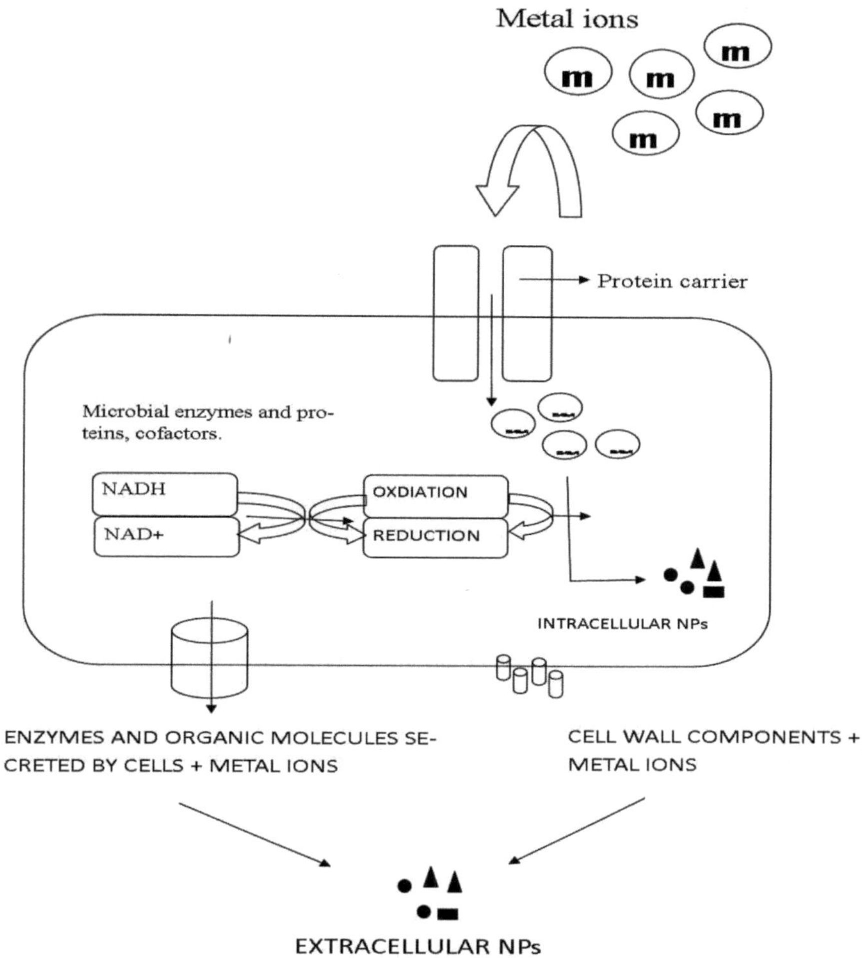

Fig. 4 Schematic representation of microbial mechanism of NPs synthesis

synthesize cadmium. The observation of the experiment demonstrates that different buffer-fortified cultures showed enhanced tolerance to cadmium. Triacin and phosphate buffer increased the tolerance of *K. aerogenes* to cadmium from 10 μM to 2 mM–5 mM. This result indicated that the growth medium composition of the bioagent is also responsible for nanoparticle synthesis. *Escherichia coli* to synthesize nanoparticle CdS showed that the production of nanoparticles was significantly affected by the growth phase of *E. coli* (Rozamond et al. 2004). In previous literature, suspension of *lactobacillus* has been utilized for the successful synthesis of TiO_2 nanoparticles. The study concluded that the enzymes generated on the cell wall of the microbes determine the reaction mechanism for the synthesis (Jha et al. 2009). Bacteria surviving in extreme climatic conditions are ideal for production of nanoparticles. Like growth phase and medium the environmental conditions also govern the morphology and characteristics of a nanoparticle produced. Thus, these are few mechanisms used by bacteria to fabricate NPs though the detailed pathway is still unknown.

Several studies have also been reported on fungi-mediated production of nanoparticles pertinent for application in waste water treatment. Utilization of fungi as a bioagent for biofabrication of nanoparticles is highly accepted among researchers. Fungi exhibit tolerance capacity for metal ions due to their cell wall structure. The yield is also higher than the bacterial cell. Fungal biosynthesis of nanoparticles is comparatively less expensive than bacteria as fungi has higher tendency to accumulate metal ion. Fungal biosynthesis is simpler in nature hence is widely used by researchers to produce variety of metal nanoparticles like gold, silver, zinc, etc. Extracellular synthesis mechanism is generally used by fungi to fabricate nanoparticles as this process avoids the usage of detergents, ultrasound, any physical factor and also doping for intracellular components such as protein, nucleic acids, and fats are not required (Gahlawat and Choudhury 2019). *Trichoderma viridae* and *Hypocrea lixii* are two fungi used by Mishra et al. to carry out extracellular NPs synthesis and study temperature and pH-dependent effect on the synthesis. The observation showed that *Trichoderma* could fabricate NPs at 30 °C and *H. lixii* could produce NPs at 100 °C (Mishra et al. 2014). *Fusarium oxysporium* released some reducing agents when incubated with metal ion as stated in literature. The fungal biomass was cultured and separated by filtration. The supernatant was next used to incubate with metal ions for few hours. The solution turned into yellowish brown which indicated the formation of silver NPs. The supernatant used was cell free; hence the observation was concluded that the reducing agents present in the solution helped in formation of the nanoparticles (Ahmad et al. 2003).

Yeast-mediated production of nanomaterial has been reported as a potential alternative of fungi-mediated nanomaterial synthesis. Yeasts have the potential to survive in a high concentration of metal ions and have the capability to deposit a high amount of metal ions from a medium (Shah et al. 2015). This feature of yeast has been used by different researchers for the green synthesis of NPs (Waghmare et al. 2015; Apte et al. 2013; Zhang et al. 2016). *Yarrowin lipolytica* is marine yeast, explored for green synthesis of AgNps, the result showed that a brown pigment named melanin is responsible for the synthesis of NPs (Apte et al. 2013). In synthesis of numerous enzymes and rapid growth with the use of simple nutrients, yeast strains possess certain benefits over bacteria and the synthesis of metallic nanoparticles employing the yeast is being considered (Kumar et al. 2011).

3.4 Nanobioremediation (NBR) Technology as Eco-Friendly Approach to Waste Water Treatment

NBR is an emerging technology used for remediating pollutants. Waste water comprises a variety of organic and inorganic compounds, heavy metals which are not biodegradable in nature. The conventional ways of treating waste water are costly and increase the carbon footprint. Recent advances have made bioremediation an attractive alternative to the researchers for treating the waste water. Bioremediation has proven to be promising and effective technology but it has its own drawbacks. The three prime approaches of biological remediation is mediated through microbes, plants, and enzymes. The waste water contaminants are hazardous to some extent for microbes and plants. Many of the organic and inorganic pollutants have become resistant to the enzyme-mediated degradation process. Thus, a combined technology was needed to remediate the calcitrants. Microbe-assisted nanotechnology and bioremediation could mitigate the problem. Recent studies have revealed that, contaminants like atrazine, molinate, and chloropyrifos, can be degraded with nanosized zerovalent ions (Zhang 2003; Ghormade et al. 2011).

Nanoencapsulated enzymes can easily degrade the complex organic compounds by microbial pathways. The combined effect of nanomaterials, microbes, and biotechnological approach provides the opportunity to remediate waste water effectively. Heavy metals are a threat to human and animal life due to its non-biodegradable nature and toxicity. Different varieties of novel materials, such as graphene derivatives (Zhao et al. 2011), carbon-based sorbents (Moreno-Castilla et al. 2004), chelates (Sun et al. 2006), activated carbons (Kobya et al. 2005), chitosan/natural zeolites (Wang et al. 2009), and clay minerals (Oubagaranadin and Murthy 2009) are being investigated with the aim of adsorbing heavy metal ions from aqueous solutions. Due to lower adsorption capabilities and efficiencies, the conventional adsorbents commonly used have been limited in heavy metal remediation (Wang et al. 2012).

Nanostructured silica, magnetic nanoparticles embedded on silica is used for heavy metal removal. Calcium, combined with Zn NPs is effective in removing lead ions from waste water. Researches have shown iron nanoparticles, nZVI, and polymer-coated nanoparticles to play a significant role in heavy metal removal like Cr (VI) and As (III). Carbon tetrachloride, pentachlorobenzene, chloro and dichlorobenzene, chloromethane, acid red, acid orange are a few variety of pollutants being released and can be treated with the Nano iron Technology or NZVI technology. Mahanty et al. (2020) biofabricated iron oxide nanoparticles from *Aspergillus tubingensis* (STSP 25) obtained from the rhizosphere of *Avicennia officinalis* in Sundarbans, India. The synthesized nanoparticles were able to remove more than 90% of heavy metals. The metals Pb, Ni (II), Cu (II), and Zn (II) were removed from waste water and also with a regeneration capacity of up to 5 cycles. He also stated that the adsorption was done chemically in endothermic reaction. A nanocomposite was made using EPS of *Chlorella vulgaris* and was co-precipitated with iron oxide NPs. The study revealed the successful modification of NPs when observed under FT-IR. The modification was because of the functional group of EPS of the microorganism.

The removal percentage of PO_4^{3-} was 91% and NH_4^+ was 85% (Govarthanan et al. 2020). A copper-resistant strain of *E. coli* sp. SINT7 was used to produce copper nanoparticles. The microbe-assisted NPs were used to degrade azo dyes and effluents produced by the textile industry. The observation of the study showed the reduction of the dyes, reactive black 5- 83.61%, congo red- 97.07%, direct blue 1- 88.42%, and malachite green- 90.55% respectively at lower con 25 mg/l. at 100 mg/l conc, this was reduced to 76.84%, 83.90%, 62.32%, 31.08%. The biogenic NP was able to degrade industrial effluents like suspended solids, chlorides, and phosphates in treated sample (Noman et al. 2020).

Carbon-based nanomaterials are used for the removal of heavy metals from waste water. The nontoxicity of the material and good absorbance capacity makes it an extensively used nanomaterial. Activated carbon was the first commonly used nanomaterial for removal of metal ions. Advances in nanotechnology and recent researches have delivered innovative carbon NMs like fullerenes, graphenes, and CNTs. The conversion of graphene oxides to graphene nano sheets can be achieved through the use of crude polysaccharides obtained from *Pleurotus flabellatus,* a study revealed by Das et al. (2014). Though activated carbon is a better adsorbent for organic and inorganic pollutants, it has some limitations in the applicability to most heavy metals, specifically to arsenic As(V) (Daus et al. 2004). CNTs are used as adsorbents for divalent heavy metals, Cu^{2+}, Zn^{2+}, Pb^{2+}, Cd^{2+}, and Co^{2+} (Tofighy and Mohammadi 2011). Experimental studies by Pyrzynska and Bystrzejewski revealed that CNTs and magnetic nanoparticles encapsulated by carbon show higher capacity of absorption studies on metals like cobalt and copper, compared to activated carbon (Pyrzynska and Bystrzejewski 2010). The absorption capacity of NMs depends on factors like metal ion concentration and pH, as stated by Stafiej and Pyrzynska (2007).

Polymers are capable of removing both heavy metals and organic compounds as it acts as good adsorbents. A case study on polymers as adsorbents chitosan biopolymers and its adsorption mechanism was studied for arsenic removal. The study showed that the mechanism involves either of the three, such as chelation of metals, formation of ion pairs, and electrostatic interaction, when adsorption is carried out by chitosan biopolymer (Guibal 2004).

3.5 Advantages of Using Nanotechnology Over Conventional Methods

Conventional methods used in waste water since ancient times are not cost-effective. The advanced nanotechnology is advantageous over conventional methods in cost reduction and lowering the carbon footprint. Traditional way of waste water remediation comprises of removal of recalcitrant by filtration, flocculation, activated charcoal, and ion exchange resins (Karthikeyan et al. 2005; Vijayaraghavan et al. 2007; Wang and Chen 2009). In general, treating waste water objectives were concerned with several crucial factors like (i) removal of suspended solids (ii) treating BODs, and (iii) elimination of harmful microorganisms.

Literature states that during early 1970s to 1990s, the vision of waste water treatment changed and focused more on environmental effects. Conventional methods involve a release of a lot of toxic elements which causes a long-term environmental problem. Removal of metals by conventional methods is becoming inadequate to meet the current status about environmental concern. The use of biological materials, including living and non-living microorganisms, to remove and recover toxic or precious metals from industrial waste waters has gained popularity over the years due to increased performance, availability, and low cost of raw materials (Ahluwalia and Goyal 2007; Benaissa and Elouchdi 2007; Bunluesin et al. 2007), microorganisms including bacteria (Ansari and Malik 2007). Algae (Mallick 2003), fungi, and yeasts (Dursun 2003) can efficiently accumulate heavy metal from their external environment (Ghimire et al. 2007; Pan et al. 2007; Ziagova et al. 2007). Nano bioremediation uses greener technology to treat the waste water. Microbe-assisted nano remediation is much more effective than only microbe-assisted bioremediation of effluents. The advantages and disadvantages of different methods used in nanoparticle synthesis are mentioned in Table 1.

Table 1 Advantages and disadvantages of different methods of synthesizing NPs

Methods	Chemicals/ microorganisms used	Advantages	Disadvantages	Reference
Physical	–	–	Low effectiveness, joined firmly	Karthikeyan et al. (2005), Vijayaraghavan et al., (2007), and Wang and Chen (2009)
Chemical	Reducing agents, such as methoxypolyethylene glycol, sodium borohydride, potassiumbitartarate, hydrazine. Stabilizing agents, such as polyvinyl pyrrolidone, sodium dodecylbenzyl sulfate	High monodispersity	Costly, low yield, use of toxic chemicals, not environment friendly	Wang and Chen (2009)
Biological	Microorganisms- bacteria, fungi, algae, plant source, and also other bio-based sources	Low mono-dispersity eco-friendly, cost-effective, synthesis process is easier than conventional		Farah and Shakoori (2007), Panchal et al. (2018), and Enmala et al. (2019)

The combined process of enzymes with nanotechnology is gaining importance. This renders NMs less harmful to environment. The enzyme molecules are responsible for decreasing the surface energy, as they reduce the cell interaction through steric hindrance (Dwevedi 2019). The eco-friendly nature of the enzymes provides additional distinctiveness of catalysis, which makes NMs more effective in bioremediation process. Conventional methods cannot be combined with any other mechanism as it would increase the cost and also the energy required. Nanotechnology is that group of technology which can be applied in combination with any biological method to make the process more hastily and effectively.

3.6 Future Prospect

Waste water remediation has become crucial technology concerned with the human health and environmental protection. To minimize and mitigate the drawbacks of conventional waste water treatment techniques, nanobioremediation has gained importance in real life implementation to serve the community. The nanoparticles possess unique quality and wide range of utility. This cost-effective technology of using nanoscience to treat waste water and industrial effluents has also minimized the environmental threats. Advancements in research have rendered nanomaterials as extensively studied material for application in remediation of effluents. Choice of microbes, optimum conditions, and choice of suitable metal ions for synthesizing nanoparticles can evolve the process and application. The major challenge in nano bioremediation process is the synthesis mechanism of the nanoparticles by microbes. The exact mechanism undergone during the synthesis process is still unknown. Studies and experiments are needed to guide the progress in this field. The study of operon systems of bacteria which is an important component for heavy metal resistance is required to make advancements in molecular level of the mechanism used. The specific control of NMs in molecular level would control the size, shape, design, increase affinity, capacity, and also selectivity of pollutants. This is another challenge in using microorganisms for biogenic production of NPs. Scale up is easy but specifications are required. Bacterial process of synthesis is slower than fungal process. This projects another challenge in utilizing bacteria as a biogenic source. This problem could be mitigated by bioengineering techniques. The genetic alteration may lead to high yield of enzyme and other organic molecules required for NPs synthesis. Enzyme-assisted nanotechnology is gaining popularity in all fields. Treating waste water using enzyme immobilization techniques may open new ways to mitigate the problems. Future nanobioremediation techniques could explore various microbial enzymes for carrying out the process. Xenobiotic degrading enzymes are protected when combined with nanoparticles. The stability also increases. Thus, treatment of waste water using nano bioremediation technology restores the environmental concern and also carries out remediation effectively.

References

Adeleye AS, Keller AA (2014) Long-term colloidal stability and metal leaching of single wall carbon nanotubes: effect of temperature and extracellular polymeric substances. Water Res 49:236–250

Adeleye AS, Conway JR, Garner K, Huang Y, Su Y, Keller AA (2016) Engineered nanomaterials for water treatment and remediation: costs, benefits, and applicability. Chem Eng J 286:640–662

Ahluwalia SS, Goyal D (2007) Microbial and plant derived biomass for removal of heavy metals from wastewater. Biores Technol 98:2243–2257

Ahmad A, Mukherjee P, Senapati S, Mandal D, Khan MI, Kumar R, Sastry M (2003) Extracellular biosynthesis of silver nanoparticles using the fungus *Fusarium oxysporum*. Colloids Surface B 28:313–318

Akoijam C, Langpoklakpam JS, Chettri B (2015) Cyanobacterial diversity in hydrocarbon-polluted sediments and their possible role in bioremediation. Int Biodeterior Biodegradation 103:97–104

Alalm MG, Tawfik A, Ookawara S (2015) Comparison of solar TiO_2 photocatalysis and solar Photo-Fenton for treatment of pesticides industry wastewater: operational conditions, kinetics, and costs. J Water Process Eng 8:55–63

Al-Dhabi NA, Mohammed Ghilan A-K, Arasu MV (2018) Characterization of silver nanomaterials derived from marine *Streptomyces* sp. Al-dhabi-87 and its in vitro application against multidrug resistant and extended-spectrum beta-lactamase clinical pathogens. Nanomaterials 8:279

Anil Kumar S, Abyaneh MK, Gosavi SW, Kulkarni SK, Pasricha R, Ahmad A et al (2007) Nitrate reductase-mediated synthesis of silver nanoparticles from $AgNO_3$. Biotech Lett 29:439–445

Ansari MI, Malik A (2007) Biosorption of nickel and cadmium by metal resistant bacterial isolates from agricultural soil irrigated with industrial wastewater. Biores Technol 98:3149–3153

Apte M, Sambre D, Gaikawad S, Joshi S, Bankar A, Kumar AR, Zinjarde S (2013) Psychrotrophic yeast yarrowia lipolytica NCYC 789 mediates the synthesis of antimicrobial silver nanoparticles via cell-associated melanin. AMB Express

Azari A, Babaei A-A, Kalantary RR, Esrafili A, Moazzen M, Kakavandi B (2014) Nitrate removal from aqueous solution using carbon nanotubes magnetized by nano zero-valent iron. J Maz Univ Med Sci 23(2):14–27

Azizi S, Ahmad MB, Namvar F, Mohamad R (2014) Green biosynthesis and characterization of zinc oxide nanoparticles using brown marine macroalga *Sargassum muticum* aqueous extract. Mater Lett 116:275–277

Behnajady MA, Modirshahla N, Hamzavi R (2006) Kinetic study on photocatalytic degradation of C.I. Acid Yellow 23 by ZnO photocatalyst. J Hazard Mater 133(1–3):226–232

Benaissa H, Elouchdi MA (2007) Removal of copper ions from aqueous solutions by dried sunflower leaves. Chem Eng Process 46:614–622

Bharagava RN, Saxena G, Mulla SI (2020) Introduction to industrial wastes containing organic and inorganic pollutants and bioremediation approaches for environmental management. In: Bioremediation of industrial waste for environmental safety Springer, Singapore, pp 1–18

Bharde A, Wani A, Shouche Y, Joy PA, Prasad BLV, Sastry M (2005) Bacterial aerobic synthesis of nanocrystalline magnetite. J Am Chem Soc 127:9326–9327

Borrego B, Lorenzo G, Mota-Morales JD, Reyes HA, Mateos F, Gill EL, LOsa N, Burmistrov VA, Pestryakov AN, Brun A, Bogdanchikova N (2016) Potential application of silver nanoparticles to control the infectivity of Rift Valley fever virus in vitro and in vivo. Nanomedicine: Nanotechnol, Biol Med 12(5)1185–1192

Brim H, Mcfarlan SC, Fredrickson JK, Minton KW, Zhai M, Wackett LP, Daly MJ (2000) Engineering *Deinococcus radiodurans* for metal remediation in radioactive mixed waste environments. Nat Biotechnol 18:85–90

Bruins MR, Kapil S, Oehme FW (2000) Microbial resistance to metals in the environment. Ecotoxicol Environ Saf 45:198–207

Bujoli B, Roussi'ere H, Montavon G, Laib S, Janvier P, Alonso B, Fayon F, Petit M, Massiot D, Bouler JM, Guicheux J, Gauther O, Lane SM, Nonglaton G, Pipelier M, Leger J, Talham DR,

Charles T (2006) Novel phosphate–phosphonate hybrid nanomaterials applied to biology. Prog Solid State Chem 34(2–4)257–266

Bunluesin S, Kruatrachue M, Pokethitiyook P, Upatham S, Lanza GR (2007) Batch and continuous packed column studies of cadmium biosorption by *Hydrilla verticillata* biomass. J Biosci Bioeng 103:509–513

Castro L, Blázquez ML, González FG, Ballester A (2014) Mechanism and applications of metal nanoparticles prepared by bio-mediated process. Rev Adv Sci Eng 3:199–216

Chen Y, Bagnall DM, Koh HJ et al (1998) Plasma assisted molecular beam epitaxy of ZnO on c-plane sapphire: growth and characterization. J Appl Phys 84(7):3912–3918

Chen ZP, Li Y, Guo M, Xu F, Wang P, Du Y, Na P (2016) One-pot synthesis of Mn-doped TiO_2 grown on graphene and the mechanism for removal of Cr(VI) and Cr(III). J Hazard Mater 310:188–198

Daneshvar N, Salari D, Khataee AR (2004) Photocatalytic degradation of azo dye acid red 14 in water on ZnO as an alternative catalyst to TiO_2. J Photochem Photobiol A: Chem 162(2–3):317–322

Das VL, Thomas R, Varghese RT, Soniya EV, Mathew J, Radhakrishnan EK (2014) Extracellular synthesis of silver nanoparticles by the *Bacillus* strain CS 11 isolated from industrialized area. 3 Biotech 4:121–126

Daus B, Wennrich R, Weiss H (2004) Sorption materials for arsenic removal from water: a comparative study. Water Res 38(12):2948–2954

Davis AS, Prakash P, Thamaraiselvi K (2017) Nanobioremediation technologies for sustainable environment. In: Bioremediation and sustainable technologies for cleaner environment. Springer, pp 13–34

De Gisi S, Minetto D, Lofrano G, Libralato G, Conte B, Todaro F, Notarnicola M (2017) Nano-scale Zero Valent Iron (nZVI) treatment of marine sediments slightly polluted by heavy metals. Chem Eng Trans 60:139–144

Dursun AY (2003) The effect of pH on the equilibrium of heavy metal biosorption by *Aspergillus niger*. Fresenius Environ Bull 12(11):1315–1322

Dwevedi A (2019) Overview of combo technology (nanotechnology and enzyme technology) and its updates. Solutions to environmental problems involving nanotechnology and enzyme technology. Academic Press, Cambridge, MA, pp 1–47

Dzionek A, Wojcieszy D, Guzik U (2016) Natural carriers in bioremediation: a review. Electron J Biotechnol 23:28–36

Enmala MK, Srithi PD, Sarkar S, Chavali M, Vasavi I, Kuppam C (2019) Nanobioremediation: a novel and sustainable biological advancement for ecological clean-up. In: Nanotechnology in biology and medicine, p 245

Farah AR, Shakoori RSAR (2007) Heavy metal resistant *Distigma proteus* (Euglenophyta) isolated from industrial effluents and its possible role in bioremediation of contaminated wastewaters. World Journal of Microbiology and Biotechnol 23:753–758

Fariq A, Khan T, Yasmin A (2017) Microbial synthesis of nanoparticles and their potential applications in biomedicine. J Appl Biomed 15:241–248

Fernández-Llamosas H, Castro L, Blázquez ML, Díaz E, Carmona M (2017) Speeding up bioproduction of selenium nanoparticles by using vibrio natriegens as microbial factory. Sci Rep 7:16046

Foster HA, Ditta IB, Varghese S, Steele A (2011) Photocatalytic disinfection using titanium dioxide: spectrum and mechanism of antimicrobial activity. Appl Microbiol Biotechnol 90(6):1847–1868

Gahlawat G, Choudhury AR (2019) A review on the biosynthesis of metal and metal salt nanoparticles by microbes. RSC Adv 9:12944–12967

Ghimire KN, Katsutoshi I, Keisuke O, Hayashida T (2007) Adsorptive separation of metallic pollutants onto waste seaweeds, *Porphyra yezoensis* and *Ulva japonica*. Sep Sci Technol 42:2003–2018

Ghormade V, Deshpande MV, Paknikar KM (2011) Perspectives for nano-biotechnology enabled protection and nutrition of plants. Biotechnol Adv 29:792–803

Govarthanan M, Jeon CH, Jeon YH, Kwon JH, Bae H, Kim W (2020) Non-toxic nano approach for wastewater treatment using *Chlorella vulgaris* exopolysaccharides immobilized in iron-magnetic nanoparticles. Int J Biol Macromol 162:1241–1249
Grasso G, Zane D, Dragone R (2020) Microbial nanotechnology: challenges and prospects for green biocatalytic synthesis of nanoscale materials for sensoristic and biomedical applications. Nanomaterial. https://doi.org/10.3390/nano10010011
Guesh K, Mayoral A, Alvarez CM, Chebude Y, D´ıaz I (2016) Enhanced photocatalytic activity of TiO_2 supported on zeolites tested in real wastewaters from the textile industry of Ethiopia. Microporous Mesoporous Mater 225:88–97
Guibal E (2004) Interactions of metal ions with chitosan-based sorbents. Sep Purif Technol 38:43–74
Guo M, Song W, Wang T, Li Y, Wang X, Du X (2015) Phenyl functionalization of titanium dioxide-nanosheets coating fabricated on a titanium wire for selective solid-phase microextraction of polycyclic aromatic hydrocarbons from environment water samples. Talanta 144:998–1006
Gupta VK, Agarwal S, Saleh TA (2011) Chromium removal by combining the magnetic properties of iron oxide with adsorption properties of carbon nanotubes. Water Res 45(6):2207–2212
He F, Zhao D (2005) Preparation and characterization of a new class of starch-stabilized bimetallic nanoparticles for degradation of chlorinated hydrocarbons in water. Environtal Sci Technol 39:3314–3320
Holmes JD, Smith PR, Evans-Gowing R, Richardson DJ, Russell DA, Sodeau JR (1995) Energy-dispersive X-ray analysis of the extracellular cadmium sulphide crystallites of *Klebsiella aerogenes*. Arch Microbiol 163:143–147
Hulkoti NI, Taranath TC (2014) Biosynthesis of nanoparticles using microbes—a review. Colloids Surf B: Biointerfaces 121:474–483
Hussain CM, Palit S (2020) Functionalization of nanomaterials for industrial applications: recent and future perspectives. In: Handbook of functionalized nanomaterials for industrial applications. Elsevier, Amsterdam, pp 3–14
Imamura K, Yoshikawa T, Hashimoto K, Kominami H (2013) Stoichiometric production of aminobenzenes and ketones by photocatalytic reduction of nitrobenzenes in secondary alcoholic suspension of titanium (IV) oxide under metal-free conditions. Appl Catal B 134–135:193–197
Janotti A, Van deWalle CG (2009) Fundamentals of zinc oxide as a semiconductor. In: Reports on progress in Physics. IOP publishing, pp 17, 29. https://doi.org/10.1088/0034-4885/72/12/126501
Jegan A, Ramasubbu A, Saravanan S, Vasanthkumar S (2011) One-pot synthesis and characterization of biopolymer-Iron oxide nanocomposite. Int J Nano Dimens 2:105–110
Jha AK, Prasad K, Kulkarni AR (2009) Synthesis of TiO_2 nanoparticles using microorganisms. Colloids Surf B 71:226–229
Kalhapure RS, Sonawane SJ, Sikwal DR, Jadhav M, Rambarose S, Mocktar C, Govender T (2015) Solid lipid nanoparticles of clotrimazole silver complex: an efficient Advances in Materials Science and Engineering 7 nano antibacterial against Staphylococcus aureus and MRSA. Colloids Surf B: Biointerfaces 136:651–658
Karthikeyan T, Rajgopal S, Miranda LR (2005) Chromium(VI) adsorption from aqueous solution by *Hevea brasiliensis* sawdust activated carbon. J Hazard Mater 124:192–199
Kaul RK, Kumar P, Burman U, Joshi P, Agrawal A, Raliya R, Tarafdar JC (2012) Magnesium and iron nanoparticles production using microorganisms and various salts. Mater Sci Pol 30:54–258
Khin MM, Nair AS, Babu VJ, Murugan R, Ramakrishna S (2012) A review on nanomaterials for environmental remediation. Energy Environ Sci 5(8):8075–8109
Kim SH, Lee SW, Lee GM, Lee BT, Yun ST, Kim SO (2016) Monitoring of TiO_2-catalytic UV-LED photo-oxidation of cyanide contained in mine wastewater and leachate. Chemosphere 143:106–114
Kobya M, Demirbas E, Senturk E, Ince M (2005) Adsorption of heavy metal ions from aqueous solutions by activated carbon prepared from apricot stone. Biores Technol 96:1518–1521

Krishnaraj C, Ramachandran R, Mohan K, Kalaichelvan PT (2012) Optimization for rapid synthesis of silver nanoparticles and its effect on phytopathogenic fungi. Spectrochim Acta Part A: Mol Biomol Spectrosc 93:95–99

Kumar D, Karthik L, Kumar G, Roa KB (2011) Pharmacologyonline 3:31100–1111

Kundu D, Hazra C, Chatterjee A, Chaudhari A, Mishra S (2014) Extracellular biosynthesis of zinc oxide nanoparticles using Rhodococcus pyridinivorans NT2: multifunctional textile finishing, biosafety evaluation and in vitro drug delivery in colon carcinoma. J Photochem Photobiol B: Biol 140:194–204

Kusior A, Klich-Kafel J, Trenczek-Zajac A, Swierczek K, Radecka M, Zakrzewska K (2013) TiO_2–SnO_2 nanomaterials for gas sensing and photocatalysis. J Eur Ceram Soc 33(12):2285–2290

Lee Y, Kim S, Venkateswaran P, Jang J, Kim H, Kim J (2008) Anion co-doped Titania for solar photocatalytic degradation of dyes. Carbon Letters. 9(2):131–136

Li YH, Wang S, Wei J et al (2002) Lead adsorption on carbon nanotubes. Chem Phys Lett 357(3–4):263–266

Li X, Xu H, Chen Z-S, Chen G (2011) Biosynthesis of nanoparticles by microorganisms and their applications. J Nanomater, 1–16

Liang XJ, Kumar A, Shi D, Cui D (2012) Nanostructures for medicine and pharmaceuticals. J Nanomater. https://doi.org/10.1155/2012/921897

Liu F, Yang JH, Zuo J et al (2014) Graphene-supported nanoscale zero-valent iron: removal of phosphorus from aqueous solution and mechanistic study. J Environ Sci 26(8):1751–1762

Lu H, Wang J, Stoller M, Wang T, Bao Y, Hao H (2016) An overview of nanomaterials for waste water treatment. In: Advances in material sciences and engineering. Hindawi Publishing Corporation. https://doi.org/10.1155/2016/4964828

Mahanty S, Chatterjee S, Ghosh S, Tudu P, Gaine T, Bakshi M et al (2020) Synergistic approach towards the sustainable management of heavy metals in wastewater using mycosynthesized iron oxide nanoparticles: biofabrication, adsorptive dynamics and chemometric modeling study. J Water Process Eng 37:101426

Mahdavi M, Namvar F, Ahmad MB, Mohamad R (2013) Green biosynthesis and characterization of magnetic iron oxide (Fe_3O_4) nanoparticles using seaweed (*Sargassum muticum*) aqueous extract. Molecules 18:5954–5964

Mallick N (2003) Biotechnological potential of *Chlorella vulgaris* for accumulation of Cu and Ni from single and binary metal solutions. World J Microbiol Biotechnol 19:695–701

Marooufpour N, Alizadeh M, Hatami M, Lajayer BA (2019) Biological synthesis of nanoparticles by different groups of bacteria. In: Microbial nanobionics. Springer, Cham, Switzerland, pp 63–85

Matheson LJ, Tratnyek PG (1994) Reductive dehalogenation of chlorinated methanes by iron metal. Environ Sci Technol 28(12):2045–2053

Mills A, Le Hunte S (1997) An overview of semiconductor photocatalysis. J Photochem Photobiol, A 108(1):1–35

Mishra A, Kumari M, Pandey S, Chaudhry V, Gupta K, Nautiyal C (2014) Biocatalytic and antimicrobial activities of gold nanoparticles synthesized by trichoderma sp. Biores Technol 166:235–242

Moghaddam AB, Moniri M, Azizi S, Rahim RA, Ariff AB, Saad WZ, Namwar F, Navaderi M (2017) Biosynthesis of ZnO nanoparticles by a new Pichia kudriavzevii yeast strain and evaluation of their antimicrobial and antioxidant activities. Molecules 22:1–18

Moon G, Kim D, Kim H, Bokare AD, Choi W (2014) Platinum-like behavior of reduced graphene oxide as a cocatalyst on TiO_2 for the efficient photocatalytic oxidation of arsenite. Environ Sci Technol Lett 1(2):185–190

Moreno-Castilla C, Alvarez-Merino MA, Lopez-Ramon MV, Rivera-Utrilla J (2004) Cadmium ion adsorption on different carbon adsorbents from aqueous solutions. Effect of surface chemistry, pore texture, ionic strength, and dissolved natural organic matter. Langmuir 20:8142–8148

Nam KT, Kim D-W, Yoo PJ, Chiang C-Y, Meethong N, Hammond PT, Chiang YM, Belcher AM (2006) Virus-enabled synthesis and assembly of nanowires for lithium ion battery electrodes. Science 312:885–888

Nguyen AT, Hsieh CT, Juang RS (2016) Substituent effects on photodegradation of phenols in binary mixtures by hybrid H_2O_2 and TiO_2 suspensions under UV irradiation. J Taiwan Inst Chem Eng 62:68–75

Noman M, Shahid M, Ahmed T, Niazi MBK, Hussain S, Song F, et al (2020) Use of biogenic copper nanoparticles synthesized from a native *Escherichia* sp. as photocatalysts for azo dye degradation and treatment of textile effluents. Environmental Pollution 257:113514

Ohsaka T, Shinozaki K, Tsuruta K, Hirano K (2008) Photoelectrochemical degradation of some chlorinated organic compounds on n-TiO_2 electrode. Chemosphere 73(8):1279–1283

Oubagaranadin JUK, Murthy ZVP (2009) Adsorption of divalent lead on a montmorillonite—Illite type of clay. Ind Eng Chem Res 48:10627–10636

Pan B, Zhang Q, Du W, Zhang W et al (2007) Selective heavy metals removal from waters by amorphous zirconium phosphate: behavior and mechanism. Water Res 41:3103–3111

Panchal A, Swientoniewski LT, Omarova M, Yu T, Zhang D, Blake DA, John V, Lvov YM (2018) Bacterial proliferation on clay nanotube pickering emulsions for oil spill bioremediation. Colloids Surf B: Biointerfaces 164:27–33

Parmon V (2008) Nanomaterials in catalysis. Mater Res Innovations 12(2):60–61

Patil S, Chandrasekaran R (2020) Biogenic nanoparticles: a comprehensive perspective in synthesis, characterization, application and its challenges. J Genet Eng Biotechnol 18:67. https://doi.org/10.1186/s43141-020-00081-3

Pavani KV, Kumar NS (2013) Adsorption of iron and synthesis of iron nanoparticles by Aspergillus species kvp 12. Am J Nanomater 1:24–26

Priyabrata M, Ahmad A, Deendayal M, Satyajyoti S, Sudhakar RS, Mohammad IK, Renu P, Ajaykumar PV, Mansoor A, Rajiv K, Murali S (2001) Fungus mediated synthesis of silver nanoparticles and their immobilization in the mycelial matrix: a novel biological approach to nanoparticle synthesis. Nano Lett 1:515–519

Pyrzynska K, Bystrzejewski M (2010) Comparative study of heavy metal ions sorption onto activated carbon, carbon nanotubes, and carbon-encapsulated magnetic nanoparticles. Colloids Surf A: physchemical Eng Asp 362:102–109

Rawal SB, Bera S, Lee D, Jang DJ, Lee WI (2013) Design of visible-light photocatalysts by coupling of narrow band gap semiconductors and TiO_2: effect of their relative energy band positions on the photocatalytic efficiency. Catal Sci Technol 3(7):1822–1830

Ray PZ, Shipley HJ (2015) Inorganic nano-adsorbents for the removal of heavy metals and arsenic: a review. RSC Adv 5(38):29885–29907

Reynolds DC, Look DC, Jogai B, Litton CW, Cantwell G, Harsch WC (1999) Valence-band ordering in ZnO. Phys Rev B: Condens Matter Mater Phys 60(4):2340–2344

Rizwan MD, Singh M, Mitra CK, Morve RK (2014) Ecofriendly Application of Nanomaterials: nanobioremediation. J Nanoparticles Res 1:1–7

Rozamond YS, Mao CB, Gao XX, Justin LB, Angela MB, George G, Brent LI (2004) Bacterial biosynthesis of cadmium sulfide nanocrystals. Chem Biol 11:1553–1559

Salem SS, Fouda A (2021) Green synthesis of metallic nanoparticles and their prospective biotechnological applications: an overview. Biol Trace Elem Res 199:344–370

Schmidt ML, MacManus DJL (2007) ZnO-nanostructures, defects, and devices. Mater Today 10:40–48

Shah M, Fawcett D, Sharma S, Tripathy SK, Poinern GEJ (2015) Green synthesis of metallic nanoparticles via biological entities. Materials 8:7278–7308

Shamsuzzaman MA, Khanam H, Aljawfi RN (2013) Biological synthesis of ZnO nanoparticles using *C. albicans* and studying their catalytic performance in the synthesis of steroidal pyrazolines. Arab J Chem 2017(10):S1530–S1536

Sharma B, Dangi AK, Shukla P (2018) Contemporary enzyme based technologies for bioremediation: a review. J Environ Manage 210:10–22

Siddiqi KS, Husen A, Rao RAK (2018) A review on biosynthesis of silver nanoparticles and their biocidal properties. J Nano Biotechnol. https://doi.org/10.1186/s12951-018-0334-5

Stafiej A, Pyrzynska K (2007) Adsorption of heavy metal ions with carbon nanotubes. Sep Purif Technol 58:49–52

Subramaniyam V, Subashchandrabose SR, Thavamani P, Megharaj M, Chen Z, Naidu R (2015) *Chlorococcum* sp. MM11—a novel phyco-nanofactory for synthesis of iron nanoparticles. J Appl Phycol 27:1861–1869

Sun S, Wang L, Wang A (2006) Adsorption properties of cross linked carboxymethyl-chitosan resin with Pb (II) as template ions. J Hazard Mater 136:930–937

Tang WW, Zeng GM, Gong JL, Liang J, Xu P, Zhang C, Huang BB (2014) Impact of humic/fulvic acid on the removal of heavy metals from aqueous solutions using nanomaterials: a review. Sci Total Environ 468–469:1014–1027

Tofighy MA, Mohammadi T (2011) Adsorption of divalent heavy metal ions from water using carbon nanotube sheets. J Hazard Mater 185:140–147

Vijayaraghavan K, Ahmad D, Abdul Aziz ME (2007) Aerobic treatment of palm oil mill effluent. J Environ Manage 82:24–31

Waghmare SR, Mulla MN, Marathe SR, Sonawane KD (2015) Ecofriendly production of silver nanoparticles using candida utilis and its mechanistic action against pathogenic microorganisms. 3 Biotech 5:33–38

Wang J, Chen C (2009) Biosorbents for heavy metals removal and their future. Biotechnol Adv 27:195–226

Wang X, Zheng Y, Wang A (2009) Fast removal of copper ions from aqueous solution by chitosan-g-poly(acrylic acid)/attapulgite composites. J Hazard Mater 168:970–977

Wang X, Guo Y, Yang L, Han M, Zhao J, Cheng X (2012) Nanomaterials as sorbents to remove heavy metal ions in wastewater treatment. J Environ Anal Toxicol 2(7):1–7

Yadav A, Kon K, Kratosova G, Duran N, Ingle AP, Rai M (2015) Fungi as an efficient mycosystem for the synthesis of metal nanoparticles: progress and key aspects of research. Biotech Lett 37:2099–2120

Yan J, Han L, Gao W, Xue S, Chen M (2015) Biochar supported nanoscale zerovalent iron composite used as persulfate activator for removing trichloroethylene. Biores Technol 175:269–274

Yogalakshmi KN, Das A, Rani G, Jaswal V, Randhawa JS (2020) Nano-bioremediation: a new age technology for the treatment of dyes in textile effluents. In: Bioremediation of industrial waste for environmental safety. Springer, Singapore, pp 313–347

Zare B, Babaie S, Setayesh N, Shahverdi AR (2013) Isolation and characterization of a fungus for extracellular synthesis of small selenium nanoparticles. Nanomedicine 11:13–19

Zhang W (2003) Nanoscale iron particles for environmental remediation: an overview. J Nanoparticles Res 5:323–332

Zhang X, Qu Y, Shen W, Wang J, Li H, Zhang Z, Li S, Zhou J (2016) Biogenic synthesis of gold nanoparticles by yeast magnusiomyces ingens lh-f1 for catalytic reduction of nitrophenols. Colloids Surf A: Physchemical Eng Asp 497:280–285

Zhao G, Li J, Ren X, Chen C, Wang X (2011) Few-layered graphene oxide nanosheets as superior sorbents for heavy metal ion pollution management. Environ Sci Technol 45:10454–10462

Ziagova M, Dimitriadis G, Aslanidou D, Papaioannou X, Litopoulou-Tzannetaki E, Liakopoulou-Kyriakides M (2007) Comparative study of Cd (II) and Cr(VI) biosorption on *Staphylococcus xylosus* and *Pseudomonas* sp. Biores Technol 98:2859–2865

Sustainable Technologies for Treatment of Industrial Wastewater and Its Potential for Reuse

Ramya Suresh, Rajivgandhi Subramaniyan, Senthil Kumar K., Naveen Kumar, and Maheswari Chenniappan

1 Introduction

Many waterways have become unsanitary and harmful to humans and other living organisms because of fast population growth, uncontrolled quick urbanization, industrial and technical advancement, energy usage, and waste creation from home and industrial sources. In Nigeria, environmental contamination is guided by little or no rules at all (Durotoye et al. 2018). As a result, many businesses pollute waterways with wastewater that has been improperly treated or not treated at all (Minhas et al. 2022). For India and the rest of the globe, water, food, and energy security are becoming increasingly critical and vital challenges. The concomitant consequences of agricultural expansion, industrialization, and urbanization have led to the closure of most watersheds in India and worldwide, resulting in mild to severe water shortages. Water usage effectiveness and demand management can help fulfill the increasing fresh water needs. As a result, with necessary treatment, wastewater and low-quality water are emerging as promising sources for demand control (Gowd et al. 2022; Kumar and Tortajada 2020).

In India's main cities, sewage production is estimated at 38,356 million liters per day (MLD), while treatment capacity is just 11,798 MLD. Similarly, just 60% of industrial effluent is treated, with the majority coming from large-scale operations. Municipal sewage treatment facilities owned and operated by the state and municipal wastewater treatment facilities used to treat industrial wastewater also fall short of the required performance criteria. Consequently, treatment plant effluent is frequently unfit for domestic use, and reuse of wastewater is typically limited to agriculture and industry. However, in underdeveloped nations, where wastewater is rarely treated

Ramya Suresh (✉) · R. Subramaniyan
Sanskrithi School of Engineering, Puttaparthi, Andhra Pradesh 515134, India
e-mail: ramya24591@gmail.com

Senthil Kumar K. · Naveen Kumar · M. Chenniappan
Kongu Engineering College, Perundurai, Erode, Tamilnadu 638060, India

M. P. Shah (ed.), *Microbial Technologies in Industrial Wastewater Treatment*,
https://doi.org/10.1007/978-981-99-2435-6_9

and significant quantities of untreated wastewater are utilized in agriculture, there is a greater danger to health and the environment (Kaur et al. 2012).

India has just 2.45% of the world's land area and only four percent of the world's water resources, but it has a population of 16%. About 1123 BCM of the country's total utilizable water supply has been calculated (690 BCM from the surface and 432 BCM from the subsurface), which accounts for just 28% of the water produced from precipitation. Seventy percent (688 BCM) of water is diverted for agriculture, and that number might rise to 1180 BCM by 2050. Water from the earth is a major source of irrigation. Irrigation accounts for around 212.5 billion gallons of the 433 billion gallons of groundwater recharged each year. Demand for home and industrial water use is expected to rise to 29.2 billion gallons per year by 2025. As a result, a reduction in irrigation water availability to 162.3 BCM is projected. If current growth trends continue, the global population will top 1.5 billion people by 2050. There is a pressing need, therefore, for effective management of water resources through improved water usage efficiency and effluent water recycling (Sato et al. 2013; Starkl et al. 2013).

With the fast growth of cities and the provision of residential water, the amount of gray/wastewater is growing in proportion to this growth. CPHEEO estimates that between 70 and 80% of the water used for household purposes is disposed of as trash. There are 234 Sewage Treatment Plants in India, according to the World Bank (STPs). This is mostly due to river action plans that have been implemented since 1978–1979 and are situated in (only about 5% of) cities and towns along important riverbanks (Sundaravadivel and Vigneswaran 2001). It is estimated that an additional 13,468 MLD of industrial sewage is produced, but only 60% of it is treated. CETPs have been set up for clusters of small-scale industries that can't afford the cost of a sewage plant. Sludge drying beds, secondary clarifiers, flash mixers, and clariflocculators are only some of the treatment processes used in these facilities to remove solids from the water (Muthukumaran and Ambujam 2003). The major procedures of screening, grit removal, and sedimentation remove large particles and settable solids. CETP-treated industrial wastewater can be used for various domestic purposes. Traditional wastewater treatment methods are time-consuming, costly, and difficult to operate and maintain. Sludge removal, treatment, and handling of India's wastewater treatment have been shown to be the most ignored areas of operation (Kamyotra and Bhardwaj 2011; Shah Maulin 2020; Yadav et al. 2019). Poor design, lack of maintenance, frequent power outages, and a lack of technical manpower have resulted in the majority of wastewater treatment plants being shut down. Wastewater disposal is a major issue due to the lack of treatment capacity and the rising amount of sewage being generated. The Sewage treatment Board is already selling a major percentage of the wastewater from STPs to local farmers for a fee, or the polluted wastewater ends up in river basins and is utilized for irrigation (Ramteke et al. 2010; Singh et al. 2008). A variety of industrial wastewaters were examined in this study to learn more about their origins, impacts, treatment options, and possible reuse.

2 Sources of Wastewater

2.1 Dairy Industry Wastewater

The dairy sector is thought to have a significant impact on water pollution since there are around 286 big- and small-scale dairy enterprises in India that are responsible for a huge amount of waste output, both solid and liquid. The dairy industry in India produces around 110–275 MT of wastewater each year, which is disposed of in landfills. Receiving stations, cheese plants, bottling plants, casein plants, butter plants, dried milk plants, condensed milk plants, and ice cream plants are some of the places where dairy sector effluent is discharged (Yonar et al. 2018; Shah Maulin 2021a, 2021b). Cleaning out containers, trucks, pipelines, and other dairy equipment generates waste as does washing and drying the product that is still within. Dairy waste can also be generated by spills caused by overflows, leaks, carelessness, and boiling over. Aside from that, dairy waste may be traced back to settling tank sludge or the containers and detergent used in washing (Shete and Shinkar 2013b).

2.2 Sugar Industry Wastewater

Wastewater from the sugar industry is mostly generated during the cleaning process. Large volumes of wastewater are generated during the cleaning of milling house floors and in the many boiling house divisions, such as heat exchangers, filtrations, suction pans, centrifugation, etc. Wastewater can also include rinse water used for cleaning the filter fabric of a rotating vacuum filter as well as periodic treatment of salt water and sulfur dioxide production lines. Using NaOH and HCl to clean heating systems and condensers on a regular basis to remove crust from the pipe wall leads to the loading of organic and inorganic pollutants into wastewater. A large portion of the wastewater generated is the result of leaks in the plumbing, pumps, and centrifuges. Other sources of wastewater include steam desolation, sprayer pool overflowing and contaminated condenser-cooled water that is discharged as wastewater (Kushwaha 2015; Sahu and Chaudhari 2015).

2.3 Tannery Wastewater

Wet tanning procedures in leather production, such as rust remover and liming as well as unhairing effluent, chrome tanning, fat-liquoring and dyeing, and rinsing, create a lot of tanning wastewater. The main sources of pollution are separated during the manufacturing operation, and then the effluent collected from these sources is mixed with effluent retrieved from other operations in most industries. Finally, further treatment is performed on all of the collected effluent. Carbon, oxygen, sulfide, nitrogen ammonia, and suspended particles are all high in integrated tannery effluent (Mani and Bharagava 2018; Nath et al. 2005).

2.4 *Textile Wastewater*

Dyes and finishes might be hampered by the presence of several types of contaminants in fibers. Materials included in natural fibers include lipids, mineral sands, and plant stuff, whereas synthetic fibers may include spun coating and weaving lubricants. When making fabric, contamination can occur, such as from lubricant being used to lubricate industrial equipment, mill debris, and transient garment marks. These contaminants are eliminated by scouring. Water scouring is chosen over solvent scouring because it is nonflammable, nontoxic, abundant, and less expensive. Polyester is easier to scour than wool or cotton. As well as common household cleaning products like dishwashing liquids and bar soap, there are a wide variety of specialty cleaners and lubricants that may be used to help in scrubbing. After scrubbing, the products are rinsed (or washed) completely. The resulting effluent is physically active and may be hazardous due to the scouring chemicals and substances emitted from the substance while washing. The wastewater has a lot of COD and solids in it. BOD may be high (Bisschops and Spanjers 2003; Pang and Abdullah 2013).

2.5 *Paper Industry Wastewater*

In paper making, water is used for a lot of different things, like making raw materials, making chemicals, making fibers, transporting them, diluting pulp, making paper webs, and cleaning equipment. Water is also utilized for a variety of other purposes, including chilling, warming in the form of vapor, and lubricating. Recycled papers are easier to pulp since they don't require a lot of chemicals, but rather water. The other way to make pulp is to use sulfites, which are chemicals that can be found in wood bark and chips. Sulfites can be used to make pulp from wood chips and bark, but they can also make a lot of sulfur-containing compounds and smelly exhaust gas molecules from reduced sulfur that make the air smell bad. The effluent from mills that use wood as a raw material contains lignin, hemicellulose, cellulose, and wood extractives degradation products. During the chloride bleaching step, the breakdown of lignin results in the formation of potentially hazardous chemicals including formic acid, acrolein, formaldehyde, chlorophenols, phthalates, and furans, among other things. However, pulping with recycled paper uses no chemicals and depends heavily on water (Elvira et al. 1997; Sonsale et al. 2021).

2.6 Pharmaceutical Industry Wastewater

Effluents in a pharmaceuticals manufacturing firm generally come from the biosynthesis and processing of the medications. The majority of Active Pharmaceutical Ingredients (APIs) that are delivered globally are created by chemical synthesis employing molecular, inorganic, and biological processes. The amount of wastewater created increases because multiproduct pharmaceutical reactors and separators are often large or utilized inefficiently. In the pharmaceutical sector, there are several subprocesses, making it difficult to describe each product waste (Martz 2012).

Pharmaceutical compounds, APIs, and medications can be delivered to rivers, underground reservoirs, and aquifers from a variety of sources. There are a variety of on-site waste treatment systems and septic tanks that can fail in the absence of rain, as can leaking sewage lines, illegal dumping, sewer-to-storm sewer cross-connections, and poorly managed pet or livestock waste. Wastewater can carry chemicals that are used in homes, businesses, and agriculture every day. Antibiotics, hormones, cleaning surfactants, cleaning agents, solvents, protective coatings, pesticides, and antioxidants are among these substances. Several investigations indicate that wastewater, groundwater, and surface water all contain high concentrations of harmful and dangerous chemicals. Inorganic and non-organic wastes, heavy metals, and possible inhibitors can all be found in wastewater, which may eventually reach a water catchment region or the underground as landfills (Patneedi and Prasadu 2015).

3 Effects of Wastewater

3.1 Effects of Dairy Industry

Anaerobic conditions and the production of pungent unpleasant smells are inevitable as dairy wastewater decomposes quickly and reduces the dissolved oxygen content in receiving streams. Malaria-carrying flies and mosquitoes thrive in the receiving water, which is then contaminated with illnesses including dengue virus, yellow fever, and chicken guniya. Some fish and algal species are said to be poisoned by increased concentrations of dairy manure. Additionally, the extremely odorous black sludge produced by the decomposition of dairy manure has been discovered to have a harmful effect on aquatic life at certain concentrations. Dairy wastewater comprises soluble and trace organics. In addition to reducing do, they encourage gas release, induce odor and taste, add color or turbidity, and promote eutrophication. Water, air, and biodiversity contamination are the primary environmental issues associated with milk production. A buildup of algae and bacteria, as well as a lack of oxygen in rivers, is a common result of these activities. As a result, fish populations begin to decline over time. As a result, dairy effluents must be treated using proper methods (Goli et al. 2019; Sinha et al. 2019).

3.2 *Effect of Sugar Industry Wastewater*

As a result of the production of wastewater, air pollutants, and solid wastes, sugar mills have an enormous environmental effect. For every ton of crushed cane, a sugar mill produces 1 m^3 of effluent. Aquatic life suffers when effluent from sugar mills is dumped into bodies of water with high TDS. During the sugar-producing period in November–April, the irrigation for majority of rabicrop acquired unsuitable water surrounding the sugar mill sites. The terrible consequences of untreated sugar industry effluents have an impact on human life as well. In fresh water systems, vast volumes of organic substances and sludge washed from mills deteriorate and consume available oxygen, resulting in a large number of fish fatalities. Additionally, owing to inappropriate wastewater treatment, the septic conditions developed which would emit foul-smelling hydrogen sulfide led to crystallization of metal and soluble salts that render the watercourses black. Using sugar industrial effluent for agriculture irrigation directly limits the germination rate without sufficient treatment and dilution. The effluent from sugar mills also enters the topsoil and seeps into the groundwater, generating polluted pools that alter the chemical composition and ultimately degrade the quality of the groundwater. Water bodies in surface and groundwater are contaminated by untreated wastewater discharged by these industries. Additionally, it harms the ecosystems of bodies of water, such as the Left Bank Outfall Drainage (LBOD) system (Kushwaha 2015).

3.3 *Tannery Wastewater*

Toxic heavy metal poisoning of agricultural soil has resulted from the nearby Jajmau, Kanpur leather industry's irrigation use. Toxic heavy metal deposition in surface soils, groundwater contamination, and plant uptake can be exacerbated by the use of sludge as an irrigation medium. This can lead to an increase in Cd, Zn, Cr, Ni, Pb, and Mn concentrations, as well as a decrease in the soil's heavy metal retention capacity. Because of the high levels of phosphorus in soil, caution should be exercised while applying phosphorus treatments. A high amount of phosphorus might result in water contamination. In nature, Cr(VI) is mostly found as the dissolvable anion, which is exceedingly hazardous at high concentrations. It is toxic, cancerous, and teratogenic at high concentrations. As a result of the oxidation of nitrogen, the end product of tannery processes is a high concentration of chlorine and nitrates. Sodium chloride as a feedstock in leather causes the soil to become alkaline and therefore, raise its pH. Concerns about the purity and color of watercourses affected by tanning effluents are a massive concern. Microbial degradation of biodegradable organic materials (such as proteins and carbohydrates) in stream waters is the primary source of stream water dissolved oxygen content decline due to tannery waste. Dissolved oxygen depletion, which is harmful to aquatic species, and anaerobic activity, which results in the generation of noxious gases, are the primary causes of this effect. There is a significant increase in reactive oxygen species (ROS) production in wastewater from tanneries, as well as phytotoxic effects and high heavy metal accumulation. This causes stress to plants (e.g., salt stress, which impacts various metabolic processes and reduces vegetative growth before eventually reducing reproductive growth) and has a significant impact on respiration, photosynthesis, germ sprouting, and mitotic activity. Accumulation is influenced by a variety of factors, including the kind of plant and its element, as well as other factors such as the amount of dissolved oxygen in solution and root secretion (Mani and Bharagava 2018; Praveena et al. 2013).

3.4 Effect of Textile Industry

People in the textile industry have a lot of problems with the way they use water, how they treat it, and how they dispose of wastewater. The discharge of a wide variety of azo dyes into industrial effluent has resulted in significant damage of the environment. Chemicals like naphthol and dyes, nitrates, and acetic acid make the effluent very poisonous as do sulfur and other metals like copper and arsenic as well as heavy metals like lead and cadmium. Other auxiliary chemicals include nickel and cobalt and certain metal salts. The temperature and pH of the mill effluent are typically often very high, which is exceedingly harmful. It also blocks sunlight from penetrating, causing an imbalance in the ecology (Singh et al. 2021).

Furthermore, allowing this effluent to run in the fields blocks soil pores, reducing soil production. Soil hardens, making root penetration more difficult. The sewage pipes are corroded and incrusted by the effluent that travels down the sewers. Drains and rivers become clogged with wastewater, rendering drinking water unsafe for human use if it is flows freely. The hue of waterways is considered an aesthetic issue rather than a danger to the environment because of the toxins it contains. People are more concerned about "non-natural" colors like red and purple than they are about blue or green rivers. Drain maintenance costs are also increased due to drain leakage, which is caused by wastewater. Air emissions, particularly those containing volatile organic compounds (VOCs), excessive noise or odor, and workplace safety are all equally important environmental concerns (Mani and Bharagava 2018).

Textile mills generate emissions into the atmosphere as a result of nearly all of their operations. Boilers at textile mills typically produce nitrogen and sulfur oxides, which are harmful to the environment. In addition to resin finishing and drying operations and printing and dyeing activities as well as fabric preparation and wastewater treatment plants, textile processes generate considerable air pollution. High-temperature drying and curing of mineral oils emit hydrocarbons. Volatile chemicals like formaldehyde and acids can be released during these processes. Dye exposure can have both acute and long-term impacts on organisms, depending on the duration of exposure and the concentration of the dye. They're also carcinogenic, which means they can harm an unborn child by causing cancer of the intestines or brain problems. Textile dyes can produce allergic reactions in the eyes, skin, mucous membranes, and upper respiratory tract (Khan and Malik 2014).

3.5 Effect of Paper and Pulp Industry

The effluents of pulp mills and bleach plants are strongly colored as a result of the breakdown products of polymeric lignin and include chlorinated aromatics. The pulp and paper industry's wastewater contains dioxin-like chemicals, which are both harmful to humans and ecologically persistent. And over seven thousand billion gallons of rich color and hazardous waste effluents are generated by the paper and pulp industry each year, mostly including large molecular mass, altered and bleached lignins. A range of reactions was seen in fish populations residing downstream of bleach kraft pulp and paper mills, according to the findings of the studies. There was a decline in secondary sexual features, reduced gonads, altered fish reproduction, and a delay in sexual maturity. Septic fungus bloomed in the river that received effluents as a direct result of the pulp and paper plant. Due to the extreme turbidity and dark hue of suspended particles, water opacity and river or lakebed blanketing might occur. Anaerobic decomposition beneath a heavy blanket may release hydrogen sulfide into aquatic habitats if the blanketing is severe. Its dark color and blanketing can inhibit photosynthetic activity (Chhonkar et al. 2000; Singh and Chandra 2019).

There are a number of elements in pulp mill effluent that can induce crop nutrient imbalance, raise soil salinity, and degrade soil structure, all of which can lead to a decrease in crop yield in the long term. Every year, the pulp and paper sector generates around 100,000 dry tons of solid waste. High chemical and biological oxygen needs have caused the effluents to become alkaline. Toxic trace elements found in paper mill wastewater can build up in soils and eventually end up in the food chain, posing serious health risks to humans and animals. The irrigation of dirty water from pulp and paper mills harms the soil, growth, quality, and yield of crops (Giri et al. 2014).

3.6 Effect of Pharmaceutical Industry

Antibiotics and other non-biodegradable organic matter present in pharmaceutical industrial wastewater have been reported to contain high levels of pollutants such as plants and animals steroids, hormone secretion, beta-lactamides, and anti-inflammatories as well as anti-depressants and cytostatic agents, as well as detergent metabolites and flame retardants. The acute toxicity of pharmaceutical compounds, including genotoxicity and mutagenesis potential, necessitates their removal from the environment. Most stages yield mother liquor that comprises unreacted ingredients, by-products, and residual organic solvent base. There may also be a variety of other chemical compounds, such as acidic or alkaline solutions and bases (Sharma et al. 2020).

Due to their inherent biological activity, APIs, or pharmaceutically active molecules, are a serious problem. A number of studies have established the hazardous effects of pharmaceutical effluents on living creatures by identifying particular chemical components. Root development was severely inhibited in experimental experiments utilizing the effluent concentration test when the v/v concentration was less than 50%. Other chromosomal abnormalities, such as micronucleated cells and the loss of chromosomes in varied concentrations of effluent showed that pharmaceutical effluent compounds were harmful and genotoxic to the cells they were found in (Martz 2012).

When the effluent is treated, superbugs are known to multiply and produce extracellular antibiotic resistant genes. In the receiving stream, the "resistome reservoirs" of these ARGs may be released and change native bacteria. DNA may be taken up, integrated, and expressed in a stable and functional manner through the natural process of transformation. Antimicrobial resistance and virulence genes have long been associated with horizontal gene transfer. Model bacteria are stimulated by the disinfection by-product bromoacetic acid (BAA), which is a controlled substance. Waste must be made non-functional by proven methods that make cell-free DNA ineffective. This can be done by reducing the size of DNA fragments below the point where they can be used by bacteria, or by changing the chemical substances and formation of the DNA so that it can be used to change microbes in watercourses (Gadipelly et al. 2014b; Patneedi and Prasadu 2015).

4 Treatment Methods

A broad range of physical, chemical, and biological processes are employed in wastewater treatment to eliminate pollutants from wastewater. There are a range of technologies that may be used in specific wastewater treatment to reduce varying degrees of contaminants. Primary, secondary, and tertiary treatments of wastewater are the three levels of treatment. There are a number of primary treatment techniques that are used to remove solids such as oil and grease, floating debris, and coagulants and coagulant aids. First, inorganic contaminants are removed from water using physical and chemical processes, which reduces the pollution burden in secondary treatment. After initial treatment, organic contaminants in the wastewater are removed using secondary treatment. For each unit process, there is a varied amount of time it takes for it to complete. tertiary treatment is an advanced treatment technique used to eliminate the residual organic and inorganic contaminants in the effluent following primary and secondary treatment (Tikariha and Sahu 2014).

4.1 Dairy Wastewater

Mechanical treatment incorporates filters, grit chambers, skimmer or sedimentation tanks, and other mechanical systems. The screens are angled towards in the flow direction, so that they are visible upward from. In order to prevent big items from floating in water sources or wastewater from clogging tiny pipelines and interfering with the operation of effluent pumps, screens are used to screen out the larger objects. Grit chambers are capable of removing larger foreign matter like grits and soil from a fluid stream. Oil, grease, fruit peels, and wood fragments may all be removed from a mixture using a skimming tank. It is possible to remove suspended particles from wastewater in a sedimentation tank by allowing it to flow slowly. The sludge that collects at the tank's bottom is treated a second time to reduce its toxicity (Shete and Shinkar 2013a).

The process of precipitation is used in chemical treatment. The procedure necessitates the addition of flocculants such as aluminum salts, ferrous salt, and limestone to the wastewater, as well as thorough mixing with the use of agitators. Because of this, inorganic phosphate forms very small particles, which eventually mix to create bigger flocks of the mineral. These flocks subsequently settle to the bottom of the tank, where they become primary sludge, while the clean effluent is discharged into the basins for biological treatment (Karadag et al. 2015).

After primary treatment, biological treatment, which is also known as secondary treatment, is utilized to eliminate any remaining materials. Dairy wastewater treatment facilities rely on it as one of their go-to options. Biodegradation of dairy wastewater is preferable because these processes use soluble chemicals and tiny colloids to remove organic pollutants from the water. There are two basic types of biological therapies, aerobic and anaerobic, based on oxygen needs. Aerobic bioremediation systems oxidize organic matter to carbon dioxide and water, while anaerobic systems are used for the treatment of high-strength dairy wastewater, where micro-organisms transfer organic matter and nutrient-rich to methane and CO_2 whereas, the rest of biofuels are used for growth and maintenance of the cells. Dairy effluent is often treated using aerobic systems these days, although this has a number of problems. Aerobic systems have a number of drawbacks, including high demand for energy, filamentous development, and acidification that occurs quickly due to the high lactose content and poor water buffer capacity. Anaerobic systems, on the other hand, are more dependable for the processing of milk wastewater treatment that has a high concentration of organic matter. This kind of processing is a cost-effective method since it does not necessitate the use of aeration or a big area, and it creates a little quantity of sludge (Sinha et al. 2019).

Chemoprecipitation is a sedimentation technique that utilizes chemicals to separate dissolved from suspended particles. The removal of phosphorus and heavy metals is possible in contemporary practice. There have been numerous substances used as substances over the years, including alum, ferrous sulfate, iron salts, and others. Alum is one of the most commonly used precipitants. Metal ions, anions, organic compounds, detergents, and greasy emulsions are the primary targets of their treatment (Sinha et al. 2019).

In wastewater treatment, coagulation/flocculation techniques are primarily used to separate suspended, colloidal, and dissolved contents, and they are directly applied to untreated water. There are two ways to go about this. Substances (coagulation inhibitors) like iron or aluminum are utilized in the first step, coagulation, to counteract the stability-promoting components of the system. Flocculation, the second phase, brings together particles that have become unstable and allows them to be readily separated via gravity settling (Loloei et al. 2014).

Dairy wastewater treatment with electrocoagulation is an advanced technology that is becoming increasingly popular. Metal plates submerged in water are used as conductors for the electric current. Electrostatic charges are responsible for the majority of the retention of inorganic pollutants, toxic substances, colloidal matter, and other contaminants in water. Because of the additional electrical charge applied to the polluted water, electrical charges that keep the particulates together are devastated, and the particles get separated from the fresh water. As they agglomerate, they create an easily destructible mass. This method produces minimal sludge. Electrocoagulation produces waste fluids that are crystal clear, tasty, odorless, and colorless. Due to the high cost of power, however, this method is prohibitively labor-intensive. Electrocoagulation successfully removed the COD from dairy wastewater of about 86.5% in a batch experiment using 3 A, pH 9, and a 75-min electrolysis duration (Akansha et al. 2020).

Many of the non-degradable organic components in wastewater can be removed using the adsorption treatment method. Adsorbents for wastewater treatment most commonly employ activated carbon. Some low-cost adsorbents are also employed for wastewater treatment, notably biochar, coal fly ash, and straw dust. Removing organic contaminants from dairy effluent was made easier with the use of husk as an adsorbent. Adsorbent dose of 5 g/L, pH 2 and temperature of 30 °C might remove as much as 92% of the contaminant from the solution. Sugarcane bagasse and rice husk were also used as adsorbents to treat dairy wastewater. Activated charcoal was utilized to remediate dairy wastewater. Dairy effluent was treated with activated charcoal to remove up to 65.34% COD and 67.23% BOD (Kushwaha et al. 2010).

Water treatment, water reclamation, and desalination applications all rely on membrane separation to perform their functions successfully. Microfiltration, ultrafiltration, nanofiltration, reverse osmosis, and electrodialysis are some of the major membrane separation technologies used to remove pollutants from dairy wastewater. Minimal temperature and lower power consumption are ideal conditions for these procedures to work. It is possible to recover a large percentage of a viable product. Fouling of the membrane might reduce the permeate flow because of the high cost of the equipment. A microfiltering system was used to remove about 90% of the COD from dairy wastewater treatment. A high-performance bioreactor, an aerobic jet loop reactor, paired with a ceramic membrane filtering unit, was utilized to test its potential for the treatment of dairy processing wastewater. At 53 kg COD m^3 d1, 97–98% of COD removal efficiency was achieved in less than three hours of hydraulic retention. Membrane biofilters accompanied by nanofiltration for the treatment of dairy wastewater with the goal of reusing the treated effluent. Organic matter and color were effectively removed from the feed effluent by membrane bioreactors (Vourch et al. 2008).

Dairy effluent may be effectively treated using ozonation to remove COD and odor. Ozone breakdown at high pH speeds up the removal of COD from dairy. It is well known that ozone is a powerful disinfectant. As a result, the company's environmental footprint and costs are reduced. Another benefit of using ozone in food processing is that it may be produced on-site and does not require shipping or storage like conventional chemical sanitizers. Cleaning and washing activities in milk processing plants account for the vast majority of the wastewater generated by this sector (Ramya Suresh et al. 2021).

4.2 Tannery Industry

Physical and chemical interactions play a major role in initial treatment as a pretreatment. Its goal is to reduce the organic load entering the biological sewer system by efficiently removing suspended particles and organics. In each situation, the wastewater quality and volume entering the biological treatment facility are likewise changed. Grille and screening processes, in particular, make for the bulk of the first treatment. Scurf and hair, as well as other dense things like sand, can be removed by these techniques. The elimination of colloidal matter is also made possible by these techniques (Zhao and Chen 2019).

The methods also entirely eradicate a number of unwanted features of wastewater, including sedimentation, flotation of certain organic products, and sedimentation and air flotation of a variety of coordination compounds and coagulation products, among others. Tanneries' oil effluent is often treated using the air flotation method. Using this approach, air is blown into the effluent through an aerating system, and the bubbles created in the water cause oil drips to rise to the top of the water, where they are collected. As a result, the oil–water dispersion is successfully separated from the water and completely obliterated. Regardless of the treatment process utilized in any facility, integrated tannery effluent is alkaline. As a general rule, the pH of a solution must be neutralized in order for further treatment to take place. Chemical neutralization, filtration neutralization, and mutual neutralization of alkaline and acidic wastewater are the three types of neutralization processes. For convenience, great efficiency, and controllability, chemical neutralization is the most used approach. HCl and NaOH are used in this procedure (Zhao and Chen 2019).

Coagulation is a typical wastewater treatment technology for tanneries because of its low cost, compact footprint, quick processing time, and appropriateness for batch treatment. In wastewater treatment, the flocculant utilized in the coagulation technique is critical. Chemical flocculants, rather than biological flocculants, are the most often employed flocculants in China's coagulation therapy nowadays. We developed a biological flocculant for treating tannery wastewater by screening bacillus in this work. Previous study suggests that this microbial flocculant is an essential part in removing total nitrogen, nitrogen-ammonia-chroma-COD from tannery waste.

Physicochemical treatment methods like adsorption are quite widespread, and it is one of the key strategies for reducing colored effluent. Most often used adsorbents are activated carbon, diatomite, Sio_2, and biosorbents made from recycled materials. This technology is not only cost-effective and environmentally friendly, but it also offers a novel means of recycling garbage. After going through a series of chemical changes, the three unique biosorbents made from agricultural residues (rice bran, wood ash, and bagasse) have great significant in pollutant removal. These biosorbents are used to clean tannery wastewater. To use these three kinds of adsorbents, researchers found that they were able to successfully remove COD and other pollutants from the water. To complicate matters further, the tanning process produces leftover metal ions that are difficult to remove from tannery effluent. By oxidizing organics and decreasing metal ions, micro-electrolysis technology cleanses water. It has been shown that using the micro-electrolysis technique for treating tannery effluent yields good results. Concentration levels of COD, suspended particles, sulfide, and total chromite may all be lowered by a significant percentage. To reclaim coolants that were previously utilized in the manufacturing of salted or pickled leather, electrolytes must be separated from the leather. An additional use for electrodialysis was the recovery of leftover tanning fluid from neutral salts like chromate during the process (Suman et al. 2021; Zhao and Chen 2019).

Large amounts of wastewater and sludge are treated using the oxidation ditch technique, a biological treatment method that can only handle a modest load. Stable operation is possible since it is very tolerant to environmental fluctuations. The oxidation ditch technique employs mechanical aeration and low-speed revolution to decrease energy usage. Because of this, tannery effluent is frequently treated using biological methods. The problem is that it takes a lot of land. Since huge tanneries use it, it is ideal. Intermittent operating processes that deform activated sludge are used in the Sequencing Batch Reactor (SBR) technology. A total of five fundamental processes are involved: inflow, reaction; sedimentation; effluent; and standby (the latter of which is optional). Massive amounts of building material are not needed in the SBR approach because there is only one reaction tank. Furthermore, the floor area required is less than the typical activated sludge process, making this technology more adaptable. The intermittent release of a considerable quantity of tannery effluent is ideal for this technology. As a result, the treatment of wastewater in medium and small tanneries can certainly benefit from the SBR approach (Suman et al. 2021).

The built wetland approach offers a number of advantages, including less investigation, lower operating costs, lower energy usage, and a high pollution removal rate. The operational costs of the built wetland approach are lower than those of the typical secondary biological treatment technology. Some researchers have already shown how this technology may be applied to tannery wastewater treatment. After treatment in a built wetland, tannery effluent had BOD and COD levels of 91.89% and 89.56%, respectively. Using an air flocculation ditch-constructed wetland with tannery wastewater, the COD was removed at a rate of more than 97.54% and ammonia nitrogen was removed at a rate of more than 89.56%. In addition to being easy and cost-effective, the operation is also steady and trustworthy (Ongen et al. 2013).

Ozone, hydrogen peroxide, and sodium hypochlorite are utilized as oxidants in the oxidation technique. The low cost and minimal floor area requirements of this approach make it an attractive treatment option, but its efficacy still has to be enhanced. The Fenton technique is the most often used oxidation process for the extensive remediation of tannery effluent, and it is the most widely used approach overall. In addition to being utilized for the ultimate extensive processing or preprocessing of recalcitrant organic wastewater, this method partly oxidizes the refractory organics present in the wastewater stream. The method alters a number of properties in tannery effluent, including biodegradability, solubility, and coagulation, among other things. All of these alterations are advantageous in terms of future therapy. Implementation of Fenton technology is straightforward due to the system's short response time and straightforward equipment (Ahmadi et al. 2005; Korpe et al. 2019).

4.3 Textile Industry

The textile discharge wastewater is treated using a biodegradation technique to eliminate the organic substrates. Microbes were first used to degrade colors about two decades ago. The breakdown of synthetic colors by bacteria is a simple process, but the mechanism behind it is complicated. There must be the right circumstances and depth of understanding to cultivate microorganisms. Organic matter, such as dyes and microbes, affects the degradation performance, as do temperature and pH, as well as the amount of dissolved oxygenation in the stream. Biological approaches include those that are aerobic, anaerobic, anoxic, or facultative, as well as any combination of these. Microorganisms are used in anaerobic processes to remove contaminants from wastewater when there is sufficient dissolved oxygen present, whereas anaerobic techniques employ microorganism without oxygen to do the same task. The biodegradation process offers several benefits over physical and chemical techniques, including environmental friendliness, low prices, minimal infrastructure and operational expenses, little solid waste generation, and mineralization into harmless end products that are completely complete. The effectiveness of biodegradation is dependent on microbe selection and enzyme activity. Microbes and proteases were identified and tested for dye removal since there are an endless number of them. An important biological feature of textile wastewater treatment is the identification and elimination of potentially pathogenic bacteria. There are a variety of microorganisms that may remove different types of colors from textile effluent water, including bacteria, molds, and algae (Holkar et al. 2016).

In the chemical wastewater treatment of textile effluent, coagulation, and flocculation procedures are often utilized because they need short incubation durations and are easy to handle. Alum, lime, and ferric salts like ferrous sulfate or ferric chloride are the most often used coagulants in wastewater treatment operations. In wastewater treatment, coagulants such as alum and oxidizing agents are used to help increase the size of the microscopic particles that form agglomerates (larger particles). The removal of dispersion colors from effluents requires the use of flocculation-coagulation-based physicochemical methods. When colors such as reactive and azo dyes are added to water, these technologies function poorly. Both flocculation and coagulation are limited by their low dye removal efficacy and large amounts of by-products sludge production, respectively (Holkar et al. 2016).

Its ability to remove a large variety of colors from wastewater has made sorbtion technology popular. Several adsorbent selection factors, including affinity, capacity, and desorption fundamental qualities are important, and they must be considered when selecting adsorbents for use in the color mitigation procedure. Due to its high surface-to-volume and adsorption capacity, commercialized activated charcoal is an effective adsorbent for the removal of dyes. A constraint of its utilization is the high cost and difficulty of its recycling or desorption. Various researchers have employed low-cost adsorbents such gypsum, silica, ash, bioenergy by-products, and resins for adsorption applications. Different biomass wastes including wheat residue, rice husk, and modified ginger wastes have also been employed as an adsorbent for the removal of dyes and pigments in textile effluent wastewater. The following investigations used low-cost adsorbents to extract dye from wastewater and were reported by several researchers. As an adsorbent, modified wheat residue (MWR) was used in one investigation (RR-24). The use of modified ginger waste allowed us to eradicate the crystal violet (CV) pigment (MGW). To eliminate Reactive Orange 84, another study used activated carbon from cotton flower agrowaste. Methyl orange and malachite green were absorbed by adsorbents in the process of dyeing (Potato plant waste). Adsorbent (activated carbon from waste tea) adsorbs acid blue 25 (AB25) (ACWT). Straw-based absorbents absorbed methylene blue. Sugarcane Bagasse Ash was used to absorb Acid Orange II. Capsicum annuum seeds absorbed Reactive Blue 49. Bagasse fly ash adsorbs Orange-G and Methyl Violet colors. Activated Prunus Dulcis absorbed the acid green 25 dye. In order to remove the cationic dye, we used an adsorbent (Clinoptilolite). While these adsorbents may be reused or discharged, dumped, and high-priced, their use has been limited by these issues. When starting with low initial concentrations of pollutants, adsorbents can be employed to adsorb them. They can also be utilized when the adsorbents are less expensive, readily available, and easy to produce or desorb (Holkar et al. 2016).

Ultrafiltration (UF), nanofiltration (NF), microfiltration (MF), and reverse osmosis (RO) have been used to remove contaminants from textile effluents. In textile wastewater treatment, filter media selection criteria and their ability to account for temperature and chemical content play an essential role. Membrane technologies are used in textile factories to minimize effluent wastewater BOD, COD, and color. When you first start investing in water-insoluble colors (like indigo dye), membrane clogging and waste generation, such as the creation of water-insoluble dye waste, all of these issues require further treatment. It is possible to remove both cation and anion contaminants from wastewater by using an ion exchange method. In most cases, synthetic resins are used for the ion procedure. The ion exchange method is often used to soften hard water. Until now, however, it has only been used to remove color from water. The benefit of this approach is that adsorbents are not destroyed. Water-soluble colors might be removed using this method. However, dispersion dyes, which are water-insoluble, are less effective (Holkar et al. 2016).

For textile effluent, electrochemical processes are frequently employed to remove dyestuff. Textile effluent may be effectively cleaned of hazardous and inorganic pollutants with this method, which can be used in either a direct or indirect fashion. To degrade dyestuffs in textile effluent, electrochemical processes such as mercury anode, graphite rode, iron, boron-doped diamond cathode, platinum foil, aluminum/titanium/platinum, and SS304 cathode are commonly used in electrochemical processes. Small amounts of chemicals are required for the process, which makes this approach highly cost-effective, and the process may be readily stabilized by controlling the electric current (Garcia-Segura et al. 2018).

Hydroxyl radicals are produced in high concentrations in advanced oxidation processes (AOP). Various oxidants, such as chlorine, oxygen, sulfur dioxide, and hydrogen peroxide, are used in the treatment of wastewater. Oxidizing chemicals can be used to attack the Chromophore. Radicals of hydroxyl are extremely potent oxidizers. Oxidative radicals readily react with most colors. Both inorganic and organic contaminants can be oxidized by them. Fenton's reactive and photocatalytic oxidation techniques can also be used in AOP procedures (use of energy source from sunlight for enhancing semiconductor catalyst). Used Fenton's reagent chemicals, which are essentially iron salts, to increase the oxidation of complex chemical contaminants (by speeding up the breakdown of H_2O_2) that were previously resistant to biological destruction. The iron sludge that is produced as a by-product of the Fenton reaction is a drawback of this method. Chemical oxidation is a process that uses oxidizing chemicals such as O_3 and H_2O_2 to break down organic matter. In order to remove synthetic colors from effluents, the ozonation technique is utilized. Azo dyes are colored by ozonation because the structured quadruple link in azo dyes is broken by the ozone gas action. There is no change in the amount of effluent and no solid trash formation as a by-product when ozone is employed in a gaseous state. In addition, ozone gas may develop harmful by-products, sometimes from disintegrating dyes in effluent water, which is the fundamental constraint of employing ozone gas. An additional drawback of the ozonation procedure is its short half-life of about ten minutes in water with a pH of 7, as well as its expensive cost. The pH, salts, and temperature of the wastewater are all factors that affect the ozonation process' stability. The breakdown of ozone gas in alkaline conditions ($pH > 8.5$) is more rapid. As a result, it is necessary to continuously check the pH of cotton discharge wastewater. Because UV light increases the creation of high quantities of hydroxyl radicals, an integrated treatment method using H_2O_2 and UV light to remove the color from the effluent is also viable. The combination of UV radiation and H_2O_2 for dye removal from dye-containing wastewater is desirable since it produces no solid waste and has an unpleasant odor. We employ UV light to accelerate H_2O_2 breakdown into hydroxyl radicals. During chemical oxidation, hydroxyl radicals convert contaminants into CO_2 and H_2O (Hutagalung et al. 2020).

4.4 Paper and Pulp Industry

A simple liquid–solid separation technique, such as gravity settling (clarifiers) or flotation, can be employed for the first treatment. Flotation is more extensively utilized. Dissolved air flotation (DAF) is the most prevalent type of flotation. Liquid infused with air is injected under pressure into a shallow tank containing pulp mill-treated water. When the pressure in the tank drops, air bubbles are made when the water that was injected into the tank is released into it. Because of their capacity to attract and carry solid particles, rising bubbles tend to accumulate these particles as they rise. The surface is then scraped to remove the particles. DAF can handle enormous volumes of water and produce a flotation concentrate with a broad range of total solids (300–5000 mg/L) at a reasonable cost. Chemical coagulation and sedimentation are the main methods for removing colloidal particles from recycled paper mill effluent because of its high solids concentration. Organic particles smaller than 0.2 m can be removed using this method. Starch and other insoluble substances can be removed by employing solid and liquid separation methods with the use of coagulants and/or flocculants. The organic components in the supernatant are subsequently removed using biological treatment. The organic load assessed by COD can be reduced by as much as 20% using typical coagulants such as aluminum sulfate, ferric chloride, aluminum-based products, bentonite or talc minerals, and organic polymer. Combining standard chemical features such as molecular mass, electron density, basicity, and even the existence or absence of silicate can help reduce the size of hybrid coagulants even more. Prior to further processing, DAF sludge and biological treatment plant sludge are frequently mixed (Mahmood and Elliott 2006).

Depending on the dosage, colloidal gold can be a successful main therapy approach. It removes suspended particles, certain toxicities, color, COD, and BOD, making further treatment more cost-effective and efficient. Polyelectrolytes, alum, and ferrous salts are among the most often utilized compounds. It was found that coagulants that have high valence electrons, such as aluminium sulfate, are typically used because the amount of metal ions needed for coagulation is lowered, which further reduces the cost of the process. Aluminum sulfate, on the other hand, has to have a number of drawbacks. The pH range of 5.5–6.5 is optimal for coagulation, so adding alkali to untreated sewage is typically needed to attain this pH. This increases the sulfate ion level in the effluent, which affects the subsequent biological treatment, and the flocculant floc generated is particularly fragile, which restricts the floc removal. An amino acid, protein, and major fatty acid are all eliminated during the coagulation-flocculation when aluminum-based coagulants are employed. This makes wastewater less biodegradable. In addition, the leftover alum and ferric-based coagulants hinder the biochemical treatment process by lowering microbe respiration and organic matter clearance. Polysilicato-iron (PSI) was created to replace the usage of aluminum coagulant. Poly-aluminium chloride (PAC) has been shown to be less efficient than PSI because of its strong bridging capability, generating flocs that settle faster, allowing for a smaller sedimentation tank. Nutrient bioavailability can be preserved by PSI, but it is rarely employed. Chemical coagulation sludge is hazardous because it contains hydroxides and has a high pH, making it caustic. Treatment, transportation, and disposal of sludge can be expensive. When it comes to treating paper effluent, one of the most prevalent methods is activated sludge (AS). But combining anaerobic therapy with an aerobic biological technique has recently been more extensively applied. It has been observed that fungus may successfully breakdown chemicals in wood. Other microorganisms like microalgae have also been employed and explored. Aerobic treatment, also known as activated sludge, is one of the first secondary treatment methods and includes aerating and recirculating a part of aerobic silt back to the system's input. A sedimentation basin and an aeration basin are the two primary components of the system (Rajeshwari et al. 2000).

Sludge, which is made up of high concentrations of cultivated microorganisms, is added to the aeration tank to generate a mixed liquor. The oxygen provided by the aerated fluid enhances biomass respiration and aids in BOD sorption, assimilation, and metabolism. Once the biomass has settled in the sedimentation tank, the treated wastewater may be clarified. An amount of the concentrated biomass at the sedimentation basin's bottom is removed in order to keep the activated sludge process' biosolids concentration constant. The aeration basin reuses most of the settled biomass. BOD and COD removal from pulp mill effluent may be reduced by up to 85% - 98% using activated sludge. In order to remove the color and stubborn chemicals contained in paper effluent, aerobic digestion alone is ineffective. However, anaerobic treatment may remove up to 90% of AOX and COD from the wastewater. In the lack of oxygen, anaerobic treatment takes place When bacteria hydrolyze molecules like carbohydrates, hemicellulose, and chlorinated chemicals, the anaerobic digestion process begins, making them more biodegradable by breaking them down into hydrophilic and simpler derivatives. In the second step, acidogenesis, commonly known as fermentation, polysaccharides, and micronutrients are transformed into CO_2, hydrogen, ammonia, and organic acids (Rajeshwari et al. 2000). The organic acids are then converted to acetic acid during the acetogenesis process, releasing more ammonia, hydrogen, and carbon dioxide (CO_2). As a final step, methanogenesis converts these compounds into methane and more CO_2. As a waste product or a resource of sustainable energy, methane, and CO_2 may be collected and utilized in many ways. Furthermore, compared to the aerated biological treatment, this method creates substantially less solid waste. Because an air pump isn't required, the system uses less energy. Consequently, in certain cases, anaerobic therapy is used in conjunction with aerobic treatment. Hydrogen sulfide is a major drawback of anaerobic treatment because sulfate-reducing bacteria convert apart sulfates into sulfides in the presence of oxygen (Kamali and Khodaparast 2015).

Both suspended particles and organic waste may be removed using an MBR, which is a hybrid-activated sludge treatment/membrane filtering system. Compared to traditional biological treatment, this treatment offers larger volume rating system, shorter hydraulic retention durations, longer solid retention times, reduced sludge generation, and chances for simultaneous nitrification/denitrification. In addition, supplementary clarifiers are no longer required with the deployment of MBR. Because of this and the shorter HRT, the plant footprint is reduced. Even though MBRs use a lot of energy, they get clogged up, and they need to be replaced every few years, they have become popular in the paper industry because of their advantages over traditional biological treatment. For MBR, fouling is a constant source of frustration. The membrane's energy efficiency and longevity are shortened, and maintenance and operation expenses rise as a result. Sludge flocs, bacteria, and cell debris are all examples of suspended particles that contribute to fouling. Membrane permeability is reduced as a result of the deposition of these materials on the membrane's surface and within its pores. A long-term MBR application is complicated by the heterogeneity of the suspended particles and active microorganisms found in fluid-phase suspended solids. In order to improve wastewater treatment and boost membrane application, fouling management has been one of the most important fields of study (Kamali and Khodaparast 2015).

Because of the high energy consumption and limitations of industrial use, membrane technology is currently not selected as a stand-alone method for the treatment of pulp and paper effluent. When it comes to wastewater treatment, the most often used technology is a force-propelled membrane process that relies on hydraulic pressure for separation. Pore sizes of approximately 0.01–10 μm are excellent for the removal of germs and other suspended particles. Smaller particles can be removed using the UF method, but the pressure needed to do so is higher because of the smaller pore diameters. There are several advantages to using low-pressure systems like those used in UF and MF: Contaminants that are bigger than the pores in the filter media are removed in both systems, resulting in a filtrate of a clean liquid. Unsurprisingly, the pulp sector has been more interested in UF. Some metals have been demonstrated to be able to be excluded by UF. It is feasible to retain iron, magnesium, and calcium compounds during the clarifying process by adding polyelectrolytes. When it comes to these membranes, RO is the most popular for separating monovalent ions such as salt and chloride from water. Treatment of wastewater and distillation has been at the forefront of reclaiming water for this country and its people. With recyclable materials as raw material and bleaching effluent, the Lucart paper mill achieved a COD of 25 ppm and 96–99.9% removal rates. MF and UF membranes remain the most popular despite RO's promising outcomes, owing to their cheaper initial and ongoing expenditures. As a result, UF is more commonly utilized than MF because of its greater color and COD removal rates (Pizzichini et al. 2005; Toczyłowska-Mamińska 2017).

Due to its strong reactivity and poor oxidation selectivity, AOP is particularly well-suited for treating wastewater containing refractory or difficult-to-remove organic chemicals, among other things. Biodegradability can be improved by using AOP in conjunction with biodegradation. AOP can be affected by a variety of parameters, including but not limited to the organic compound's rate of reaction, pH, and temperature. As an example, ozone, hydrogen peroxide, ultraviolet radiation, mixed ferrous and ferric salts, and radiation from ultrasonic, solar, and thermal energy may all be used to accelerate the formation of OH radicals. Hydrogen peroxide, Fenton reagent, photo Fenton and ozone combinations are all believed to be useful for the oxidation of pulp and paper industry effluents, as are UV/H_2O_2 and $UV/H_2O_2/Fe^{2+}$ combinations. Fenton's reagent is the most successful of these procedures when it comes to removing the material at a rapid rate. To generate radicals, Fenton reagents make use of metal salts (often iron) as homogeneous catalysts. Ecofriendly, they don't require sophisticated apparatus or high pressure operation, enabling Fenton procedure suitable for applicability on any scale. Fenton process. As a simple pre-treatment for biotreatment, AOP is also beneficial in the removal of organic contaminants and phenolic compounds from various bleaching effluents, which are commonly present in bleaching waste. This is because the oxidation of dissolved compounds is not possible with H_2O_2, therefore hydroxide ions ($HO^•$) are formed when H_2O_2 reacts with another chemical, such as Fe^{2+}. As a result of the interaction of Fe^{2+} with hydrogen peroxide, Fe^{3+} can also be reacted with. Decomposition is attributed to quick Fenton reaction cycles since Fe(III) is reduced to Fe(ii) immediately after each cycle. The pace at which radicals are produced and the concentration of the oxidizing agent created during the Fenton reaction determine the process' efficiency. The supply of ferrous catalyst, peroxide concentration, its mix, pH, tempature, and treatment duration all have a direct influence on the Fenton reaction's efficiency. It is possible to minimize the breakdown of peroxide into non-oxidizing species by carefully controlling the intake and transformation of the Fenton reagent (Haq et al. 2020; Suresh et al. 2021).

4.5 *Pharmaceutical Industry*

The pharmaceutical industry uses a wide range of methods to treat and dispose of wastewater. Due to the raw materials and techniques used in the manufacture of various medications, wastewaters created by these sectors might vary in composition and volume, by facility, season, and even time. Pharmaceutical wastewater has typically been treated using biological techniques. The bioremediation of pharmaceutical wastewater involves both aerobic or anaerobic treatment techniques. The authors found that Pseudomonas aeruginosa (P. aeruginosa) performed better than P. pseudomallei in degrading phenolic effluents from fermentation operations. An example of an aerobic treatment technology is the Activated Sludge (AS) processing, which includes the prolonged oxygenation activated sludge process, the AS process with activated charcoal, and membrane bioreactors. The most popular aerobic treatment, activated sludge, has been shown to be effective for a variety of types of pharmaceutical wastewater. Traditional activated sludge treatment, a low-cost approach, relies on temperature and hydraulic retention time as two of the most important variables. Besides these, the existence of organic compounds, COD, BOD, pH, and the presence of non-biodegradable materials all have an impact on the AS method's performance (Shah and Shah 2020).

Biological processes are frequently utilized as a secondary therapy after these procedures as a main treatment. These procedures are carried out in order to lessen the organic burden on the next biological process. Adsorption with activated carbon, ozone therapy, and electrocoagulation have all seen widespread application. They are commonly employed in combination with biological approaches as well. Due to the antibiotic molecule's complicated structure, persulfate was unable to successfully cleanse the wastewater. Two EC-PS processes were coupled on the sample, resulting in an efficacy of 96% and a peak activity of 75 min. When persulfate was activated by electrocoagulation, the process became more efficient, allowing the clearance of antibiotics more quickly (Li and Yang 2018).

Activated carbon (AC) is a good adsorbent for organic content because of its high specific surface area and well-developed micropores and surface chemical characteristics. Recent results showed that waste-derived activated carbon can remove ibuprofen. Since, it is so easy to get raw materials for carbon generation, the AC method offers a significant benefit. Powder form reactive charcoal or granulated activated charcoal is used in the AC process. In comparison to granular charcoal, powder charcoal is often used, but GAC is typically reused in packed-bed columns. Because of the long sludge retention period that can be obtained in a small reactor capacity by using MBR for pharmacy wastewater treatment, MBRs have received a great deal of interest in the last decade. Microorganism concentrations in the MBR might rise to 20 mg/L. Because of the high biomass content, bigger organic molecules have a greater chance of breakdown. Membrane treatment also has the benefit of not being constrained by the sludge's settling properties because it separates suspended materials using membranes. An MBR paired with an activated sludge process reactor achieved removal efficiency of 98.56% for TSS and 89.34% for total COD in research on analgesic and anti-inflammatory wastewaters. Micro- or ultrafiltration membrane retention can generally be ignored, however, biodegradation plays a crucial role, since better percentage removal was obtained for larger SRTs, whereas membrane retention can be omitted. However, the MBR treatment utilizing ultrafiltration only partially succeeded in eliminating persistent medicines and their transformation products, and hence wastewater was released into the environment with tiny amounts of these substances. Additional treatment stages utilizing sophisticated treatment techniques, such as chemical activation, oxygen oxidative, advanced oxidation processes (AOP), NF, or membrane filtration, might lessen this waste discharge (Gadipelly et al. 2014b).

When compared to the standard immunosorbent, the Molecularly Imprinting Technology (MIP) exhibits higher selectivity and affinity, greater stability, and less effort in the manufacturing process. Compared to biological receptors, MIPs have great mechanical strength and are resistant to strong chemical conditions, heat, and pressure. Tetracycline breakdown products and molecularly imprinted polymers are used to make affinity membranes for removing tetracycline from water (Gadipelly et al. 2014a).

Fenton and photo-Fenton processes, wet air oxidation, and ultrasonic irradiation and microwave radiation, which generally operate at 2450 MHz in either monomode or multimode vessels are the most common AOPs. AOPs can be used alone or in combination with the other physico-chemical and biological processes, depending on the type of medicinal effluent and the treatment target of destruction or transformation. Oxidizing agents, such as ozone, are very powerful oxidizing agents, which may either break down in water to create more powerful hydroxyl radicals, or target functional groups of organic molecules by an electrophilic process. Toluene, phenols, nitrophenols, nitroanilines, trichloromethylpropanol (TCMP), and other biodegradable organic contaminants are found in pharmaceutical wastewater. Ozonation has been widely used to remove antibiotics. In other cases, ozonation may not be the best option since molecules having amide bonds are resistant to ozone, which makes ozonation ineffective. Hydrogen peroxide reacts with ferrous or ferric ions in Fenton's reagent, resulting in hydroxyl radicals through a radical chain reaction. Fe functions as the catalyst in a heterogeneous chemical process. This procedure is the best option for wastewater treatment because ferrous is an abundant element. Using Fenton oxidation, researchers have demonstrated that refractory effluents may be made less harmful and more quickly treated biologically (Akmehmet Balcıoğlu and Ötker 2003).

5 Potential Use of Wastewater

As dairy product consumption rises, the quantity of wastewater generated by the sector is expected to follow suit. Chemical materials are increasingly being used in this business on an annual basis. Failure to create new, better technologies at a rapid pace might have serious ramifications for future generations. The dairy sector has the ability to reuse wastewater since it is both a major user and a large source of it. Purified wastewater may be used in a variety of applications, including boilers, cooling towers, and even plant cleaning. Additionally, the dairy sector will directly profit from in-house wastewater treatment due to a significant reduction in the amount of money it will have to pay in charges for wastewater collection and treatment. For instance, in the UK, decreased discharge costs account for 70% of overall savings from AD. In addition, farms that use effluents to irrigate pastures will indirectly assist the dairy business. As a result, effective wastewater treatment in the dairy sector is critical (Li and Yang 2018).

Various water streams are used in the sugar industry, including imbibition water, cooling water, boiler make-up liquid, scrubber feed and scrubber make-up water, condenser feed, and condenser make-up water. One ton of cane crushed requires an average of 11 m^3 of water in a single day for these actions. As a result, a sugar refinery with a production capacity of 2,500 TCD requires 27,500 m^3 of water per day. If treated wastewater can be reused for various water-consuming activities, fresh water consumption can be reduced and zero water discharge can be achieved. The conventional use for cleaned sugar industry effluent is irrigation, on the other hand. Plant growth and agricultural output were observed to be influenced by residual contaminants in treated wastewater. There have also been reports of compromised soil health.

It is critical for pharmaceutical manufacturing facilities to prevent waste streams from contaminating the environment with hazardous materials such as acids, acids with heavy metals, and a variety of key active pharmaceutical ingredients. Mycelia and solvents are present in considerable quantities in the fermentation broth in the fermentation plants. The solvents have extremely high BOD concentrations, and some of them are not biodegradable. In addition, recovering the drug can significantly minimize waste disposal costs associated with the primary unit process, as well as raw water requirements associated with the secondary unit operation, resulting in a rapid offset of waste-treatment overheads and an improvement in the process's economics. Reduced water consumption and operational expenses can be achieved by repurposing the recovered by-product and reusing the water for boilers' feed or cooling systems and other processes. Because of heat pinching and boiler feed, water and energy costs can be reduced by using hot waste streams as heat exchangers (heat pinching) (Gadipelly et al. 2014a).

Pollutant removal from paper and pulp industry manufacturing facility wastewater presents a number of obstacles, the complexity of which is greatly influenced by the type of wastewater being treated. The difficulty will be determined by how pure the treated water has to be. When it comes to managing wastewater, the first step is figuring out exactly what the system requires and what the water quality requirements are for each usage. Then, the water may be reused to the best of its ability. As a second step, treating the final effluent is necessary in order to meet the discharge restrictions. Both the pulp and papermaking process, as well as the wastewater treatment system, must be robust and trustworthy. In recent years, the importance of reducing the quantity of solid sludge created as a by-product of wastewater treatment operations has become more widely recognized. The initial sludge volume might be absorbed by various types of organisms residing in different zones of wastewater treatment operations if the treatment is carried out in phases. Sludge dewatering and maximal reduction of soluble and colloidal particles from the water circuit are also critical to maintaining a somewhat clean water supply. Including the development of methods to reduce sludge production, opportunities to use produced sludge for positive uses have been investigated as opposed to only incurring expenditures for waste treatment and disposal. Wastewater treatment is primarily concerned with removing contaminants, but it will also need to include ways to reduce the generation of other types of pollution in the future, like gaseous pollutants (Gadipelly et al. 2014a).

6 Conclusion

In underdeveloped nations like India, the challenges related to wastewater reuse originate from its inadequate treatment. The problem thus is to create suitable low-cost, low-tech, user-pleasant solutions, which on the one hand prevent endangering our large wastewater-dependent livelihood and then, on the other hand, safeguard deterioration of our priceless natural resources. Until the advent of modern oxidation technologies, only aerobic and anaerobic biotechnological techniques were utilized to remediate wastewater together with traditional physicochemical approaches. The mechanical, biological, and chemical treatment approaches were confronted with constraints such as poor efficacy, hazardous by-products, and high expenses of operation. Solid waste generation has also been regarded as a potential for improving physicochemical technologies like coagulation. Due to the noteworthy waste handling costs and associated environmental downsides, the development of technologies with a low solid waste output is of considerable relevance. The processes used to remove and degrade contaminants from industrial effluents clearly influence the amount of solid waste created. Simultaneously, the growing severity of pollution produced by industrial wastewater emphasizes the importance of a sustainable treatment method. To attain the discharge criteria, a combination of several treatment technologies requires implementation. However, further study should focus on the effectiveness and affordability of these possible industrial wastewater procedures.

References

Ahmadi M, Vahabzadeh F, Bonakdarpour B, Mofarrah E, Mehranian M (2005) Application of the central composite design and response surface methodology to the advanced treatment of olive oil processing wastewater using Fenton's peroxidation. J Hazard Mater 123:187–195

Akansha J, Nidheesh PV, Gopinath A, Anupama KV, Suresh Kumar M (2020) Treatment of dairy industry wastewater by combined aerated electrocoagulation and phytoremediation process. Chemosphere 253:126652

Akmehmet Balcıoğlu I, Ötker M (2003) Treatment of pharmaceutical wastewater containing antibiotics by O_3 and O_3/H_2O_2 processes. Chemosphere 50:85–95

Bisschops I, Spanjers H (2003) Literature review on textile wastewater characterisation. Environ Technol 24:1399–1411

Chhonkar P, Datta S, Joshi H, Pathak H (2000) Impact of industrial effluents on soil health and agriculture-Indian experience: Part I-Distillery and paper mill effluents

Durotoye TO, Adeyemi AA, Omole DO, Onakunle O (2018) Impact assessment of wastewater discharge from a textile industry in Lagos, Nigeria. Cogent Eng 5:1531687

Elvira C, Sampedro L, Dominguez J, Mato S (1997) Vermicomposting of wastewater sludge from paper-pulp industry with nitrogen rich materials. Soil Biol Biochem 29:759–762

Gadipelly C, Pérez-González A, Yadav GD, Ortiz I, Ibáñez R, Rathod VK, Marathe KV (2014) Pharmaceutical Industry Wastewater: Review of the Technologies for Water Treatment and Reuse. Ind Eng Chem Res 53:11571–11592

Gadipelly C, et al (2014b) Pharmaceutical industry wastewater: review of the technologies for water treatment and reuse. Ind Eng Chem Res 53:11571–11592

Garcia-Segura S, Ocon JD, Chong MN (2018) Electrochemical oxidation remediation of real wastewater effluents—a review. Process Saf Environ Prot 113:48–67

Giri J, Srivastava A, Pachauri SP, Srivastava PC (2014) Effluents from paper and pulp industries and their impact on soil properties and chemical composition of plants in Uttarakhand, India. J Environ Waste Manag 1:26–30

Goli A, Shamiri A, Khosroyar S, Talaiekhozani A, Sanaye R, Azizi K (2019) A review on different aerobic and anaerobic treatment methods in dairy industry wastewater. J Environ Treat Tech 6: 113

Gowd SC, Ramakrishna S, Rajendran K (2022) Wastewater in India: an untapped and under-tapped resource for nutrient recovery towards attaining a sustainable circular economy. Chemosphere 291:132753

Haq I, Mazumder P, Kalamdhad AS (2020) Recent advances in removal of lignin from paper industry wastewater and its industrial applications—a review. Biores Technol 312:123636

Holkar CR, Jadhav AJ, Pinjari DV, Mahamuni NM, Pandit AB (2016) A critical review on textile wastewater treatments: possible approaches. J Environ Manage 182:351–366

Hutagalung SS, Muchlis I, Khotimah K (2020) Textile wastewater treatment using advanced oxidation process (AOP). IOP Conference Series: Materials Science and Engineering 722:012032

Kamali M, Khodaparast Z (2015) Review on recent developments on pulp and paper mill wastewater treatment. Ecotoxicol Environ Saf 114:326–342

Kamyotra J, Bhardwaj RM (2011) Municipal wastewater management in India, India Infrastructure Report, p 299

Karadag D, Köroğlu OE, Ozkaya B, Cakmakci M (2015) A review on anaerobic biofilm reactors for the treatment of dairy industry wastewater. Process Biochem 50:262–271

Kaur R, Wani S, Singh A, Lal K (2012) Wastewater production, treatment and use in India. National report presented at the 2nd regional workshop on safe use of wastewater in agriculture, pp 1–13
Khan S, Malik A (2014) Environmental and health effects of textile industry wastewater. In: Environmental deterioration and human health. Springer, pp 55–71
Korpe S, Bethi B, Sonawane SH, Jayakumar KV (2019) Tannery wastewater treatment by cavitation combined with advanced oxidation process (AOP). Ultrason Sonochem 59:104723
Kumar MD, Tortajada C (2020) Assessing wastewater management in India. Springer
Kushwaha JP, Srivastava VC, Mall ID (2010) Treatment of dairy wastewater by commercial activated carbon and bagasse fly ash: Parametric, kinetic and equilibrium modelling, disposal studies. Biores Technol 101:3474–3483
Kushwaha, JP (2015) A review on sugar industry wastewater: sources, treatment technologies, and reuse. Desalination Water Treat 53: 309–318
Li Z, Yang P (2018) Review on physicochemical, chemical, and biological processes for pharmaceutical wastewater. IOP Conference Series: Earth and Environmental Science 113:012185
Loloei M, Alidadi H, Nekonam G, Kor Y (2014) Study of the coagulation process in wastewater treatment of dairy industries. Int J Env Health Eng 3:12
Mahmood T, Elliott A (2006) A review of secondary sludge reduction technologies for the pulp and paper industry. Water Res 40:2093–2112
Shah Maulin P (2021a) Removal of refractory pollutants from wastewater treatment plants. CRC Press
Shah Maulin P (2021b) Removal of emerging contaminants through microbial processes. Springer
Mani S, Bharagava RN (2018). Textile industry wastewater: environmental and health hazards and treatment approaches. In: Recent advances in environmental management. CRC Press, pp 47–69
Martz M (2012) Effective wastewater treatment in the pharmaceutical industry. Pharm Eng 32:48–62
Minhas PS, Saha JK, Dotaniya M, Sarkar A, Saha M (2022) Wastewater irrigation in India: current status, impacts and response options. Sci Total Environ 808:152001
Muthukumaran N, Ambujam N (2003) Wastewater treatment and management in urban areas-a case study of Tiruchirappalli city, Tamil Nadu, India. Proceedings of the Third International Conference on Environment and Health, Chennai, India, pp 15–17
Nath K, Saini S, Sharma YK (2005) Chromium in tannery industry effluent and its effect on plant metabolism and growth. J Environ Biol 26:197–204
Ongen A, Kurtulus Ozcan H, Arayıcı S (2013) An evaluation of tannery industry wastewater treatment sludge gasification by artificial neural network modeling. J Hazard Mater 263:361–366
Pang YL, Abdullah AZ (2013) Current status of textile industry wastewater management and research progress in Malaysia: a review. Clean Soil Air Water 41:751–764
Patneedi CB, Prasadu KD (2015) Impact of pharmaceutical wastes on human life and environment. Rasayan J Chem 8:67–70
Pizzichini M, Russo C, Meo CD (2005) Purification of pulp and paper wastewater, with membrane technology, for water reuse in a closed loop. Desalination 178:351–359
Praveena M, Sandeep V, Kavitha N, Jayantha RK (2013) Impact of tannery effluent, chromium on hematological parameters in a fresh water fish, Labeo Rohita (Hamilton). Res J Animal, Veterinary and Fishery Sci 1:1–5
Rajeshwari KV, Balakrishnan M, Kansal A, Lata K, Kishore VVN (2000) State-of-the-art of anaerobic digestion technology for industrial wastewater treatment. Renew Sustain Energy Rev 4:135–156
Ramteke PW, Awasthi S, Srinath T, Joseph B (2010) Efficiency assessment of common effluent treatment plant (CETP) treating tannery effluents. Environ Monit Assess 169:125–131
Ramya Suresh, Baskar Rajoo, Maheswari Chenniappan, Manikandan Palanichamy (2021) Experimental analysis on the synergistic effect of combined use of ozone and UV radiation for the treatment of dairy industry wastewater. Environ Eng Res
Sahu OP, Chaudhari PK (2015) Electrochemical treatment of sugar industry wastewater: COD and color removal. J Electroanal Chem 739:122–129

Sato T, Qadir M, Yamamoto S, Endo T, Zahoor A (2013) Global, regional, and country level need for data on wastewater generation, treatment, and use. Agric Water Manag 130:1–13

Shah A, Shah M (2020) Characterisation and bioremediation of wastewater: A review exploring bioremediation as a sustainable technique for pharmaceutical wastewater. Groundw Sustain Dev 11:100383

Sharma K, Kaushik G, Thotakura N, Raza K, Sharma N, Nimesh S (2020) Enhancement effects of process optimization technique while elucidating the degradation pathways of drugs present in pharmaceutical industry wastewater using Micrococcus yunnanensis. Chemosphere 238:124689

Shah Maulin P (2020) Microbial bioremediation & biodegradation. Springer

Shete BS, Shinkar NP (2013a) Comparative study of various treatments for dairy industry wastewater. IOSRJEN 3:42–47

Shete BS, Shinkar NP (2013b) Dairy industry wastewater sources, characteristics & its effects on environment. Int J Curr Eng Technol 3:1611–1615

Singh A, Gautam R, Sharma R (2008) Performance evaluation of a common effluent treatment plant (CETP) treating textile wastewaters in India. J Ind Pollut Control 24:111–121

Singh AK, Chandra R (2019) Pollutants released from the pulp paper industry: Aquatic toxicity and their health hazards. Aquat Toxicol 211:202–216

Singh J, Gupta P, Das A (2021) Dyes from textile industry wastewater as emerging contaminants in agricultural fields. In: Sustainable agriculture reviews 50. Springer, pp 109–129

Sinha S, Srivastava A, Mehrotra T, Singh R (2019) A review on the dairy industry wastewater characteristics, its impact on environment and treatment possibilities. In Emerging issues in ecology and environmental science. Springer, pp 73–84

Sonsale AN, Purohit J, Pohekar SD (2021) Renewable & alternative energy sources for strategic energy management in recycled paper & pulp industry. Bioresour Technol Rep 16:100857

Starkl M, Stenström T, Roma E, Phansalkar M, Srinivasan RK (2013) Evaluation of sanitation and wastewater treatment technologies: case studies from India. J Water, SanitIon Hyg Dev 3:1–11

Suman H, Sangal VK, Vashishtha M (2021) Treatment of tannery industry effluent by electrochemical methods: A review. Mater Today: Proc 47:1438–1444

Sundaravadivel M, Vigneswaran S (2001) Wastewater collection and treatment technologies for semi-urban areas of India: a case study. Water Sci Technol 43:329–336

Suresh R, Rajoo B, Chenniappan M, Palanisamy M (2021) Feasibility of applying nonthermal plasma for dairy effluent treatment and optimization of process parameters. Water Environ J 35:1038–1050

Tikariha A, Sahu O (2014) Study of characteristics and treatments of dairy industry wastewater. J Appl & Environ Microbiol 2:16–22

Toczyłowska-Mamińska R (2017) Limits and perspectives of pulp and paper industry wastewater treatment—a review. Renew Sustain Energy Rev 78:764–772

Vourch M, Balannec B, Chaufer B, Dorange GD (2008) Treatment of dairy industry wastewater by reverse osmosis for water reuse. Desalination 219:190–202

Yadav A, Raj A, Purchase D, Ferreira LFR, Saratale GD, Bharagava RN (2019) Phytotoxicity, cytotoxicity and genotoxicity evaluation of organic and inorganic pollutants rich tannery wastewater from a Common Effluent Treatment Plant (CETP) in Unnao district, India using Vigna radiata and Allium cepa. Chemosphere 224:324–332

Yonar T, Sivrioğlu Ö, Özengin N (2018) Physico-chemical treatment of dairy industry wastewaters: a review. In: Technological approaches for novel applications in dairy processing, p 179

Zhao C, Chen W (2019) A review for tannery wastewater treatment: some thoughts under stricter discharge requirements. Environ Sci Pollut Res 26:26102–26111

An Introduction to Bioelectrochemical System (BES) for Microbial Electro Remediation

Senthil Kumar K., Naveen Kumar, C. Anantharaj, N. Pooja, and Ramya Suresh

1 Preamble

Microbial Fuel Cell is a bioelectrochemical cell in which microorganisms are employed to utilize the carbon sources for power generation (Tardast A et al. 2012). The main principle behind this process is the production of electrons alongside carbon dioxide and protons, when a carbon source is utilized by a microorganism anaerobically. The conversion of acetic acid by a microorganism Shewanella putrefaciens under anaerobic conditions is given as follows:

$$C_2H_4O_2 + 2H_2O \xrightarrow{\textit{Anaerobic condition}} 2CO_2 + 8e^- + 8H^+$$

The organic matter which acts as the feed is fed along with the microorganisms in the anodic compartment which consists of an anode. This anodic compartment is maintained in anaerobic conditions. The microorganisms employed in Microbial Fuel Cell mostly belong to the exoelectrogens class. The exoelectrogenic microorganisms release the electron produced during their metabolic processes. These electrons get shuttled from the outer cell of the microorganisms to the anode. In MFC, the anodic compartment and the cathodic compartment are separated by a Proton Exchange Membrane (Ghasemi M et al. 2012). Oxygen is supplied at the cathodic chamber

Senthil Kumar K. (✉)
Department of Chemical Engineering, Kongu Engineering College, Erode, India
e-mail: senthilkumar.chem@kongu.edu

Naveen Kumar · C. Anantharaj
Department of Civil Engineering, Easwari Engineering College, Chennai, India

N. Pooja
Department of Chemical Engineering, Central Institute of Petrochemicals Engineering & Technology, Chennai, India

Ramya Suresh
Department of Civil Engineering, Sanskrithi School of Engineering, Puttaparthi, India

M. P. Shah (ed.), *Microbial Technologies in Industrial Wastewater Treatment*,
https://doi.org/10.1007/978-981-99-2435-6_10

where the electron from the anodic chamber, proton which diffuses from PEM reacts with it to form water.

$$2O_2 + 8H^+ + 8e^- \longrightarrow 4H_2O$$

There are several advantages such as the application of MFC in effluent, biosensor, biohydrogen production in using the Microbial Fuel Cell other than energy generation.

2 Various Designs

A sequence of MFC may be utilized for efficient bioenergy production with a series of MFC. This model is called a stacked-type MFC (Logan BE et al. 2006). Various designs of MFCs along with its power density are represented in Table 1. Also, some unique changes like coupling a Photo Bio Reactor with MFC have also proved to generate power along with wastewater treatment. Such MFCs are called Photosynthetic Microbial Fuel Cell (PMFC). Also, using PMFC, biodiesel can also be produced as a valuable by-product along with power generation.

3 Types of MFC

The two common types of MFC based on the type of the microorganism employed are Mediated and Direct Electron Transfer (DET) MFC shown in Fig. 1. In the former, MFC utilizes a moderator to bring electron from the cell cytoplasm and to shuttle it toward the anode (Li Huang et al. 2018). In MET, electrons are transferred through the base of electrochemical, which could produce metabolite by microbes or an endogenous redox mediator (Reguera G et al. 2005; Evelyn et al. 2014). The commonly used mediators are thionin (Thurston et al. 1985), sulphate (Park et al. 1997), and natural red (Park et al. 1999). These reduced mediators then deposit the electron they obtained from the cytoplasm to the anode. On electron deposition, the mediator gets oxidized again and the same process continues again. These mediators are needed to be fed into the MFC at frequent intervals due to the high instability of these mediator compounds. Frequent addition of mediators adds up the cost of operation and also the toxicity of the chemical mixture in the anodic compartment. But some mediators can be produced by the microorganism itself which shuttles the electron to the anode. Some of them include pyocyanin, 2-amino-3-carboxy-1,4naphthoquinone, and ACNQ (Rabaey et al. 2004). In Direct Electron Transfer MFC, a unique class of microorganisms called exoelectrogens are employed (Gorby YA et al. 2006; Shah MP 2020). They are gram-negative microorganisms which release the electrons extracellularly into the anodic chamber. Hence, mediators are not required here to extract the electrons within the microbial cell. Some of the exoelectrogens are Desulfuromonas

Table 1 Different designs of MFCs along with power density

Type of MFC	Fuel	Power Density (mW/ m^2)	Reference
Single compartment	Glucose	766	(Cheng S et al. 2006a)
Single compartment	Domestic wastewater	464	(Cheng S et al. 2006a)
Double compartment chamber	Glucose	860	(Liu H et al. 2005)
Double compartment	Acetate	480	(Cheng S et al. 2006b)
Up flow	Sucrose	560	(Bond and Lovely 2003)
Single chamber	Complex substrate	600	(Zhang T et al. 2007)
Single chamber	Glucose	355.5	(Bond DR et al. 2002)
II chamber H type	Acetate	13	(Chaudhuri and Lovely 2003)
II chamber H type	Glucose	33.4	(Bond and Lovely 2003)
Single chamber	Sewage sludge	6000	(Franks AE and Nevin K 2010)
Two-chamber air cathode MFC	Glucose	283	(Rahimnejad M et al. 2011)
two chamber	Marine sediment (acetate)	14	(Zhou M et al. 2013)
Two-chamber H type	Lactose	17.2	(Antonopoulou G et al. 2010)

acetoxidans, Geobacter sulfurreducens, Shewanella putrefaciens, etc. The absence of mediators proves to be very advantageous as they require only less cost and non-toxic.

4 Components of MFC

The MFC primarily comprising of a cathodic comportment, an anodic comportment, and PEM along with the electrodes. Table 2 reviews the various materials used as electrodes in Microbial Fuel Cell.

4.1 Anodic Chamber

The anodic chamber consists of an anode, the organic source, and the microorganism. In aerobic conditions, the electrons released by the microbe will be attracted toward highly electro-negative components like oxygen. So, the anodic chamber has to be

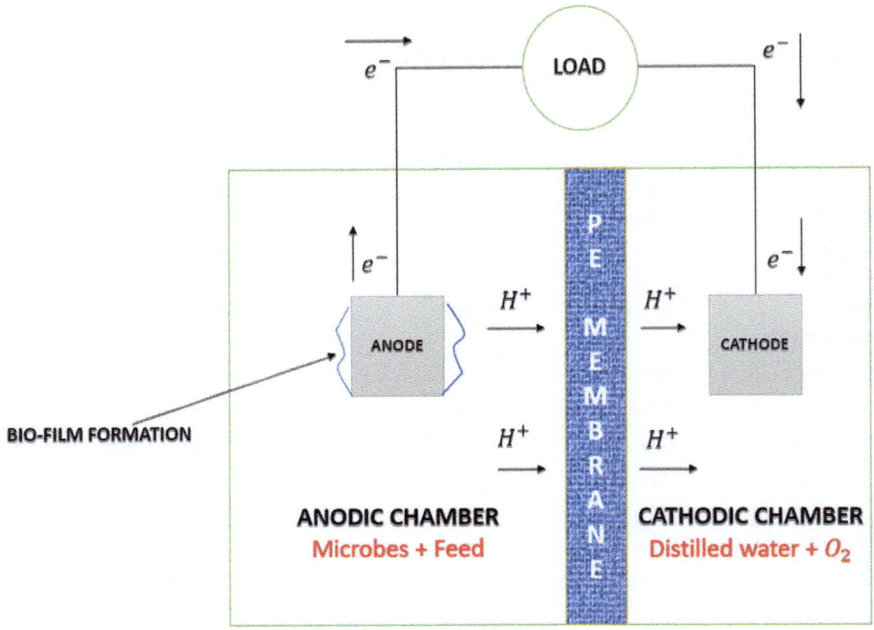

Fig. 1 Schematic representation of MFC

Table 2 Basic components of Microbial Fuel Cell

Item	Materials
Anode	Graphite granules bed, graphite fiber brush, conductive polymers, carbon paper, carbon cloth, reticulated vitreous carbon, graphite rod, graphite felt
Cathode	Carbon-paper, cloth, reticulated vitreous carbon, graphite-rod, felt, granules bed, fiber brush, conductive polymers
Anode compartment	Glass (Borosil/acrylic), polycarbonate, plexiglass
Cathode compartment	Glass (Borosil/acrylic), polycarbonate, plexiglass
Membrane materials	Salt bridge, ultrex, polyethylene.poly, Nafion, porcelain septum
Microorganisms	Aerobic or anaerobic or Facultative groups
Electron catalyst	Platinum, platinum black, Manganese dioxide, iron (Fe^{3+})

maintained in an anaerobic condition. The properties such as conductivity, corrosion resistance, hardiness, resistance, more surface area, biocompatibility, strength, etc., are to be considered during the selection of anode materials. Several MFC studies have been carried out with the carbon electrode material. Anodes synthesized from carbon can be reused in numerous structures such as carbon cloth, fiber and paper, and fiber brush, (Ishii SI et al. 2008; Jayapriya J et al. 2012). The mainly employed carbon

Table 3 List of materials used as electrodes with advantages and disadvantages

Materials	Merits	Demerits	References
Stainless steel	High conductivity, Low cost	Poor bacteria attachment, low power production	(Kim JR et al. 2007)
Carbon paper	High conductivity	easily broken, minimum specific surface area, costly	(Ishii SI et al. 2008)
Carbon cloth	More conductivity and specific surface area	Expensive	(He z et al. 2005)
Reticulated vitreous carbon	More conductivity and porosity, huge specific surface area	easily broken	(Liu H et al. 2005)
Graphite rod	More conductivity and surface area	small specific surface area, costly	(Kim HJ et al. 2002)
Graphite felt	Huge conductivity and porosity, more specific surface area, flexible	Low strength	(MirellaDi Lorenzo et al. 2009; Shah MP 2021)
Graphite granules bed	Low cost, more porosity, and surface area	More contact resistance	(Ahn Y et al. 2014)
Conductive polymers	High surface area	Low conductivity	(Yu EH et al. 2007)

material is graphite rod due to its good conductivity and cheap price. Carbon papers and carbon clothes were used in H_2 fuel cells during the initial periods (Park DH and Zeikus JG 1999; Patil SA et al. 2009). Then, synthesized anode materials were later applied in MFC to reduce the inner resistance and get better performances (Dumas C et al. 2007; Fabian Fisher 2018; Jung S and Regan JM 2007). The efficiency of anodic electrodes made from stainless steel is less in comparison with the graphite anode (Dumas C et al. 2008; Kim JR et al. 2007).

Table 3 describes various materials employed as electrodes with their advantages and disadvantages. The organic matter is fed inside the anodic chamber along with the microorganisms. The organic matter is used by the microorganisms to generate electrons and protons. An important function of the anode is to conduct the electrons generated through an external circuit.

4.2 Cathodic Chamber

In a typical MFC, the cathodic chamber consists of a cathode, distilled water, and an aerator. Nevertheless, the probable cathode materials should have excellent electrical conductivity, excellent strength, excellent catalytic environment (Om Prakash et al. 2018). Generally, MFC will be worked in a pH of 7–8 at ambient temperature conditions. In this circumstance, oxygen's rate of reduction is extremely less which in turn minimizes the performance of MFC (Ahn Y et al. 2014).

In cathodic compartment of MFC, carbon objects must be modified with supplementary catalysts for some reactions (Yu EH et al. 2007). For the majority of MFC operations, platinum is placed due to its excellent O_2 reduction rate performance. Usages of costly metals as cathodes are limited during the scale-up of the MFC concept. Though the platinum cathode is costly, it gets fouled easily when low-quality water is used in the cathodic chamber. Numerous researchers attempted the minimization of expenditure of cathode materials using effective but inexpensive materials. An effort has been made for cathode materials which were prepared with metal porphyrins and phthalocyanines carried on Ketjenblack carbon to increase the O_2 rate and catalytic activity in MFC. Iron phthalocyanine as a cathode has resulted in more oxidation rates at neutral pH than Pt catalyst. An optimum power density of 634 mW/m^2 has resulted in iron phthalocyanine–Ketjenblack carbon at pH of 7–8, which is more expensive compared to Pt catalyst (593 mW/m^2) at identical conditions. Since macrocyclic catalyst is inexpensive, the utilization of these materials is most widely preferred for applications of MFC (Xu Y et al. 2012). The aerator is used in the cathodic chamber to facilitate the flow of oxygen in the compartment.

4.3 Proton Exchange Membrane

The PEM is a vital component in the MFC. The main objectives of PEM are (i) maintaining the anaerobic environment in the anodic compartment, (ii) transferring protons between chambers (iii) minimizing back diffusion of oxygen in the anodic chamber, and (iv) maintaining long-time operating conditions. Most of the MFC operations employ Nafion membrane due to its proton conductivity. Figure 2 shows the mechanism of proton transfer by Nafion membrane. Nafion membrane is made of chemically stabilized perfluoro sulfonic acid polymer. The drawbacks in using Nafion as a membrane are, they can spread the cause of contamination thereby reducing power generation and depreciating MFC efficiency (Park DH and Zeikus JG 2003). The Nafion membrane is also costly. Several investigations are being conducted for finding an alternative PEM. A few examples are Salt Bridge (Park DH and Zeikus JG 2003), porcelain septum, interpolymer cations exchange membrane (Grzebyk and Poźniak 2005), microporous filter (Biffinger et al. 2007), physical barriers and SPEEK (Ayyaru S and Dharmalingam S 2011). All the abovementioned membranes are permeable to protons. However, intense investigation is needed to increase the potential of the membrane and its long-time strength (Cheng S et al. 2011; Rozendal RA et al. 2006).

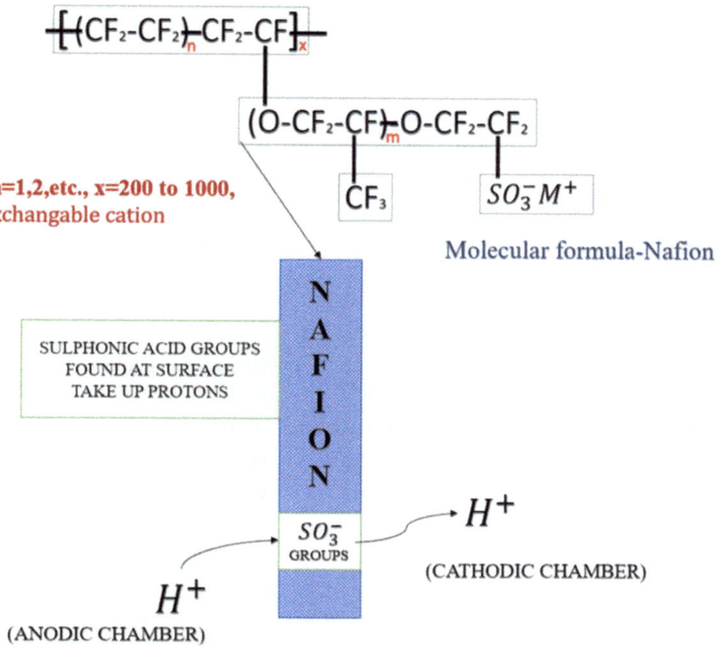

Fig. 2 Schematic representation of proton transfer mechanism in Nafion membrane

5 Selection of Substrate

The biological process primarily depends on substrate factors as it provides carbon (nutrient) and energy source. In MFC, acetate and glucose are the substrates investigated by most of the investigators in various compositions (Antonopoulou G et al. 2010; Jadhav GS and Ghangrekar MM 2009; Liu H et al. 2005). Several types of substrates like non-fermentable substrates (for example, acetate and butyrate), the fermentable substrates (glucose and sucrose), and compound substrates (effluent from domestic and food process) may be added into the anode compartment (Sun M et al. 2008, 2009; Tender LM et al. 2002). But it is very much complicated to assess and analyze the performance of MFC based on data available in literature as it depends upon different operating conditions like temperature, type of Microbial Fuel Cell, surface area, electrode material, and different respiration species (microorganisms) for increased electricity production (Rahimnejad M et al. 2011). The substrate is also crucial in influencing the bioenergy production in MFC (Bahareh Aesfi et al. 2019; Li He et al. 2017; Venkata Mohan S et al. 2014). Table 4 reviews various substrates and microorganisms used in MFCs (Venkatesh Chaturvedi and Pradeep Verma 2016).

Table 4 Details of substrates and microorganisms used in MFCs (Venkatesh Chaturvedi and Pradeep Verma 2016)

Substrates	Concentration and current density	Microorganisms	Reference
Artificial/Synthetic wastewater	510 mg/L, 0.008 mA/ cm^2	Anaerobic culture from a pre-existing MFC	(Jadhav and ghangrekar 2009)
Food firm wastes	8169 CO mg/L, 0.025 mA/cm^2	Aerobic sludge	(Quezada BC et al. 2010)
Swine wastewater	60 CO gm/L, 0.700 mA/cm^2	paddy field soil	(Ichihashi and hirooka 2012)
Slaughterhouse wastewater	900 COD mg/L, 0.130 mA/cm^2	Granular anaerobic sludge inoculum	(Katuri KP et al. 2012)
Food waste	16 g/L, 0.045 mA/cm^2	Anaerobic culture	(Choi J et al. 2011)
Rice straw hydrolysate	400 mg/mL, 137.6 mA/cm^2	Desulfobulbus and Clostridium	(Wang Z 2014)
Sucrose	2674 mg/L, 0.19 mA/ cm^2	Anaerobic sludge from septic tank	(Behera and Ghangrekar 2009)
Brewery wastewater	600 mg/L, 0.18 mA/ cm^2	Anaerobic mixed consortia	(Wen Q et al. 2009)
Chocolate industry wastes	1459 mg/L COD, 0.302 mA/cm^2	Activated sludge	(Patil SA et al. 2009)
Cellulose	4 g/L, 0.02	Pure culture of Enterobacter cloacae	(Rezaei F et al. 2009)

6 Selection of Microorganisms

The microorganism is employed in the anodic chamber along with the substrate. Only, a certain class of microorganism which has an external cellular layer is used in MFC (Logan BE 2009). These microbes are gram-negative microbes and come under the class electricigens. Initially, the microbial activity will be aerobic when it is introduced into the anodic chamber. Later, with the depletion of oxygen initially present in the chamber, the microbe starts to act anaerobically and releases electrons and protons. The exoelectrogenic capability may also be induced by providing a shock load in the anodic chamber (Cuijie Feng et al. 2014). The effective power generation at the anode depends on the presence of microbes. The microbial interaction with the anode can be easily observed with the formation of biofilm over the anode. The biofilm enhances electron transport. The microbes can be of the same species or can be of a mixed culture. The important factors to be considered during the microbial selection depend on the nature of the substrate, substrate protiens like pH, temperature, etc. Table 5 describes various microbes used in MFCs.

Table 5 Microbes used in MFCs (Senthilkumar K et al. 2022)

Mediator electricity-producing bacteria (Zhuwei Du et al. 2007)		
Microorganisms	Note	Reference
Klebsiella pneumoniae	HNQ as mediator	(Logan BE 2009)
Proteus mirabilis	Thionin as mediator	(Rhoads A et al. 2005)
Gluconobacter oxydans	Mediator (HNQ, resazurin, or thionine) needed	(Choi Y et al. 2003)
Desulfovibrio desulfuricans	Sulfate/sulfide as mediator	(Lee SA et al., 2002)
Streptococcus lactis	Ferric chelate complex as mediators	(Park DH et al. 1997)
Proteus mirabilis	Thionin as mediator	(Vega CA 1987)
Escherichia coli	Mediators such as methylene blue needed	(Thurston CF et al. 1985)
Actinobacillus succinogenes	Neutral red or thionin as electron mediator	(Schroder U et al. 2003)
Mediator-less electricity-producing bacteria (Senthilkumar K et al. 2022)		
Desulfuromonas acetoxidans	Deltaproteobacteria identified from a sediment MFC	(Park DH and Zeikus JG 1999)
Geobacter sulfurreducens	generated current without poised electrode	(Bond DR et al. 2002)
Aeromonas hydrophila	Deltaproteobacteria	(Bond DR and Lovley DR 2003)
Pichia anomala	Current generation by yeast (kingdom Fungi)	(Pham CA 2003)
Acidiphilium sp. 3.2 Sup5 Power	production at low pH	(Prasad D et al. 2006)
Thermincola sp. strain JR	Phylum Firmicutes	(Borole AP et al. 2008)
Desulfobulbus propionicus	Deltaproteobacteria	(Wrighton KC et al. 2008)

7 Parameters Affecting Current Generation

Several parameters interact among themselves to determine the operation of MFC which is a complex variable. Table 6 explains various parameters affecting bioenergy production in Microbial Fuel Cells.

7.1 Parameters from Anodic Chamber

The main parameter in an anodic compartment is the kind of substrate (Sharma Y and Li B 2010). The substrate plays an important role as the electron donor. The feed also affects the bacterial growth. Also, the bacterial concentration plays a vital role in converting organic matter to electrons, protons, and carbon dioxide. At high bacterial

Table 6 Parameters influencing current production in MFCs

Component	Parameters	Effects	References
Anodic chamber	Nature of Substrate	Determines the number of electrons to be released	[Sharma Y and Li B 2010]
	Microbe used	Selection is based on feed	[Dávila D et al. 2008]
	Volume of chamber	At constant microbial concentration, it is inversely proportional to the current generation	[Dávila D et al. 2008]
	Microbial concentration	Directly proportional to power generation	[Dávila D et al. 2008]
	Anode used	Determines effective electron transport	[MirellaDi Lorenzo et al. 2009]
	Surface area of anode	Directly proportional to power generation	[Hyung Soo Park et al. 2001]
	pH & temperature	Optimum condition varies with microbe; Affects bacterial growth	[Venkata Mohan et al. 2014]
Cathodic chamber	pH of distilled water	Higher generation of current at pH 6–7	[Venkata Mohan et al. 2014]
	Cathode used	Determines effective electron transport	[MirellaDi Lorenzo et al. 2009]
	Surface area of cathode	Directly proportional to power generation	[Hyung Soo Park et al. 2001]
	Flow rate of Oxygen	Determines DO content in cathodic chamber	[Rago L et al. 2017]
PEM	Proton permeability	Directly proportional to current generation	[Dharmalingam S et al. 2019]

concentrations, the reaction rate will be faster (Dávila D et al. 2008). Higher power density is found in mixed cultures. The nature of an electrode also has a minimal role. The type of microorganism also plays a major role in power generation. Other parameters of feed like BOD, COD, pH, and temperature also have a considerable effect on power generation (Venkata Mohan et al. 2014).

7.2 *Parameters from Cathodic Chamber*

The electrode material should be successfully employed to dissolve the electrons in cathodic chamber from anodic chamber. The maximum power density is achieved with neutral pH (6–7). The Dissolved Oxygen content also plays an important role in MFC function (Rago L et al. 2017). The dissolved oxygen is introduced by the aeration. Also, with an increase in electrode area, power generation is increased. The selection of materials is important as the electrons should not be transported to the wall of the MFC.

7.3 *Parameters from PEM*

The Proton Exchange Membrane prevents the short-circuiting of the electrons with protons in anodic chamber while maintaining anaerobic environment at the cathode side (Wen-Juan Hu et al. 2011). The rate of transportation of protons depends on the resistance offered by the PEM. The rate of transportation of protons will be high if the resistance offered by the PEM is low.

8 Applications of MFC

Recently, MFC plays a massive role in environmental applications. The applications of the Microbial Fuel Cell in different areas in our society, helping to create a sustainable environment are electricity generation, bio-hydrogen, wastewater treatment, biosensor, and desalination plants.

9 Conclusion

This review paper concludes that MFC will prove to be an effective non-conventional energy. Further, focus should be made on all the operational parameters associated with power generation along with its optimization. Several scientific studies are being made in making the MFC a scalable product. Also, the environment is mostly polluted by wastewater when compared with other foulness. Some of the drawbacks are not restricted by recent effluent treatment technologies. Achieving a sustainable environment and fulfilling their future needs is quite difficult with conventional treatment technologies. MFCs have been investigated and are now being accepted as a novel technology, which has more merits, particularly in wastewater treatment and bioelectricity generation. MFC produces more energy and produces less sludge. For the past few years, research on MFC has increased and the technology has improved at least in lab-scale studies. However, commercialization of MFC is a difficult one because of its complications in various parts of the reactor. Besides the high cost of materials, current fluctuation and more interior resistance are the barriers to electricity generation and hold down the application fields. Therefore, upcoming research should give more importance to new effective costing of MFC materials to treat wastewater efficiently. It is more essential to realize the character in nature and role of MFC electrodes. The multiplicity of wastewater can be radically degraded by them or along with additional processes. Wastewater treatment technology mainly focuses on cost and energy demands. Practically, MFC is a promising technology for removing pollutants from industrial effluent in an effective manner.

References

Ahn Y, Ivanov I, Nagaiah TC, Bordoloi A, Logan BE (2014) Mesoporous nitrogen-rich carbon materials as cathode catalysts in microbial fuel cells. J Power Sources 269:212–215

Tardast A, Rahimnejad M, Najafpour G, Ghoreyshi AA, Hossein Z (2012) Fabrication and operation of a novel membrane-less microbial fuel cell as a bioelectricity generator. Iran. J Energy Environ 1–5

Antonopoulou G, Stamatelatou K, Bebelis S, Lyberatos G (2010) Electricity generation from synthetic substrates and cheese whey using a two chamber microbial fuel cell. Biochem Eng J 50:10–15

Ayyaru S, Dharmalingam S (2011) Development of MFC using sulphonated polyether ether ketone (SPEEK) membrane for electricity generation from wastewater. Bioresour Technol 102:11167–11171

Bahareh A, Li Shiue-Lin, Henry AM, Sanchez-Torres V, Hu Anyi, Li Jiangwei, Yu Chang-Ping (2019) Characterization of electricity production and microbial community of food waste-fed microbial fuel cells. Process Saf Environ Prot 125:83–91

Balat H, Kirtay E (2010) Hydrogen from biomass–present scenario and future prospects. Int J Hydrogen Energ 35(14):7416–7426

Behera M, Ghangrekar MM (2009) Performance of microbial fuel cell in response to change in sludge loading rate at different anodic feed pH. Bioresour Technol 100:5114–5121

Biffinger JC, Ray R, Little B, Ringeisen BR (2007) Diversifying biological fuel cell designs by use of nanoporous filters. Environ Sci Technol 41:1444–1449

Bond DR, Holmes DE, Tender LM, Lovley DR (2002) Electrode-reducing microorganisms that harvest energy from marine sediments. Science 295:483–485

Bond DR, Lovley DR (2003) Electricity production by Geobacter sulfurreducens attached to electrodes. Appl Environ Microbiol 69:1548–1555

Borole AP, O'Neill H, Tsouris C, Cesar S (2008) A microbial fuel cell operating at low pH using the acidophile Acidiphilium cryptum. Biotechnol Lett 30:1367–1372

Chaudhuri SK, Lovely DR (2003) Electricity generation by direct oxidation of glucose in microbial fuel cells. Nat Biotechnol 21:1229–1232

Cheng S, Liu H, Logan BE (2006a) Increased performance of single chamber microbial fuel cells using an improved cathode structure. Electrochem Commun 8:489–494

Cheng S, Liu H, Logan BE (2006b) Power densities using different cathode catalysts (Pt and CoTMPP) and polymer binders (Nafion and PTFE) in single chamber microbial fuel cells. Environ Sci Technol. 40:364–369

Cheng S, Member S, Jin Y, Rao Y, Arnold DP, Member S (2011) An active voltage doubling AC/DC converter for low-voltage energy harvesting applications. Power 26:2258–2265

Choi J, Chang HN, Han JI (2011) Performance of microbial fuel cell with volatile fatty acids from food wastes. Biotechnol Lett 33:705–714

Choi Y, Kim N, Kim S, Jung S (2003) Dynamic behaviors of redox mediators within the hydrophobic layers as an important factor for effective microbial fuel cell operation. Bull Korean Chem Soc 24:437–440

Cuijie F, Jiangwei Li, Dan Q, Lixiang C, Feng Z, Shaohua C, Hongbo Hu, Chang-Ping Yu (2014) Characterization of exoelectrogenic bacteria enterobacter strains isolated from a microbial fuel cell exposed to copper shock load. PLoS ONE 9(11):e113379

Davila D, Esquivel JP, Vigues N, Sanchez O, Garrido L, Tomas N, Sabate N, Del Campo FJ, Munoz FJ, Mas J (2008) Development and optimization of microbial fuel cells. J New Mater Electrochem Syst 11:99–103

Dumas C, Basseguy R, Bergel A (2008) Electrochemical activity of geobacter sulfurreducens biofilms on stainless steel anodes. Electrochim Acta 53:5235–5241

Dumas C, Mollica A, Féron D, Basséguy R, Etcheverry L, Bergel A (2007) Marine microbial fuel cell: use of stainless steel electrodes as anode and cathode materials. Electrochim Acta 53:468–473

Evelyn E, Yan Li, Aaron TM, Peter, A. Gostomski (2014) Gaseous pollutant treatment and electricity generation in microbial fuel cells (MFCs) utilising redox mediators. Rev Environ Sci Biotechnol 13:35–51

Fabian F (2018) Photoelectrode, photovoltaic and photosynthetic microbial fuel cells. Renew Sustain Energ Rev 90:16–27

Franks AE, Nevin K (2010) Microbial fuel cells a current review. Energies 3:899–919

Gorby YA, Svetlana Y, Jeffrey SM, Kevin MR, Dianne M, Alice D, Terry JB, Chang IS, Kim BH, Kim KS, Culley DE, Reed SB, Romine MF, Saffarin DA, Hill EA, Liang S, Elias DA, Kennedy DW, Grigoriy P, Kazuya W, Ishii S, Logan B, Nealson KH, Fredrickson JK (2006) Electrically conductive bacterial nanowires produced by shewanella oneidensis strain mr-1 and other microorganisms. Proc Natl Acad Sci USA 103:11358–11363

Grzebyk M, Poźniak G (2005) Microbial fuel cells (MFCs) with interpolymer cation exchange membranes. Sep Purif Technol 41:321–328

He Z, Minteer SD, Angenent LT (2005) Electricity generation from artificial wastewater using an upflow microbial fuel cell. Environ Sci Technol 39:5262–5267

Park HS, Kim BH, Kim HS, Kim HJ, Kim GT, Mia K, In Seop C, Yong Keun P, Hyo Ihl C, (2001) A novel electrochemically active and Fe (III)-reducing bacterium phylogenetically related to clostridium butyricum isolated from a Microbial Fuel Cell. Anaerobe 7:297–306

Ichihashi O, Hirooka K (2012) Removal and recovery of phosphorus as struvite from swine wastewater using microbial fuel cell. Bioresour Technol 114:303–307

Ishii SI, Watanabe K, Yabuki S, Logan BE, Sekiguchi Y (2008) Comparison of electrode reduction activities of Geobacter sulfurreducens and an enriched consortium in an air cathode microbial fuel cell. Appl Environ Microbiol 74:7348–7355

Jadhav GS, Ghangrekar MM (2009) Performance of microbial fuel cell subjected to variation in pH, temperature, external load and substrate concentration. Bioresour Technol 100:717–723

Jayapriya J, Judy G, Ramamurthy V, Mudali KU, Raj B (2012) Preparation and characterization of biocompatible carbon electrodes. Compos: Part B Eng 43:1329–1335

Jung S, Regan JM (2007) Comparison of anode bacterial communities and performance in microbial fuel cells with different electron donors. Appl Microbiol Biot 77:393–402

Katuri KP, Enright AM, O'Flaherty V, Leech D (2012) Microbial analysis of anodic biofilm in a microbial fuel cell using slaughter house wastewater. Bioelectrochemistry 87:164–711

Kim HJ, Park HS, Hyun MS, Chang IS, Kim M, Kim BH (2002) A mediator-less microbial fuel cell using a metal reducing bacterium, Shewanella putrefaciens. Enzym Microb Technol 30:145–152

Kim JR, Jung SH, Regan JM, Logan BE (2007) Electricity generation and microbial community analysis of alcohol powered microbial fuel cells. Bioresour Technol 98:2568–2577

Korneel R, Nico B, Siciliano SD, Marc V, Willy V (2004) Biofuel cells select for microbial consortia that self-mediate electron transfer. Appl Environ Microbiol 70:5373–5382

Laura R, Pierangela C, Federica V, Sarah Z, Alessandra C, Lucia C, Andrea S (2017) Influences of dissolved oxygen concentration on biocathodic microbial communities in microbial fuel cells. Bioelectrochemistry 116:39–51

Lee SA, Choi Y, Jung S, Kim S. (2002) Effect of initial carbon sources on the electrochemical detection of glucose by Gluconobacter oxydans. Bioelectrochemistry 57:173–178.

Li He, Peng Du, Yizhong C, Hongwei Lu, Xi C, Bei C, Zheng W (2017) Advances in microbial fuel cells for wastewater treatment. Renew Sustain Energy Rev 71:388–403

Li H, Li X, Teng C, Manhong H (2018) Electrochemical performance and community structure in three microbial fuel cells treating landfill leachate. Process Saf Environ 113:378–387

Li J, Zheng G, He J, Chang S, Qin Z (2009) Hydrogen-producing capability of anaerobic activated sludge in three types of fermentations in a continuous stirred-tank reactor. Biotechnol Adv 27:573–577

Liu H, Grot S, Logan BE (2005) Electrochemically assisted microbial production of hydrogen from acetate. Environ Sci Technol 39:4317–4320

Logan BE (2009) Exoelectrogenic bacteria that power microbial fuel cells. Nat Rev Microbiol 7:375–381

Logan BE, Hamelers B, Rozendal R, Schröder U, Keller J, Freguia S, Aelterman P, Verstraete W, Rabaey K (2006) Microbial fuel cells: methodology and technology. Environ Sci Technol 40:5181–5192

MirellaDi L, Tom PC, Ian MH, Keith S (2009) A single-chamber microbial fuel cell as a biosensor for wastewaters. Wat Res 43:3145–3154

Mostafa G, Samaneh S, Manal I, Zahira Y, Daud WRW (2012) New generation of carbon nanocomposite proton exchange membranes in microbial fuel cell systems. Chem Eng J 184:82–89

Om P, Alka M, Kailasa SK, Shobhana C, Kumar A, Mungray (2018) Comparison of different electrode materials and modification for power enhancement in benthic microbial fuel cells (BMFCs). Process Saf Environ 117:11–21

Park DH, Kim BH, Moore B, Hill HAO, Song MK, Rhee HW (1997) Electrode reaction of Desulfovibrio desulfuricans modified with organic conductive compounds. Biotechnol Tech 11:145–158

Park DH, Zeikus JG (2003) Improved fuel cell and electrode designs for producing electricity from microbial degradation. Biotechnol Bioeng 81:348–355

Park DH, Zeikus JG (1999) Utilization of electrically reduced neutral red by Actinobacillus succinogenes: physiological function of neutral red in membrane-driven fumarate reduction and energy conservation. J Bacteriol 181:2403–2410

Patil SA, Surakasi VP, Koul S, Ijmulwar S, Vivek A, Shouche YS, Kapadnis BP (2009) Electricity generation using chocolate industry wastewater and its treatment in activated sludge based microbial fuel cell and analysis of developed microbial community in the anode chamber. Bioresour Technol 100:5132–5139

Patil S, Mulla AKS (2013) A single chamber microbial fuel cell as power supply for implantable medical devices. Energy Tech. 01–04.

Shah MP (2021) Removal of refractory pollutants from wastewater treatment plants. CRC Press

Pham CA, Jung SJ, Phung NT, Lee J, Chang IS, Kim BH, Yi H, Chun J (2003) A novel electrochemically active and Fe (III)-reducing bacterium phylogenetically related to Aeromonas hydrophila, isolated from a microbial fuel cell. FEMS Microbiol Lett 223:129–134

Prasad D, Sivaram TK, Berchmans S, Yegnaraman V. (2006) Microbial fuel cell constructed with a micro-rganism isolated from sugar industry effluent. J Power Sources 160:991–996

Quezada BC, Delia ML, Bergel A (2010) Testing various food-industry wastes for electricity production in microbial fuel cell. Bioresour Technol 101:2748–2754

Rahimnejad M, Ghoreyshi AA, Najafpour G, Jafary T (2011) Power generation from organic substrate in batch and continuous flow microbial fuel cell operations. Appl Energy 88:3999–4004

Reguera G, McCarthy KD, Mehta T, Nicoll JS, Tuominen MT, Lovley DR (2005) Extracellular electron transfer via microbial nanowires. Nature 435:1098–1101

Rezaei F, Xing D, Wagner R, Regan JM, Richard TL, Logan BE (2009) Simultaneous cellulose degradation and electricity production by Enterobacter cloacae in a microbial fuel cell. Appl Environ Microbiol 75:3673–3678

Rhoads A, Beyenal H, Lewandowshi Z. (2005) Microbial fuel cell using anaerobic respiration as an anodic reaction and biomineralized manganese as a cathodic reactant. Environ Sci Technol 39:4666–4671

Rozendal RA, Hamelers HVM, Buisman CJN (2006) Effects of membrane cation transport on pH and microbial fuel cell performance. Environ Sci Technol 40:5206–5211

Shah MP (2020) Microbial Bioremediation & Biodegradation. Springer

Schroder U, Nieben J, Scholz F (2003) A generation of microbial fuel cells with current outputs boosted by more than on e order of magnitude. Angew Chem Int Ed Engl 42:2880–2883

Senthilkumar K, Chitra Devi V, Mothil S, Naveen Kumar M (2018) Adsorption studies on treatment of textile wastewater using low-cost adsorbent. Desalin Water Treat 123:90–100

Senthilkumar K, Naveenkumar M, Samraj S (2022) Reaping of bio-energy from waste using microbial fuel cell technology, Wiley

Sharma Y, Li B (2010) Optimizing energy harvest in wastewater treatment by combining anaerobic hydrogen producing biofermentor (hpb) and microbial fuel cell (mfc). Int J Hydrogen Energy 35:3789–3797

Sun M, Sheng GP, Zhang L, Xia CR, Mu, ZXl (2008) An MEC-MFC coupled system for biohydrogen production from acetate. Environ Sci Technol 42:8095–8100

Sun M, Sheng G, Mu Z, Liu X, Chen Y (2009) Manipulating the hydrogen production from acetate in a microbial electrolysis cell-microbial fuel cell coupled system. Power Sources 19:338–343

Tender LM, Reimers CE, Stecher HA, Holmes DE, Bond DR (2002) Harnessing microbial generated power on the seafloor. Nat Biotechnol 20:821–825

Thurston CF, Bennetto HP, Delaney GM, Mason JR, Roller SD, Stirling JL (1985) Glucose metabolism in a microbial fuel cell. Stoichiometry of product formation in a thionine-mediated Proteus vulgaris fuel cell and its relation to Coulombic yields. J Gen Microbiol 131:1393–1401

Vega CA, Fernandez I (1987) Mediating effect of ferric chelate compounds in microbial fuel cells with Lactobacillus plantarum, Streptococcus lactis, and Erwinia dissolvens. Bioelectrochem Bioenerg 17:217–222

Venkata MS, Velvizhi G, Annie Modestra J, Srikanth S (2014) Microbial fuel cell: critical factors regulating bio-catalyzed electrochemical process and recent advancements. Renew Sustain Energy Rev 40:779–797

Venkatesh C, Pradeep V (2016) Microbial fuel cell: a green approach for the utilization of waste for the generation of bioelectricity, Bioresour Bioprocess

Wang Z, Lee T, Lim B, Choi C, Park J (2014) Microbial community structures differentiated in a single-chamber air-cathode microbial fuel cell fuelled with rice straw hydrolysate. Biotechnol Biofuels 7:9

Wen Q, Wu Y, Cao D, Zhao L, Sun Q (2009) Electricity generation and modeling of microbial fuel cell from continuous beer brewery wastewater. Bioresour Technol 100:4171–4175

Hu, Wen-Juan, Cheng-Gang N, Ying W, Guang-Ming Z, Wu, Zhen (2011) Nitrogenous heterocyclic compounds degradation in the microbial fuel cells Process Saf. Process Saf Environ Prot 89:133–140

Wrighton KC, Agbo P, Warnecke F, Weber KA, Brodie EL, DeSantis TZ, Hugenholtz P, Andersen GL, Coates JD (2008) A novel ecological role of the Firmicutes identified in thermophilic microbial fuel cells. ISME J 2:1146–1156

Xu Y, Rojas-cessa R, Grebel H, (2012) Allocation of discrete energy on a cloud-computing data enter using a digital power grid. Proceedings of IEEE international Conference on Green Computing and Communications, pp 615–618

Yu EH, Cheng S, Scott K, Logan BE (2007) Microbial fuel cell performance with non-Pt cathode catalysts. J Power Sources 171:275–281

Zhang T, Zeng Y, Chen S, Ai X, Yang H (2007) Improved performances of E.coli-catalyzed microbial fuel cells with composite graphite/PTFE anodes. Electrochem Commun 9:349–353

Zhou M, Yang J, Wang H, Jin T, Hassett DJ, Gu T (2013) Bioelectrochemistry of microbial fuel cells and their potential applications in bioenergy, Bioenergy Res: Adv Appl 131–153

Zhuwei Du, Haoran Li, Tingyue Gu (2007) A state of the art review on microbial fuel cells: A promising technology for wastewater treatment and bioenergy. Biotechnol Adv 25:464–482

Phytoremediation of Metals and Radionuclides

Anitha Thulasisingh, Sathishkumar Kannaiyan, Vishal Amit Kannan, and Srivarshini Govindarajan

1 Introduction

As we are observing the drastic increase in the population, the needs of the people also increase, which in turn increases the utilization of resources and disposal of waste. Thus, resulting in the accumulation of garbages and wastes which pollute the land, water, and air. Countries across the planet are confronting the same issues as the result of increased urbanization. So, the waste accumulated should be disposed of in a proper period in order to rescue the land and restore pollution-free environment. If mechanical methods and chemical methods are adopted, it might cause a further increase in pollution and wastages, hence to overcome this issue the method of bioremediation, i.e., the method of degrading and decomposing all wastes irrespective of the sources is being put to use. This technique is very effective such that, even if there is a metallic contaminant, it can be taken up, reclaimed, and treated by the help of microbes, and plant and animal species through the biological processes. In specific, this bioremediation can undergo the process called phytoremediation, which has been adopted to treat waste and disposals via the plant species application. There are various mechanisms under phytoremediation that are almost similar and only change slightly in their usage and application. The various mechanisms of phytoremediation are phytostabilization, phytodegradation, phytovolatalization, phytoextraction, phytoimmobilization, chelate-enhanced phytoremediation, rhizodegradation, and rhizofiltration. These techniques are used in the up taking of metals or radionuclides from the soil or water and then making the area contaminant free. There are various advantages and is proving to be successful in bioremediating

A. Thulasisingh · V. A. Kannan · S. Govindarajan
Department of Biotechnology, Rajalakshmi Engineering College, Chennai 602105, India

S. Kannaiyan (✉)
Department of Chemical Engineering, Sri Sivasubramaniya Nadar College of Engineering, Chennai 603110, India
e-mail: sathishkumark@ssn.edu.in

M. P. Shah (ed.), *Microbial Technologies in Industrial Wastewater Treatment*,
https://doi.org/10.1007/978-981-99-2435-6_11

the metal-contaminated site and yet, not used widely since there is still research and studies going on.

2 Soil Pollution

Soil contamination occurs when materials enter the soil, disrupting its constitution and morphology, decreasing its viability, making it more sensitive to drought, and rendering it unfit for cultivation. It is being contaminated with heavy metals, acids from industrial residues, etc.. Other than biomolecules, heavy metals in the soil and rock bed, are primarily non-biodegradable and hence retain in the environment. Heavy metals include Hg, Cd, Pb, Ar, Cu, Sn, Zn, and Cr are the predominant toxic substances found in the environment that causes Pollution (Knox et al. 1999). The Institute of Soil Science and Plant Cultivation has made a regulation due to the presence of heavy elements tolerance in the surface of the soil to a level of 0.2 m into six categories, which are shown in Table 1.

2.1 Sources of Soil Pollution

Metals are known one, which is one of the most harmful emerging pollutants since they do not degrade via physiological mechanisms and so persist for extended periods of time. The sources of the heavy metal aspects can be divided into two categories AT missile installations (TNT, elements, and explosive), agricultural farmland (rodenticides, weed killers, metals, and lead), industrial locations (organic metals, As), electronic wastes, sites for timber processing and extraction discharges (commodities), polluted soils, and sediments have been found as the chief source for the soil pollution and contamination. Based on the source deposition, there are several ways for soil to get infected, and they can be classified as point or nonpoint sources. Chemical discharges are distinct, confined, and generally, easily measured, and are included as the point source contaminant. Industrial outfall pipes, unprotected surface runoff exit sewer lines, and grey water outfalls are all forms of point sources. The contamination

Table 1 Lists out the tolerance levels of heavy elements on the surface of the soil

Degree of level	Level
0	Normal level
I	Slightly increased level
II	Initial contamination
III	Midlevel contamination
IV	High-level contamination
V	Very high-level contamination

based on the location and resource, they are being classified as radionuclides and non-radionuclides pollutants. The non-radionuclides include missiles, e-wastes from electronics and digital applications, pesticides and herbicides, food waste contaminated with heavy metals, medical waste from hospitals and dispensaries.

2.2 *Non-Radionuclide Pollutants*

2.2.1 Missiles

Some of the explosives like 2, 4, 6-tolite, picronitric acid, carbazotic acid, hexahydro-1, 3, 5-trinitro-1, 3, 5-triazine and cyclonite hexogen accumulate in the soil and when food intake of the contaminated soil is consumed, it results in severe health issues including headache vomiting, petit mal, and chronic generalized tonic–clonic disease, memory loss and head trauma that is permanent and pervasive (Thomas 2004).

2.2.2 Insecticides

Pesticides can be hazardous to a range of plants, including raptors, fisheries, beneficial insects, and other species also, along with weeds and pest controlling. Even insecticides and herbicides are the most highly hazardous type of pesticide; herbicides can also harm wanted plants and various species. The activities of pesticides are generally influenced by pH, temperature, and climatic situations. Some common examples are DDT, Captan 2, 4-D- iso-octyl ester, and other chlorinated carbon compounds. As pesticides have the ability to penetrate the core of the soil from the top soil to the bed rocks, they contaminate the soil to a major extent (Chowdhury et al. 2008; Kasassi et al. 2008).

2.2.3 Food Wastes

Nowadays, people use a number of plastic bottles, and non-degradable things and dispose them as our daily waste concerned with food products. They are generally made of toxic polymers, asbestos, and we also dump waste into soil. The metal cans and bottles are made of cyanides, heterocyclic compounds which are dangerous to the soil. The spontaneous digging of surface soil can lead to exposure of methane and other hazardous hydrocarbon, and gases.

2.2.4 E-Wastes

The main source of e-waste is selenium, nickel, arsenic, cadmium, lead, etc., about 13% of the *e*-waste contains metals and non-metals, which serves as the cause for soil contamination. India has an annual e-waste of about 380,000 tons. Now-a-days, magnetic properties consisting of metals are isolated from e-waste using a strong field of magnetism associated with Foucault's current filters. Eddy current sites it's applicable to extract Cu, Al, and other metal alloys arising out of commercial electronic-waste (Islam et al. 2020; Pant et al. 2012, Shah MP., 2020).

2.2.5 Medical Wastes

One of the methods adopted for the disposal of clinical and dispensaries waste is incineration, which is the igniting of the compounds and things excreted as residues in order to avoid the pandemic of diseases and microbes through which they spread. This reduces the spread of disease and sterilizes, but the burnt products releases flue, gas mixture consisting of heavy metal gets mixed with the atmosphere and also sediments in the soil causing contamination with some hazardous metals and some wastes which are not treated properly are mixed with degradable waste. This contains heavy metals like zinc, mercury, silver, chromium, and few amount of cadmium (Adama et al. 2016).

2.3 Heavy Metals

Cadmium, chromium, tin, mercury, lead, zinc, iron, manganese, nickel, copper, and other heavy metals are very poisonous when accumulated, resulting in a variety of negative consequences on the environment and living organisms as it causes severe health issues to humans, plant and all other living organisms.

2.3.1 Cadmium

Cadmium is a whitish metal, water-insoluble and flammable in an amorphous form. It usually occurs with zinc and its mineral ores. It has anti-corrosive property. There is only one primary ore for the metal cadmium, it is Greenockite. The main source of cadmium in the soil as a contaminant is from the phosphorous fertilizers (PO_4^{3-}) (Karaca et al. 2002). It is one of the major hazardous heavy metals relaying a crucial character in soil contamination leading to adverse effects of human and animal immune systems (Bergkvist et al. 2003; Wang et al. 2012). The sources of cadmium are from nuclear reactors, where cadmium is being used for controlling the metal rod which absorbs the neutrons; nickel–cadmium batteries, also in fertilizers, and also from plastic industries where it is used as an anti-corrosive coating, contact switches, ancient cathode ray tubes, cable manufacturing firms, from coal industries and mining. The limited levels of cadmium does not affect the soil activity whereas beyond 2 mg/kg of soil is said to be pollutant. It adversely affects the dehydrogenase enzyme activity and inhibiting the activity of the enzymes phosphomonoesterase and urease, which are significantly playing a vital role by catalyzing the reaction of organic phosphorous to the inorganic form of phosphorous that is assimilated by the plants. It also affects negatively by the enzymatic action of arylsulfatase, which is significant in the process of mineralization by the hydrolysis of the aryl esters and breaking O-S linkage. Cadmium is carried by transport proteins and intracellular pathways, where it binds to sulfur groups in macromolecules to form complexes. Cadmium at higher concentrations (30 and 40 mg Cd per kg) contributed to the reduction of microbial species. It highly affects the fungal species as compared to bacteria and actinomycetes (Williams and Wollum 1981, Shah MP., 2021). In general, with accordance to humans, it causes renal failure, cancer risk in breast, prostate glands, skeletal deformations, hypertension, high cholesterol levels, and heart disorders along with some blood circulation malfunctioning. It also affects the MAP Kinase pathway, by binding to the alpha estrogenic receptor. It is considered as the first category carcinogenic element as represented by the International Agency for research on Cancer (Tran and Popova 2013).

2.3.2 Mercury

Mercury (Hg) is one of the naturally occurring metals which is said to be the most toxic element in the world. The main ore of the Hg is cinnabar. Mercury is silvery white d-block metal and liquid at room temperature. It is a moderate conductor of current. We might find its application in thermistors, thermocouples, mercury amalgam, alloys, lamps, dental tools, and clinical applications. Besides such an application, it is also responsible for many poisoning and contaminations. Mercury poisoning is one of the life-threatening element, which affects marine, birds, humans as well as plants. It generally occurs as $HgCl_,$ methyl mercury, mercury sulfate, etc., the natural sources of mercury include the combustion of fossil fuels, clinical and medical wastes, metal wastes, pesticides, herbicides, fertilizers, gold mining, coal mining, volcano eruption lava and batteries (Higueras et al. 2014). The sources of the contamination are seldom man made. It reaches humans through marine foods which have methyl mercury leads to serious health issues. In general, mercury has a strong binding energy so it occurs in a combined state. But it is significant to remove it in a free mobile state. The presence of mercury escalates the pH and acidity of the texture of soil and enters the plant and human cycle. Due to various massive amounts of insoluble dissolved organic substances/clay minerals, Hg is extensively bonded to soil particles. Mercury-contaminated soil is usually seen in temperate and tropical regions (Hallal et al. 2009). Humic acid inhibited Hg from being carried into plants or wiped out of the soil by lowering the quantity of accessible Hg in soil. In some climatic situations, runoff might result in Hg seeping into the natural water system. It affects the genetic makeup and leads to mutations and depresses the germination capacity of succeeding generations of plants (Wang et al. 1997). In genetic alterations, the mercury easily binds to the disulfide bond, thus resulting in sulfate–mercury sulfate bond formation. Then, it affects the S-phase of cell cycle, by reducing DNA synthesis and suppressing the amount of genetic materials. They also tend to depress the presence of other important micronutrients such as K, Mg, Ca, etc., which are up taken by xylem along with mercury as they are required for other mechanisms. Even in minimum concentration, they affect the light and dark photosynthesis (Patra and Sharma 2000).

In humans, mercury poisoning exists in different forms of mercury such as methyl mercury, metal, and mercury chloride. Mercury poisoning generally affects the brain. Methyl mercury causes nervous system damages including peripheral vision damages, numbness in the regions of fingers, wrist, mouth and certain regions, muscle fatigues and muscle pains, lagging in narrative skills in adults. In children, memory loss, eye vision damages, motor and sensory nerve repairs and their impact on the skill and co-ordination. The metal mercury has been consumed directly through the nasal and mouth in the form of vapor. It results in chronic symptoms such as shivering, modifications in feelings like mood alterations, lack of sleep, muscle atrophy and muscle twitching and brain damages, negative effects in nerve signaling, and irritations under low concentrations of mercury. When the concentration of mercury increases, it impacts heavy brain damages, renal failures, respiratory and inhaling issues, cardiovascular disorders. In the other forms of mercury, either in organic or inorganic forms of mercury, they inhibit the alimentary tracts or respiratory tracts, muscle weakness and muscle cell death, skin irritations, and amnesia. Some primary precautions to avoid mercury vapor contamination can be undertaken. They are safely disposing the CFL lamps, clinical instruments consisting of mercury. According to National Agency for Food and Drugs Administration and Control, the maximum acceptable amount of mercury is 0.001 mg per litre.

Minamata Disease

The most toxic form of mercury is methyl mercury. The mercury poisoning was found in shellfish and other fishes of the water sources in a small town of Japan named Minamata. As the fishes and shellfish were the staple food of the town, it was consumed by the dwellings of the town as not only the human but also animals were found to be infected with severe symptoms of numbers in the body parts such as hands, legs, memory loss, eye vision lost and heavier damages, and central nervous system damages. There was highest measure of about 920 parts per million of methyl mercury poisoning was observed (Harada 1982).

2.3.3 Lead

Lead is a grey color, soft, malleable element. This element is comparatively least reactive. It has both acid and basic properties. It has more resistance to corrosion because when the layer of lead has been coated the substance on which becomes passive. They are less reactive toward oxygen. In particular, lead poisoning promotes haemoglobin synthesis suppression, triiodothyronine synthesis inhibits renal, musculoskeletal, and endocrine glands malfunction, vascular endothelial abnormality, hypertension, high blood pressure, high sensitivity, short-term and long-term damage to the central nervous system such as brain, spinal cord, and peripheral nervous systems such as motor nerves, sensory neurons, and hippocampus neurons, and reproductive problems to both men and women gender. Lead is, in general, used in batteries, cells, ammunition and shrapnel, balances, army personnel, pewters, molten compositions, fair-skinned acrylics, cosmetics and facial products, pipelines and linkages, fossil fuel combustion, and protective coatings are just a few of the items used in the military (Effron et al. 2004). Industrial sources and hazardous sites, for instance, defunct lead smelters, and may potentially release lead into the environment. The natural lead levels in soil range from 50–400 parts per million. Ore, metallurgical, and processing operations have resulted in significant increases in lead levels in the environment, particularly around industrial and mining sites (Nas and Ali 2018). When lead is discharged into the air from anthropogenic processes or aeroplanes with gasoline engines, it may travel great distances before landing on the ground, where it generally adheres to the soil particles and it depends on the kind of chemical entities and the soil's properties, lead can migrate from the soil into ground water. In general, lead is being more taken by children and foetus up to the age of 0 months to 10 years as a result of their growing phase since there is absence of growth phase in elders and older people. Lead has been taken in many ways such as in cases of foods consumed, water, and personal uses like in direct contact with skin, nose, mouth such as chemicals used, lead-contaminated paints, and acrylics. Lower consumption levels of lead causes anxiety, less haemoglobin, growth hormone secretion lesser, ear damages, and intelligence quotient levels lower, so as a result of this hyperactivity in kids are being screened. According to the World Health Organization, the maximum acceptable amount of lead is advised as 0.01 mg per liter. In plants, lead poisoning restricts the activity of enzymes of dehydrogenase. Chloramphenicol acetyl transferase is a type II enzyme, which is involved in the oxidation of hydrogen peroxidases for the conversion to water. Lead in the soil reduces seed germination and slows seedling growth, lowering germination percentage, gestation factor, root/shoot elongation, threshold indicator, and root and shoot total matter.

2.3.4 Copper

Copper is an adaptable brownish-red metal which is a good conductor of electricity. It comes under the category of native metals. The predominant ores are the malachite, azurite, and also turquoise. The range of copper found in the body is 1.4–2.1 mg of total body weight. Copper and many associated elements react and complexes are been formed which are used as fungicide, antimicrobials. It is a unique metal, as any microbe cannot grow on the surface of the metal. But it has some common issues like plant infertility and also seedling growth along with germination of seeds are been prohibited by the addition of copper in its native form. It also reduces the iron intake by the plant. It has earlier yellowing in leaves and plant parts. And also, some enzymes like tyrosinase, acid phosphatases, ureases, alkaline phosphatases, and proteases are being affected minimally. It has been revealed that the accumulation and contamination of copper in the form of ions such as cu^{2+} and cu^{3+} ions, complexes like copper zinc sulfate, cupric chloride and with combination of copper–chromium–zinc–lead co-ordination compounds. These accumulations result in the inhibition of many enzymes. The strongest inhibition is seen in the enzyme kinetics of the enzyme dehydrogenase. It also had significant inhibitory action on the enzyme urease, when the copper is in the pure form. When the copper occurs in a mixed form, there is less inhibition as compared to that of its pure form. Cu resistance of the individual plant is based on the dependency of the plant on the ATP production efficiency. Some specific plant taxonomic families are found to be more resistance and forbearance toward Cu contamination such as brassica, lamiaceae, gramineae, and malvaceae. In humans, the toxicity of copper results in food poisoning, an abundant amount of DNA damage. It also plays a vital role in the later stages of the Alzheimer's disease in older people. An amount of elevation of 3 ppm (mg / kg) results in severe effects of copper contamination. But copper also has a positive effect as it aids by acting as cofactor for cytochrome oxidase, it reduces lipid content and alters the structures of thylakoids of the chloroplast, thereby, decreasing the rate of the photosynthesis process (Bouazizi et al. 2010). In humans, copper consumption may be taken up via mouth, or nose while inhaling or while having soil freshly consumed leaves and vegetables without washing the soil surface. These consumption and accumulation result in impaired nervous disorders, destructions in reproduction organs, renal and adrenal endocrine damages, memory and tissue damages in cases of new-born. The symptoms of copper toxicity include nausea feeling with irritable bowel syndromes, vomiting, head pain, abdominal and apo thorax pains and damages. The abnormal raised quantity of copper results in severe issues like cystic fibrosis, hypothyroidism, migraine head damages, diabetes mellitus, cholesterol, tumors, heart attacks, urinary infections, etc.. The amount of copper is variable according to the plant family. In general, it ranges from 1.5 to 3 kg per 5 acres. In other terms, the quantity of copper is around three–six pounds. In soil, copper is usually found in the form of copper sulfate and copper oxide (Fernandes et al. 1991).

2.3.5 Zinc

It is considered to be one of the old and primordial elements. Generally, occurs in oxidative form as zinc oxide. According to Tran 2013, there exist 23 different forms of isotopes of zinc occurring in the Earth's crust. In human body, zinc is abundantly found in blood cells. It is the cofactor of the carbonic anhydrase enzyme, which is very predominant in the metabolism of carbon dioxide and also cofactor of RNA polymerase. It helps in the degradation and absorption of many proteins and other macromolecules. They also help in insulin storage and many biochemical reactions, gene copying, DNA replications. It is also extensively used in pharmaceuticals. Zinc sulfate is mostly used as weedicides and in dying. Zinc is known to be such an important trace element, but when it becomes elevated then, its regular amount has an adverse effect. This is due to its interaction with other heavy metals in the soil it reacts and forms compounds that interrupt the metabolic cycles of soil and nature. In general, zinc has the ability to suppress the absorption and accumulation of other metals and minerals. The zinc contamination and toxicities can be viewed by the naked eyes as the pigmentation of yellow changes from yellow to purple in later periods, due to phosphate deficiency. The excessive up taking of zinc shows inhibited development, delayed or absence of germination, transpiration rate is been decreased, root systems are been affected in consuming metals, and water (Tran and Popova 2013). It inhibits the survival of oligotrophic bacteria, nitrogen-immobilizing bacteria, actinomycetes, copiotrophic bacteria. A sequence of heavy elements relevant to their inhibition and retardation on genera Streptomyces actinomycetes as (mercury > cadmium > zinc > copper > nickel > lead > cobalt). The sources of zinc to soil include drainages impairing, fertilizers, wastes of galvanization residues, semiconductor wastes, fuels extracting dusts (Hershfinkel et al. 2007). The highest record of zinc occurrence is 12400 mg per kilogram of soil.

2.3.6 Nickel

Nickel is known to be heavy toxic metal as well as necessary metal. It has a significance in the metabolism of the enzyme urease, which is very prerequisite for the health of mortal beings. It is found in excess amounts in meta-igneous rocks which are formed as a result of metamorphosis of the evolutionary changes (Nagajyoti 2010). It increases cytotoxic effects in the human body and leads to human tumors and blood cancer. Its occurrence in the surface of the soil may range from 200 to 26,000 mg of Ni per kg of rock soil. The pavement for the accumulation of nickel is by the mechanisms of translocation and absorption. The uptake of nickel is stored in the epidermis and vascular tissues of the leaves. Whereas, the occurrence in normal topsoil is around 10–1000 mg of Ni per kg of soil. H^+-ATPase utilizes adenosine triphosphate energy to generate hydrogen ion protons to the extracellular matrix out of the cell. Nickel (+2) metal in the plants acts as the best marker for the water content uptake reduction by the plants as a result of the heavy metal toxicities. It also reduces cell division and cell differentiation activities in plants. When such plants are being consumed, it leads to cancer of the individual by the mutation caused due to the chromatin condensation (Nagajyoti 2010). It activates the HIF–1 (Hypoxia inducing factor). This causes respiratory problems. General complexes associated with nickel intake in the case of humans involve dermatitis, pompholyx, epigenetic effects, GI tract issues, heart diseases, etc. It reduces the uptake of the minerals and nutrients such as iron, magnesium, etc., which are been considered essential for major functioning. It results in the shutting of stomata and leads to the prohibition of the Calvin cycle and affects photosynthesis and also inhibits the common enzyme in the Calvin cycle such as rubisco-3-phosphoglycerate kinase, (F-1, 6 BP) fructose 1, 6 bi phosphatases, NADP dehydrogenase, etc., (Seregin and Kozhevnikova 2006).

2.4 Radionuclides Resources of Pollutants

During U–238 decay, Pb–210 is being released. The other derivatives include radium-226, radon-222, cesium-137, cesium-134, strontium-90, tellurium-129, iodide-131. This is used as a marker for conquering soil erosion (Walling 1998). The ratio of 79% of total radionuclides pollutants are from the nature and about 10% is human-made and the rest is from industries, mining, leaching, etc., (Zhu and Shaw 2000). The radionuclides behave as inactive radio-isotopes in the soil, then it is been taken up by the plants for nutrition purposes. Later, it is being consumed by human and other organisms in this manner. The radionuclides join the local ecosystem and again retain in the soil.

Radioisotopes are radioactive elements with uncertainty which tend to change to its original form from its isotopic form. It can be of two forms:

- Natural radionuclide
- Artificial radionuclide

2.4.1 Natural Radionuclide

Native radionuclides may be found throughout all diverse habitats; sediment deposition, groundwater, atmosphere, nutrition, and including our own cells contain naturally occurring radioactive elements. It includes the alpha, beta, gamma, and cosmic rays. They include 80 percent of the total available radionuclide in the Earth's crust. They are native to the Earth. It includes potassium-40, cesium-59, U-238, and Th-232 which are the main source of gamma rays origin. To analyze the environmental threat posed by radioisotope, biotechnologists have established some flora and fauna which are genetically modified that may be employed as biochemical markers of radionuclide toxicity (Kirkham and Corey 1997). Elements and radioactive isotopes might impose constant stemming mostly on microbiota. Gamma spectrophotometry was used to determine the total amounts of nine radionuclides (Cs^{137} Pb^{210}, Ra^{228}, Co^{60}, Ra^{226}, Am^{241}, Th^{228}, U^{238}, and Th^{228}) and nine metals (Ni, Co, and Cu, Zn, As, Cd, Hg, Pb, and Cr).

2.4.2 Artificial Radionuclide

Artificial radionuclides are being synthetically produced by bombardment, and other explosive processes. They account for almost 20% of the total available radionuclide in the Earth's crust. Chernobyl accident in Ukraine is a significant example for the artificial radionuclide pollution over the Earth, which occurred in 1986. Its precipitation was depicted on the flora such as potatoes, fodder, top soil, honey nectar of the flowers, mushrooms, berries, and many root tubers and root products. It includes Sr-90, I-131, Po-240, Cs-134, Po-241, and iridium-192. The levels of 0.2 Bq per meter square of area to 10 Bq per meter square of area is the normal level of the cesium-137 is to be found in specific cases of honey, if this exceeds, it results in contamination and very pandemic effects. The levels of 0.3 Bq per meter square of area to 1000 Bq per meter square of area is the normal level of the cesium-137 is to be found in specific cases of honey, if this exceeds it results in contamination and very pandemic effects. The source of this pollution is anthropogenic and industrial. The anthropogenic source includes consumer products, nuclear-based pharmaceuticals, radionuclide therapies, industrial outlets, etc., Lead-238 tends to be the most stable artificial radionuclide. The occurrence of the Fukushima nuclear crisis in 2011, mostly 131-Iodine, 134-Cesium, as well as 137-Cesium was discharged. A portion of the radiation was exposed directly into the atmosphere (Talerko et al. 2021).

2.4.3 Effects of the Radionuclide

This in general results in many disasters of blood sheds with thousands of deaths. These physiological consequences can be created directly, but they are usually induced unintentionally by free radicals, which can readily break crosslinks and produce a variety of distinct DNA damages. Free radicals are not only formed in reaction to radiation, but also in response to a variety of stresses such as exposure to ultraviolet rays from sun or susceptibility to other pollutants. This also causes skin cancer, blood cancer, and other serious health issues.

2.5 *Remediation Using Plant Kingdom*

Bioremediation based on plants is known to be a low-cost approach, non-invasive substitute, or supplement to bioscience treatment techniques that utilize plants and their accompanying microbiota to clear away the environment. This technique makes use of the mechanisms by which plants and its symbiotic rhizosphere vegetation, (the organism in the root with a symbiotic relationship) breakdown and retain organic-based and inorganic-based contaminants in the natural world. Phytoremediation is a cost-effective method of removing organic and inorganic contaminants from the environment. Generally, the plants which are used in bioremediation belong to the taxonomic plants of cruciferous family and Labiatae family.

2.5.1 Advantages of Phytoremediation

- It can be done both in the site of contamination and external transport mode of operation.
- It is a very effective method to replenish the soil and reduce the soil contamination.
- It is a cost-effective method of soil fertility replenishing.
- It has the major advantage of not spreading contamination from a particular location to nearby water bodies and other locations and atmospheres.

- It is ecofriendly and adds nutritive values to the soil.
- It's a green technology if put to a proper use can be ecofriendly and aesthetically pleasing.
- It has the capability to treat sites contaminated with more than one type of pollutant.

2.5.2 Limitations of Phytoremediation

Although phytoremediation is a very useful technology, it does have a few drawbacks:

- It depends on plants having roots that grow long and which can reach depths into the ground.
- The use of invasive and non-native species can affect the biodiversity.
- It is a slow process, since using plants takes time for its growth which follow-up to the clean-up of the contaminated site that may take up to several years or more.
- The harvested plants have to be handled well and should be disposed properly and not be left carelessly as wildlife could consume it.
- Climate plays another major factor (Maheswari and Rajeswari 2016).

There exist two kinds of phytoremediation.

Ex situ Phytoremediation
In situ Phytoremediation

2.5.3 In Situ Phytoremediation

Inside the site of contamination, phytoremediation is the low cost and environmentally friendly elimination of contaminants through plant absorption. It is generally used to clean phreatic water and topsoil. Two widely used in situ phytoremediation techniques are phytoextraction and phytostabilization. It depends on solar energy. It is used for remediation and degradation of both non-radionuclide and radionuclide resources of contaminants. In situ phytoremediation is the most preferable technique adopted for remediation. It doesn't require any transportation charges. The main advantages of in situ phytoremediation includes—it uses much less expenditure than any other physical methods using engineering techniques or any chemical methods which uses some procedures like leaching, incineration, well making, etc., they are climate and season effective, it is highly sustainable and exhibits least damages to the top soil as compared to the ex situ—phytoremediation, it also supports better microbial growth, easy to use, and aids their metabolic activities, provides carbon sequestration, and also used to extract the essential metals as in molecular levels and nanoparticles sample of gold (Au), nickel (Ni), chloride (Cl). For example, a small area of two acres is enough to clean up a waste of around 5000 tons (Shmaefsky 2020). The application of in situ phytoremediation ensures:

1) Biodegradation of the pollutants, and also not disturbing the other microbes which reserve soil fertility.

2) Existence of hydrogen acceptor by oxidation and reduction reactions
3) Ecofriendly and also enriching the essential elements in the soil (Mohanty et al. 2011).

The procedure is done as the land which consists of soil contaminated has been phytoremediated by using the more applicable plants which have biodegradation properties include rose plant, oak tree, maple tree, willow tree, tulip flower, alder tree, dahlia plant, peony plant, etc. The land to be phytoremediated is divided into four parts. The first- and third-part division of the land is further divided into two divisions as 1(A), 1(B), 3(A), and 3(B). And the remaining second- and fourth-part divisions are not subdivided and they serve as a single unit. Then, different sorts of seeds desired are cultivated. In the first part of the land, 1(A), lolium perenne, clover, festuca are been used as herbs for the crops. These plants are obliterated from the land for every 30 days. Then, the second part also followed the same procedure but effaced for every 90 days. In 3(A), it is seeded with weeds, and is effaced for every 30 days. Part 4 is being kept as arable land. The part 1(B) is flourished by using phytoremediation plants (rose, tulip, dahlia, and peony). The 3(B) parts of the land is harvested with the help of phytoremediating trees (maple, oak, and willow) (Denys et al. 2006). After regular cultivation of plants for around three years, the lands get its own normal texture and properties and can be used for better cultivation.

2.5.4 Ex Situ Phytoremediation

Externally done plant remediation necessitates the extraction and treating of contaminated soil, making it a costly cleaning technique. In the case of moderately polluted soils, phytoremediation using ex situ remediation plants appears to have the ability to replace existing techniques. But the requirement to blend sediments with soil and/or sand to facilitate the establishment of the most often-used species results in a 30% increase in the volume of the synthesized substrate. The duration for the ex situ phytoremediation process ranges from 60 to 90 days. This type requires some basic optimization parameters involving temperature, oxygen, space, pH, etc., Hence, it is also called "OFF—SITE" phytoremediation. Generally, ex situ type of remediation requires large-scale bioreactors for the cleaning process. It is a time-consuming process. It is preferred for land with comparably fewer heavy metals and pollutant concentrations. It includes bio augmentation, rhizofiltration, biocomposting. The method is not deeply limited. It avoids the spread of the contaminated soil over the uncontaminated soil (Shmaefsky 2020).

2.6 *Mechanisms Under Phytoremediation*

Phytoremediation is immensely a wide-ranging technique under which there are various mechanisms adopted and applied for treating the heavy metals and radionuclides which are present in the soil. The various mechanisms include phytostabilization, phytodegradation, phytovolatalization, phytoextraction, phytoimmobilization, chelate-enhanced phytoremediation, rhizodegradation, and rhizofiltration. The mentioned mechanisms and the type of plant opted are based on the amount of metal or radionuclide present in the site (Shackira and Puthur 2019).

2.6.1 Phytostabilization

Phytostabilization is the technique in which the contaminants, i.e., the metals which are present in the soil are immobilized with the help of plants. The metal contaminants are entrapped and accumulated near the roots and in the rhizosphere region. This method is an inexpensive phytotechnology and also less invasive. At present, this technique has been widely recognised and has been put to use to decontaminate the soil and thereby, restore its physical, chemical, and biological characteristics.

Physical removal of the contaminant metal is not possible with the help of phytostabilization, it just deactivates and immobilizes the metal ions and prevents it from mixing in ground water or dust thereby blocking the movement of these ions to the food chain (Muthusaravanan et al. 2018). Organic contaminants like pesticides and hydrocarbons can also be possibly degraded with the help of this technology with the assistance of microbes that are associated with the roots of the green plants.

The objectives of phytostabilization are:

a. To change the trace element speciation in the soil aiming to reduce the solubility and exchangeable fractions of these elements.
b. To reduce the direct exposure of soil-heterotrophic living organisms to pollutants.
c. To limit the metal mobility thereby increasing the quality of the environment.
d. To stabilize the vegetation cover and limit trace element uptake by crops (Vassilev et al. 2004).

Phytostabilization is applicable in the sites where the contamination levels are varied and with different textures of soil such as its pH, salinity, and the level of metal content. Based on all these criteria and through specifically selecting the appropriate species of plant as well as adding the required amendments for the field chosen the efficiency of the process can be improved and can be carried out well. Hence, these are the two most important basis for the phytostabilization process (Fig. 1).

Basis for the Selection of the Plant Species

Selection of the plant species is the most important step in the phytostabilization process. Only the correct species of plant can be helpful in getting the process done

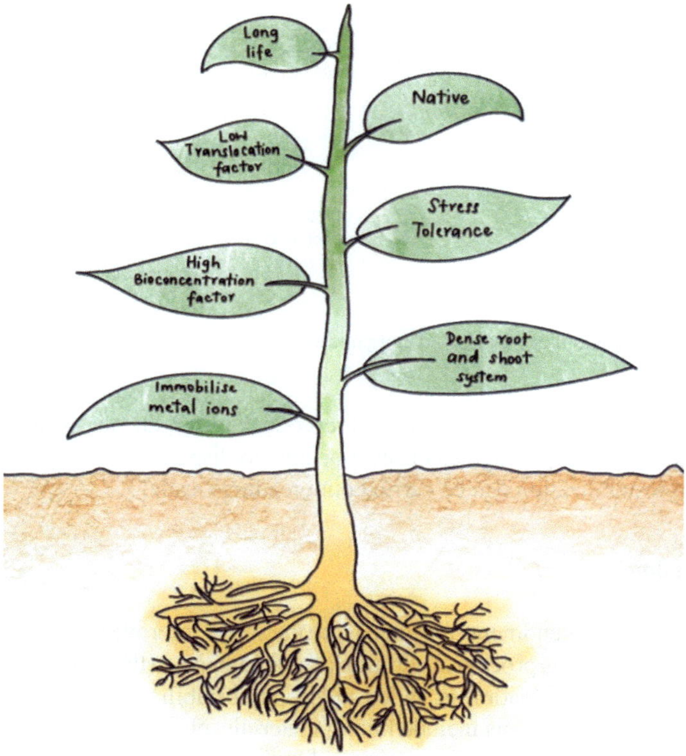

Fig. 1 Criteria adopted for the selection of phytostabilization

effectively. Thus, for selecting the plant species for phytostabilization, the main criterias which are to be considered are:

- The plant which is chosen for a contaminated field should be native to the place, and this is due to the fact that when a native plant is chosen, it is already familiar and most suited to the environmental factors (rainfall, availability of sunlight) of that place and will be able to grow quicker and cover the ground faster.
- The plant species which is to be selected should have good tolerance to other factors that can cause stress to it such as drought and salinity of soil.
- The preferred plant should only immobilize the metal ions, they shouldn't take up the metals, i.e., exclude or limit the translocation to shoots.
- The plant species must have a high bioconcentration factor and low translocation factor.
- The plants should have the capability to tolerate multiple metals and metalloids that are present in the soil.
- The root and shoot system of the plant should be dense enough for phytostabilization.

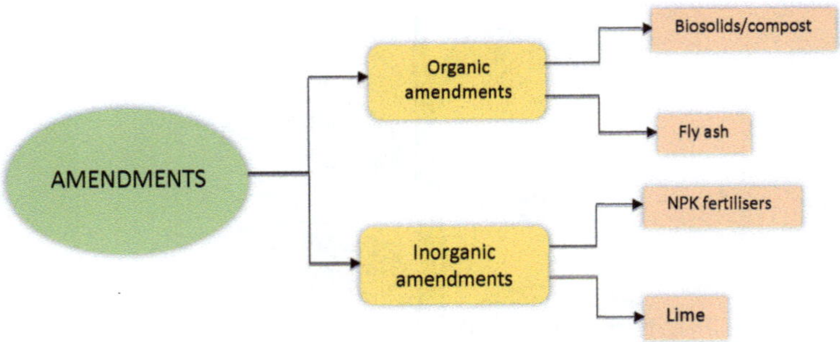

Fig. 2 Specific functions of organic and inorganic amendments used in phytostabilization process

- The plants should have a long life and reproduce faster, i.e., the plants must have large quantities of propagules or must be able to disperse its seeds effectively so that phytostabilization can be carried out on a large scale.

Soil Amendments

Different kinds of amendments are added into the soil to improve the rate of phytostabilization process. This process of addition of amendments in the soil so as to increase the rate of phytostabilization is called aided phytostabilization (Alvarenga et al. 2009). General amendments include liming materials, phosphate, compost, zeolites, aluminosilicates, etc. There are two types of amendments put to use to achieve the process that is organic and inorganic materials as amendments.

The organic amendments include biosolids/compost and fly ash and the inorganic amendments include lime and NPK fertilizers. The organic amendments bring chemical changes (like pH, soil solution composition, and organic acid content) and biological properties (like microbial diversity) in the soil. The changes that happen in the soil are that the soil becomes more stable by which the biological availability decreases. Hence, the nutritional quality of the soil improves, helps plant growth better, and the leaching of metal is reduced. Figure 2 represents the specific functions of organic and inorganic amendments used in phytostabilization process.

Process of Phytostabilization

Sorption, precipitation, complexation, or metal valence reduction are the basic steps in phytostabilization (Ghosh and Singh 2005).

a. Eliminating the contaminants

The contaminated metal or metalloid ions are not exactly taken up by the plant for eliminating it but the plants play a major role by just immobilizing them and reducing the bioavailability. For example, the normally grown agricultural crops can

Table 2 Examples of plant species and the metal which they phytostabilize along with its soil conditions

S.No	Metal contaminant	Plant species	Soil condition
1	Arsenic (As)	*Agrostis capillaris*	Wetlands
2	Cadmium (Cd)	*Quercus ilex* subsp. *ballota*	Mine tailings
3	Mercury (Mg)	*Triglochin maritima*	Wetlands
4	Manganese (Mn)	*Typha latifolia*	Industrial Sludge
5	Zinc (Zn)	*Festuca rubra*	Mine tailings

escalate the phytostabilization process of a metal-contaminated soil, and at the same time, they produce biomass that are rich in zinc (Zn) content, which can then be used as stock fodder. Therefore, the primary role of the plant was stabilizing the field and the biomass which was produced its value was also increased.

In the previous paragraph, the example which was discussed was for a normal plant which does a satisfactory accumulation of the contaminants. This process can be further improved by using hyperaccumulating plants. Hyperaccumulating plants are plants that can accumulate the metals at increased levels, i.e., a hundred or thousand levels higher than normal plants. The roots of these plants secrete exudates which in a sense play a pivotal role in phytostabilization (Colzi et al. 2012). The exudates help by boosting the accumulation, stabilization, or volatilization of the contaminants from soil (Anamika Kushwaha et al. 2016). The metal ions are not uptaken by roots but are stabilized in the rhizosphere with the help of root exudates. The exudates can be classified as high-molecular-weight and low-molecular-weight compounds. These exudates can be either sugars, polysaccharides, organic and amino acids, peptides, and proteins depending on the plants. Table 2 represents some plant species and the metals, which they phytostabilize with their conditions of the soil.

b. Cell wall Binding

Cell wall is an important part of a plant root for this process. There are two ways the roots uptake the metals—by simple passive diffusion through the plasma membrane and through special metal transporters (active metal uptake).

The pectin compound from the cell walls bind to divalent and trivalent metal cations, specifically homogalacturonans (HGA) domain of the pectin compound bind with the metal ions. The HGA inactivates the absorbed metal ion and thereby trapping them in the apoplast region of the roots itself and, hence, reducing the toxicity. In conclusion, if there is a higher level of pectin in the roots, the binding of the metal ions is greater and thereby restricting the metal ions accumulation within the roots itself.

c. Rhizosphere modification

The phytostabilization of the contaminated sites is directly linked with the rhizosphere conditions of the soil. The changes regulate the soil's biochemical properties, the transformation, mobility, and bioavailability of metal/metalloids.

The H^+/OH^- fluxes and the pH of the rhizosphere are regulated by the cations and anions uptake by plant roots. As the acidity of the soil increases, the adsorption of the metal decreases. The major cause of the soil acidification is during the nitrogen cycle. The acidification factor of the soil is important because it affects the solubility and speciation of metal(loid) ions in several ways, mainly (a) metal(loids)s speciation alteration; (b) surface charge tempering in variable charge soil; and (c) influence in the redox reaction of the metal(loid)s. The consequences faced when the pH decreases is that the surface negative charge decreases which results in lower cation adsorption, it also causes the decrease in hydroxy species of metal cations and lastly it dissolute the metal(loid) compounds.

Even the bacteria that are present in the rhizosphere play a role in the metal complexes formations. Few prokaryotes that are present like archaea and eukaryotes like algae and fungi excrete extracellular polymeric substances (EPS) such as polysaccharides, glycoproteins, lipopolysaccharides, soluble peptide, etc. These substances influence the anion functional groups that can absorb metal(loid) ions. For example, organisms that are involved in EPS production include *Bacillus megaterium, Acinetobacter*, sulfate-reducing bacteria, and cyanobacteria. Not only the EPS plays a role even the cell wall of these bacteria play an important role in metal(loid)s adsorption and redox reactions.

d. Hydraulic control

This includes the use of plants to control the water migration that are present at the subsurface through the transpiration process. The plants act as the hydraulic pump since they have a dense root system, and this invariably helps in phytostabilization. The movement of contaminants is regulated through leaching and surface runoff by controlling the flow of water in soils.

Factors Affecting Phytostabilization

To reform and revegetate the contaminated soil, the phytostabilization technique has to be carried out efficiently. The factors that can affect the process and reduce its efficiency are soil, plant, contaminant, and environmental factors.

(i) *Plant factors*

Plants are the prime factor for the phytostabilization process to carry out properly, and the transformation of metal(loid)s and binding of soil particles are managed. For sustainable management of soil remediation, proper type of plant must be selected with beneficial phenotypic and genotypic characteristics. The other critical and most important feature of the plants is that, the roots must be morphologically dense and should be able to penetrate deep into the soil. Willow species that have deep roots are being recognized for accumulating significant amounts of Cd and Zn from the soil (Utmazian et al. 2007).

Enhancing the root biomass and morphology helps in more efficient phytostabilization. Genetic engineering can help and may be useful for enhancement. It helps in modifying the root morphology and/or for identifying and cloning the relevant enzyme-controlling genes into more deep-rooted plants. The changes can affect the microbial associations in the rhizosphere. Then, there are plant species known as hyperaccumulators which are defined as the plant species that are capable of accumulating metal(loid)s above the threshold concentrations. For instance, they can hold up to 10,000 mg kg^{-1} dry weight of shoots for Zn and Mn, 1000 mg kg^{-1} for Co, Cu, Ni, As, and Se, and 100 mg kg^{-1} for Cd. Examples of hyperaccumulators are the *Brassicaceae* and *Fabaceae* families and many more species of up to 400 species (Bolan et al. 2011).

(ii) *Soil Factor*

As we have discussed that the plant selection was a very important factor for the phytostabilization, it also must be noted that the various properties of soil also play a vital role in it since plants grow on it. The physical, chemical, and biological properties of soil control plant growth and determines the successful outcome of this technology. The dynamics of the metal(loids) are also regulated by the soil property thereby affecting their stabilization in the soil. The soil pH is a major factor that influences the process which aids in the sorption of both inorganic and ionizable organic contaminants and controls the speciation of contaminants and they also provide the amendment to assist in plant growth.

(iii) *Contaminant Factor*

The bioavailability and mobility of the contaminants are affected by the reaction (adsorption, complexation, precipitation, and reduction) in these contaminants which in turn affects the phytostabilization. Sorption and complexation with inorganic and organic ligands by soil colloids are the chemical interactions that contribute to metal(loid) retention. Contaminants also have the ability to change the plant growth and the microbes associated with it which has great effects on phytostabilization.

(iv) *Environmental Factor*

The two main environmental factors that affect phytostabilization are rainfall and temperature. The sites where the plants grow will have less availability of water due to the contamination and hence depend on rainfall in the area, which helps in controlling contaminants leaching and erosion of soil and its sediments. Whereas, the temperature affects both the plant growth and characteristics of the soil surface such as cracking and crust formation.

Advantages and Limitations of Phytostabilization

When compared to other techniques followed, the phytostabilization is less expensive, and do not destroy the soil in any way. As we use the plants for the phytostabilization there is a chance of revegetation which then builds up a proper ecosystem restoring and rendering the site, even making the soil physically stable. Valuable products can be gained from biomass of the vegetation used in the phytostabilization in turn generating income from it.

Inspite of having advantages and being one of the effective methods, it also has some limitations. It is only useful at sites where the contaminants are near the surface level and will not be otherwise much effective. All the parameters and factors must be monitored at all times to prevent an increase in metal solubility and leaching. The contaminants remain in the same site. The vegetation and soil may require long-term maintenance to prevent rerelease of the contaminants and future leaching.

2.6.2 Phytovolatilization

This technique is a little different from the phytostabilization method. In phytovolatilization, the contaminants are not only immobilized but are also absorbed by the plant and are transported and released into the atmosphere in reduced toxicity. The examples of the types of toxic metals which tend to convert easily to volatile forms are selenium, arsenic, and mercury. The volatile forms that are released into the atmosphere are dimethyl selenide and mercuric oxide. Although, it is released into atmosphere, it will be toxic to the living system and hence this technique remains as a debatable one. It can be considered as a permanent solution for contamination sites because the gaseous volatilized products are highly unlikely to redeposit on or near the site. Phytovolatilization shouldn't be followed in places having high population and places with unpredictable weather (Muthusaravanan et al. 2018).

There are two different forms of phytovolatilization, namely, direct and indirect phytovolatilization.

Direct Phytovolatilization

This form of phytovolatilization can be simply called as the normal phytovolatilization since it follows the same process of up taking the contaminants and transpiring, it into the atmosphere. This form is innate and very well-studied and well-known method. In this form, the contaminants are immobilized in the soil, absorbed by the roots, translocated through the stems, and finally, volatilize the compound leaves. This method differs slightly from the standard transpiration process of water through the vascular pathway because the contaminants are hydrophobic and tend to diffuse through the hydrophobic membranes of cutin into the epidermis and dermal tissues.

Indirect Phytovolatilization

This form of phytovolatilization wholly depends on the root activities of the plant and its increase in the volatile contamination flux from underground. The roots grow and traverse different areas and depths of the underground, and cause changes in the chemical properties and transport in the soil which in turn cause an increase in the flux of the volatile compounds in the soil.

The mechanisms that the root follows are:

Rising of soil permeability
Bulk motion of water toward the surface
Water table level reduction
Chemical transport via hydraulic redistribution
Bulk motion with gas fluxes

Among the various mentioned mechanisms, the most well-known method is the water table level reduction. In this, the plant removes underground water thereby increasing the width of vadose zone, which helps the volatile contaminants to diffuse directly through air resulting in an increase in the volatile contaminant flux instead of diffusing through the water which is a lot harder. The root activities interfere in the contaminant transport in the underground because the soil texture at macro and micro scales during growth and senescence are modified. Hydraulic redistribution (HR) moves water from the saturated areas of the soil to the other parts through which the volatile contaminants move and when they come close to the surface of the soil volatilization takes place effectively.

Gases Transportation in Plants

Once the VOCs are absorbed by the plant, various plants release them differently. They get transported through the plant parts similar to water, CO_2, O_2 transportation. In this, the transpiration rate and direct phytovolatilization fluxes are related, i.e., low direct phytovolatilization takes place during night time.

The gases reach the leaves by the process of simple diffusion. This diffusion of VOCs takes place dominantly in highly volatile species of wetland plants. In herbaceous plants, VOCs escape through the primary tissues of the plant such as from stems and leaves. The VOCs transport in the epidermis with the help of two pathways: diffusion through the stomata and cuticle. Whereas, in case of woody species, direct phytovolatilization from tree trunks and branches involve the diffusion of VOCs through the secondary xylem, secondary phloem and periderm, thereby to the atmosphere (Limmer and Burken 2016).

Study and experiment on phytovolatilization of arsenic from As-contaminated soil using a particular species of plant named *Pteris vittate* was conducted by Masayuki Sakakibara and his fellow mates. Arsenic is one of the great environmental concerns because of its extensive contamination and toxicity which can cause great harm when it is present in the soil. Hence, by using the phytovolatilization for As-contaminated soil, they have given results of a decrease in the arsenic level in the soil (Sakakibara et al. 2010).

2.6.3 Phytodegradation

This technique of phytoremediation is a little different from other techniques. In phytoremediation, the toxic contaminants are broken down into simpler, less toxic forms. Many contaminants like chlorinated solvents, pesticides, and other organic compounds, and various other inorganic compounds are being degraded by phytodegradation.

The breaking of the toxic contaminants takes place through two pathways:

a) with the help of enzymes like dehalogenase and oxygenase in the plant.
b) with the help of the metabolic processes.

It is not dependent on the microorganisms that are present in the rhizosphere region. Plants accumulate organic xenobiotics like PCBs, PAHs, trinitrotoluene, etc., from the polluted site and degrade them through their metabolic activities. Some studies suggest that class 4 types of plants have better phytodegrading abilities. The factors affecting phytodegradation process include:

a) Concentration of contaminants present in the soil—If the contaminants' concentration is comparatively higher in the upper surface of the soil, the efficiency of the process is greater.
b) Uptake efficiency of pollutants—Depends on the phytochemical properties of the plant.

Similar studies suggest that it is possible and viable to use GM plants such as transgenic poplars for phytodegradation (Ali et al. 2013).

2.6.4 Phytoextraction

It is one of the phytoremediation techniques that incorporate plants to remove the heavy metals like cadmium (Cd) from the soil. It is also called phytoaccumulation or phytomining and considered as a more advanced form of phytoremediation. It is a classical method for pollutant removal from the soil and hence, cause no adverse effects on the properties of the soil. The contaminant from the soil is absorbed by the roots of the plant and is translocated to the shoots and accumulated there. The translocation and accumulation of the metal contaminants are pivotal biochemical process and is helpful for effective phytoextraction. The plant's shoot is then harvested and subjected to various steps like drying, ashing, and composting which helps us to recover the metals that are collected in the plant's shoot (Ali et al. 2013; Ahmadpour et al. 2012).

There have been few experiments carried out on the field and few commercial exercises stated by Robinson et al. 2006 which have given successful results of phytoextraction. A phytoextraction process is considered to be successful only when the level of contaminants in the soil of an area is lowered and detoxified to a certain extent which is under environmental rules and regulations. Various literatures have reported the potential of various species for the phytoextraction of heavy metals from the contaminated area (Ahmadpour et al. 2012). Some studies show experiments that are based on hydroponics, which have given an insight on the metal uptake by hyperaccumulators, accumulators, and tolerant plants. These stats help in determining the uptake efficiency and metal tolerance of probable phytoremediation species (Marchiol et al. 2004).

For a proper and successful implementation of phytoextraction as a technology for cleansing the environment, it depends on factors such as the metal availability for uptake and the plant species that has the capability to absorb and accumulate the metals in its shoot system.

The traits and characteristics a plant species should possess in order to work efficiently in the phytoextraction are:

a) ability to grow outside their area of collection
b) it should be able to grow fast
c) should have high biomass
d) should be easily harvested
e) the harvestable parts should be capable of accumulating a wide range of heavy metals

Even though these are the criteria for a plant to carry out phytoextraction, there is no such plant available that satisfies all these criteria. Although, we could use a non-accumulator plant that grows rapidly and modify genetically so that it can achieve almost all the mentioned criteria (Bhargava et al. 2012). Phytoextraction adopts two approaches:

i. Natural or continuous phytoextraction—utilizes natural metal hyperaccumulating plants. The plants used here have a natural tendency to isolate and extract metals in higher amounts from the soil and have systematic mechanism to translocate these metals to the shoot and accumulate them. These plants can tolerate high metal concentration but are slow growers and produces less biomass yield.
ii. Chemically induced phytoextraction—metal hyperaccumulation in plants is activated with the help of chemical amendments. In this, the metal up taking by roots, its translocation, is carried out with the help of chemical amendments that are present in the soil. These plants have low tolerance levels toward high metal concentration and are fast-growing plants with high yields of biomass (Mahmood 2010).

A field experiment carried out by Zhuang et al. 2007 wherein, they used eight different species of plants which is suitable to extract specific heavy metals brought a result that the phytoremediation potential of certain plant species depends on three important factors—biomass of plant, bioconcentration factor, and soil mass. And it also provided data regarding the amount of biomass production of hyperaccumulators was less than the non-accumulating plants. Bioconcentration factor is an important feature that a plant should have for phytoremediation, i.e., the metal uptake, their mobilization through plant tissues, and their accumulating ability.

2.6.5 Chelate-Enhanced Phytoremediation

Chelates are a class of coordination or complex compounds that is made of a central atom to which ligands are bonded, this technique is mainly implemented in the phytoextraction. Chelate-enhanced phytoextraction is used to overcome the limitations of the process like low metal solubility, bioavailability, and low metal translocation from root to shoot. Various chelating agents are tested and used according to their effectiveness in the process of enhancement in the phytoextraction of soil metals (Song et al. 2005). Examples of various chelating agents used in the phytoremediation are given as Ethylene Diamine Tetra Acetic acid (EDTA), N- Hydroxyethyl Ethylene Diamine Tetra Acetic acid (HEDTA), Ethylene Diamine Di-Succinate (EDDS), Nitrilotriacetic acid (NTA) and Diethylene Triamino Penta Acetic acid (DTPAA).

Even though the hyperaccumulator plants have substantial metal-accumulating capacity when compared to non-hyperaccumulator plants, they are still not able to acquire the metals completely from the soil. From recent studies, it has come to know that this is due to the restrictions on the metal bioavailability in the contaminated soil. Therefore, in these cases, the application or the implementation of the synthetic chelates and organic acids (e.g., citric acid, malic acid, oxalic acid) have been put to use to overcome the situation. Generally, organic acids are less effective than synthetic chelates in this process, among the acids. Citric acid has been proven to be much efficient in inducing plant hyperaccumulation of uranium.

All the four major steps involved in this technique—metal desorption from the soil, transport of the metals in the root through diffusion, translocation of metals to the shoot—will be affected by the addition of the chelating agents (Song et al. 2005).

Chelating Agents' Effects on Metals in Soil

The chelators solubilize the metals in the soil solution and mobilize them from the solid phase by forming soluble complexes (Alkorta et al. 2004). This happens by ligand exchange reaction and remobilizes the adsorbed metal from the solid phase. These effects of chelating agents are different for different metals and soils.

Factors that affect solubilizations are:

a) Type of metal and its distribution in the soil
b) Metal: ligand ratio
c) Metal–ligand complexes formation constant
d) Soil pH
e) Presence of competing cations
f) Free and complex metals adsorption onto the soil particles.

Example: the usage of chelating agents such as EDTA when added in a considerable amount to the soil proved to solubilize $PbCO_3$ effectively in the soil.

Up Taking of Chelated Metal Complexes by Plants

After the metal gets solubilized by the chelates, the metal–ligand complexes are absorbed by the roots of the plant and transported through the roots and to the shoots with the help of xylem. Recent studies proved that usage of both apoplastic and symplastic transport pathways are being used by the plant tissues to translocate the metals to the shoots. The two pathways depend on the plant species and the chelate used.

Example: Pb chelated with H-EDTA makes use of apoplastic pathway whereas Pb chelated with EDTA uses a symplastic pathway.

Factors that Determine Accumulation of Chelates-Induced Metal in Plants

- High concentrations of chelates can physiologically damage the root membranes and can affect the uptake of the complexes.
- Synthetic chelates can disrupt the cell membranes and its functions.
- Can lead to phytotoxicity and probably cause risks in the environment.

2.6.6 Rhizodegradation

Rhizodegradation is the process in which the organic pollutants which are present in the soil are broken down by the microorganisms that are present in the rhizosphere region soil. The contaminants get easily and actively get degraded in rhizosphere region because, there is an increase in the metabolic activities of microbes. The rhizosphere region is about 1 mm around the roots and the plant has direct influence in it. Plants secrete exudates which consist of carbohydrates, amino acids, etc. The carbon sources for the microbes are sugars and organic acids and the other nutrient-containing exudates acts as the nitrogen sources. Another factor that helps in increase in the microbial activity is the loosening of the soil and transport of water in the rhizosphere by the plant root. This technique is also called enhanced rhizosphere biodegradation, phytostimulation (Maheswari and Rajeswari 2016; Jeevanantham et al. 2019; Ramin and Ramamurthy 2012).

Advantages and Disadvantages of Rhizodegradation

Advantages of rhizodegradation are.

- Complete mineralization of the contaminant takes place.
- Low expenditure for initiation and maintenance.

Disadvantages of rhizodegradation are

- Long time period required for broad rhizosphere formation.

- If the microbes utilize the tissues from the plant rather than the contaminant for the source, it can diminish the degradation of the contaminant.

2.6.7 Rhizofiltration

Rhizofiltration is the process that uses both terrestrial and aquatic plants to absorb, concentrate the contaminants from the soil and water. It is also called as phytofiltration. The plants used here have longer and fibrous root system, which in turn helps in efficient removal of contaminants. It is present in all the other methods of phytoremediation as it is the main mechanism that operates phytoremediation. It has the capacity to remove a wide range of contaminants both organic and inorganic. Rhizofiltration reduces the mobility of the contaminant and blocks the migration of the contaminants to the ground water thereby reducing the bioavailability.

Points to remember before starting rhizofiltration:

- Selecting the proper and suitable species of plant. Hydraulic detention time and sorption by the plant roots must be considered for successful rhizofiltration.
- The type, range, and the level of metal contaminants that is present in the soil.
- Depth of the contaminants that is present in the soil.

Principle of Rhizofiltration

Rhizofiltration depends on the physical and biochemical factors of the plant root. The roots must be efficient in the synthesis of various important chemicals which will help in transporting the heavy metals upwards in the plant. The precipitation of the metals to the root surface occurs where the root exudates, pH, and other changes in the rhizosphere assist the process. Once the plant gets saturated with the metal contaminant it is harvested and disposed. The tolerance of the plant to hold the metal is helped by glutathione and other organic acid metabolism.

Process of Rhizofiltration

The plants are initially grown hydroponically, i.e., they are grown in greenhouses with their roots immersed in water not in soil, and then planting them in metal-polluted water once their roots are long and well developed. As the plants are introduced into the contaminated area the plants absorb and concentrate the metals in the roots and later to shoots. The root system of the plant provides a good amount of surface area that can absorb and accumulate nutrients, water, and other materials. Later the plants are harvested once they become saturated with the metal.

Importance of Rhizofiltration

This technique is applied in the treatment of the water that is present on the surface and in the ground, industrial and residential effluents, acid mine drainages, radionuclide-contaminated solutions, etc. The plants are capable of removing metals like copper, cadmium, nickel, lead, and zinc from soil and aqueous solutions. The radionuclides are also removed at a very small level from the streams. It is also a cost-effective technology and uses less labor cost or operational or maintenance costs. It also does not produce any secondary waste.

Constraints in Rhizofiltration

- Plants have to be grown separately in a greenhouse before planting it in a rhizofiltration system.
- A properly engineered system is necessary to control influent concentration and flow rate.
- Periodic harvesting and plant and its disposal is required.
- pH of the influent should be monitored continuously for optimal metal uptake

3 Conclusion

Heavy metals and radionuclides are one of the most crucial threats to the soil thereby to the land and the water which later does affect the human health. These contaminants released by various human activities like mining, industrial emissions, smelting of ores, and usage of various pesticides and fertilizers are the main causative of the contamination. Phytoremediation one such technology is becoming very important in the bioremediation of metal and radionuclide contaminants from the soil. The usage of hyper-accumulators to accumulate toxic metals from the soil is proving to be helpful and does no destructive effect to the soil. Still, there are researches that need to be done to make use of these plants more efficiently. Since there are reports of the loss of the agricultural soil resources from the metal contamination, if this technology is put to use there is a great chance of saving the agricultural lands. Researchers are still trying to figure out various other plants that can help in this process.

References

Adama M, Esena R, Fosu-Mensah B, Yirenya-Tawiah D (2016) Heavy metal contamination of soils around a hospital waste incinerator bottom ash dumps site. J Environ Public Health 1–6

Ahmadpour P, Ahmadpour F, Mahmud TMM, Abdu A, Soleimani M, Tayefeh FH (2012) Phytoremediation of heavy metals: a green technology. Afr J Biotechnol 11:14036–14043

Ali H, Khan E, Sajad MA (2013) Phytoremediation of heavy metals—concepts and applications. Chemosphere 91:869–881

Alkorta I, Hernandez-Allica J, Becerril J, Amezaga I, Albizu I, Onaindia M, Garbisu C (2004) Chelate-enhanced phytoremediation of soils polluted with heavy metals. Rev Environ Sci Biotechnol 3:55–70

Alvarenga P, Gonçalves AP, Fernandes RM, de Varennes A, Vallini G, Duarte E, Cunha-Queda AC (2009) Organic residues as immobilizing agents in aided phytostabilization: (I) Effects on soil chemical characteristics. Chemosphere 74:1292–1300

Azevedo R, Rodriguez E (2012) Phytotoxicity of mercury in plants: a review. J Bot 1–6

Bergkvist P, Jarvis N, Berggren D, Carlgren K (2003) Long-term effects of sewage sludge applications on soil properties, cadmium availability and distribution in arable soil. Agric Ecosyst Environ 97:167–179

Bhargava A, Carmona FF, Bhargava M, Srivastava S (2012) Approaches for enhanced phytoextraction of heavy metals. J Environ Manage 105:103–120

Bolan NS, Park JH, Robinson B, Naidu R, Huh KY (2011) Phytostabilization: a green approach to contaminant containment In: Sparks DL (ed) Advances in agronomy, Academic Press

Bouazizi H, Jouili H, Geitmann A, Ferjani EE (2010) Copper toxicity in expanding leaves of Phaseolus vulgaris L.: antioxidant enzyme response and nutrient element uptake. Ecotoxicol Environ Saf 73:1304–1308

Chowdhury A, Pradhan S, Saha M, Sanyal N (2008) Impact of pesticides on soil microbiological parameters and possible bioremediation strategies. Indian J Microbiol 48:114–127

Colzi I, Arnetoli M, Gallo A, Doumett S, Bubba MD, Pignattelli S, Gabbrielli R, Gonnelli C (2012) Copper tolerance strategies involving the root cell wall pectins in Silene paradoxa L. Environ Exp Bot 78:91–98

Denys S, Rollin C, Guillot F, Baroudi H (2006) In-situ phytoremediation of PAHs contaminated soils following a bioremediation treatment. Water Air Soil Pollut 6:299–315

Effron D, de la Horra AM, Defrieri RL, Fontanive V, Palma RM (2004) Effect of cadmium, copper, and lead on different enzyme activities in a native forest soil. Commun Soil Sci Plant Anal 35:1309–1321

Fernandes JC, Henriques FS (1991) Biochemical, physiological, and structural effects of excess copper in plants. Bot Rev 57:246–273

Ghosh M, Singh SP (2005) A review on phytoremediation of heavy metals and utilization of its by-products. Appl Ecol Environ Res 3:1–18

Harada M (ed) (1982) Minamata disease. In: Jelliffe EFP, Jelliffe DB (ed) Adverse effects of foods, Springer, Boston

Hellal J, Vallaeys T, Garnier-Zarli E, Bousserrhine N (2009) Effects of mercury on soil microbial communities in tropical soils of French Guyana. Appl Soil Ecol 41:59–68

Hershfinkel M, Sekler SWF, I, (2007) The zinc sensing receptor, a link between zinc and cell signaling. Mol Med 13:331–336

Higueras P, Fernandez Martinez R, Esbri JM, Rucandio I, Loredo J, Ordonez, A, Alvarez R (2014) Mercury soil pollution in Spain: a review. In: Jimenez E, Cabanas B, Lefebvre G (ed) The handbook of environmental chemistry, Springer, Cham

Islam A, Ahmed T, Awual MR, Rahman A, Sultana M, Aziz AA, Monir MU, Teo SH, Hasan M (2020) Advances in sustainable approaches to recover metals from e-waste-A review. J Clean Prod 244:118815

Jeevanantham S, Saravanan A, Hemavathy RV, Kumar PS, Yaashikaa PR, Yuvaraj D (2019) Removal of toxic pollutants from water environment by phytoremediation: a survey on application and future prospects. Environ Technol Innov 13:264–276

Karaca A, Naseby DC, Lynch JM (2002) Effect of cadmium contamination with sewage sludge and phosphate fertilizer amendments on soil enzyme activities, microbial structure and available cadmium. Biol Fertil Soils 35:428–434

Kasassi A, Rakimbei P, Karagiannidis A, Zabaniotou A, Tsiouvaras K, Nastis A, Tzafeiropoulou K (2008) Soil contamination by heavy metals: measurements from a closed unlined landfill. Bioresour Technol 99:8578–8584

Kirkham MB, Corey JC (1977) Pollen as indicator of radionuclide pollution. J Nucl Agric Biol 6:71–74

Knox AS, Gamerdinger AP, Adriano DC, Kolka RK, Kaplan DI (1999) Sources and practices contributing to soil contamination. In: Adriano JM, Bollag WT, Frankenberger Jr, Sims RC (ed) Bioremediation of contaiminated soils, 37th edn. Ch. 4

Kushwaha A, Rani R, Kumar S, Gautam A (2016) Heavy metal detoxification and tolerance mechanisms in plants: implications for phytoremediation. Environ Rev 24:39–51

Limmer M, Burken J (2016) Phytovolatilization of organic contaminants. Environ Sci Technol 50:6632–6643

Maheswari UK, Rajeswari K (2016) Toxicity of heavy metals-phytoremediation techniques. In: Shankar KS, Kumar RN, Pushpanjali, Nagasree K, Nirmala G, Raju NS (ed) Reshaping agriculture and nutrition linkages for food and nutrition security, ICAR—Central Research Institute for Dryland Agriculture, Hyderabad, India

Mahmood T (2010) Phytoextraction of heavy metals-the process and scope for remediation of contaminated soils. Plant Soil Environ 29:91–109

Marchiol L, Assolari S, Sacco P, Zerbi G (2004) Phytoextraction of heavy metals by canola (Brassica napus) and radish (Raphanus sativus) grown on multicontaminated soil. Environ Pollut 132:21–27

Mohanty M, Pattnaik MM, Mishra AK, Patra HK (2011) Bio-concentration of chromium—an in situ phytoremediation study at South Kaliapani chromite mining area of Orissa, India. Environ Monit Assess 184:1015–1024

Muthusaravanan S, Sivarajasekar N, Vivek JS, Paramasivan T, Mu Naushad, J Prakashmaran, Gayathiri V (2018) Suenhancements. Environ Chem Lett 16:1339–1359

Nagajyoti PC, Lee KD, Sreekanth TVM (2010) Heavy metals, occurrence and toxicity for plants: a review. Environ Chem Lett 8:199–216

Nas FS, Ali M (2018) The effect of lead on plants in terms of growing and biochemical parameters: a review. Ecol Environ Sci 3:265–268

Pant D, Joshi D, Upreti MK, Kotnala RK (2012) Chemical and biological extraction of metals present in E waste: a hybrid technology. Waste Manag 32:979–990

Patra M, Sharma A (2000) Mercury toxicity in plants. Bot Rev 66:379–422

Ramin M, Ramamurthy AS (2012) Effects of surfactants on rhizodegradation of oil in a contaminated soil. J Environ Sci Health Part A 47:1486–1490

Robinson BH, Schulin R, Nowack B, Roulier S, Menon M, Clothier BE, Green S, Mills T (2006) Phytoremediation for the management of metal flux in contaminated sites. Forest Snow Landsc Res 80:221–224

Rogiers T, Merroun ML, Williamson A, Leys N, Houdt RV, Boon N, Mijnendonckx K (2021) Cupriavidus metallidurans NA4 actively forms polyhydroxybutyrate—associated uranium-phosphate precipitates. J Hazard Mater 421:126737

Sakakibara, M, Watanabe A, Inoue M, Sano S, Kaise T (2010) Phytoextraction and phytovolatilization of arsenic from as-contaminated soils by Pteris vittata. In: Proceedings of the Annual International Conference on Soils, Sediments, Water and Energy, Oct 2006, pp 16–19

Seregin IV, Kozhevnikova AD (2006) Physiological role of nickel and its toxic effects on higher plants. Russ J Plant Physiol 53:257–277

Shah Maulin P (2020) Microbial bioremediation & biodegradation. Springer

Shackira AM, Puthur JT (2019) Phytostabilization of heavy metals: understanding of principles and practices. In: Srivastava S, Srivastava A, Suprasanna P (ed) Plant-metal interactions. Springer, Cham

Shmaefsky BR (2020) Principles of phytoremediation. In: Shmaefsky BR (ed) Phytoremediation In-situ applications, concepts and strategies in plant sciences, 1st edn. Springer, Switzerland

Shah Maulin P (2021) Removal of refractory pollutants from wastewater treatment plants. CRC Press

Song J, Luo YM, Wu LH (2005) Chelate-enhanced phytoremediation of heavy metal contaminated soil. In: Van Briesen JM (ed) Nowack B. Biogeochemistry of chelating agents, Am Chem Soc, pp 366–382

Talerko M, Kovalets I, Lev TD, Igarashi Y, Romanenko O (2021) Simulation study of radionuclide atmospheric transport after wild land fires in the Chernobyl exclusion zone in April 2020. Atmos Pollut Res Published Online

Thomas AL, David AN, Ronald LC (2004) Bioremediation of soils contaminated with explosives. J Environ Manage 70(4):291–307

Tran TA, Popova LP (2013) Functions and toxicity of cadmium in plants: recent advances and future prospects. Turk J Bot 37:1–13

Utmazian MNDS, Wieshammer G, Vega R, Wenzel WW (2007) Hydroponic screening for metal resistance and accumulation of cadmium and zinc in twenty clones of willows and poplars. Environ Pollut 148:155–165

Vassilev A, Schwitzguebel JP, Thewys T, van der Lelie D, Vangronsveld J (2004) The use of plants for remediation of metal-contaminated soils. Sci World J 4:9–34

Walling DE (1998) Use of ^{137}Cs and other fallout radionuclide in soil erosion investigations: progress, problems and prospects. ISSN 1011–4289 International Atomic Energy Agency (IAEA).Vienna, Austria

Wang C, Ji J, Yang Z, Chen L, Browne P, Yu R (2012) Effects of soil properties on the transfer of cadmium from soil to wheat in the Yangtze River delta region, China-a typical industry-agriculture transition area. Biol Trace Elem Res 148:264–274

Wang DY, Qing CL, Guo TY, Guo YJ (1997) Effects of humic acid on transport and transformation of mercury in soil-plant systems. Water Air Soil Pollut 95:35–43

Williams SE, Wollum AG (1981) Effect of cadmium on soil bacteria and actinomycetes. J Environ Qual 10:142

Zhu YZ, Shaw G (2000) Soil contamination with radionuclides and potential remediation. Chemosphere 41:121–128

Zhuang P, Yang QW, Wang HB, Shu WS (2007) Phytoextraction of heavy metals by eight plant species in the field. Water Air Soil Pollut 184:235–242

Phyto- & Microbial- Remediation of Radioactive Waste

Raksha Anand, Lalit Mohan, and Navneeta Bharadvaja

1 Introduction

The various terrestrial activities pitch in to the heavy metal and radionuclide composition of the soil. However, some of the anthropogenic factors like industrialization, mining, e-wastes, and nuclear activities are extensively contributing to its upsurge. Most of the polluted water deposits contaminants to the surface layer of the soil, which is readily being absorbed by the roots. These contaminations find their way from polluted water and soil into the food chains, thereby getting bio-accumulated and posing serious health hazards. This not only affects the ecosystem but also the physiological development of plants as well as other consumers (Shehata et al. 2019).

With the accelerated international competition regarding nuclear power and mineral legacy through mining, the overall index of radioactive waste has increased. There are operational industries with unreported leakages, untraced escape of detrimental radionuclides into water and soil. The waste from mining, the ore-wash-offs, nuclear testing, energy-generating operations, and inadvertent release are some of the general causes contributing toward radionuclide contamination (Roh et al. 2015). The other extreme contributors to radio-contamination are the episodic disasters at atomic reactors as witnessed in 1986 and 2011 at Chernobyl Nuclear Power Plant and Fukushima Daiichi nuclear power plant respectively (Prakash et al. 2013a). Today, we not only have to treat the already existing radio waste but also must research new ways to safe disposal of the forthcoming waste. These strategies must be effective and safe to both the environment as well as handlers.

Due to a longer half-life, radionuclides have an extended impact on the environment. This is the reason behind radio waste being a threat to the environment and lives. Radionuclides sometimes are precursors, contributing to the emergence

R. Anand · L. Mohan · N. Bharadvaja (✉)
Department of Biotechnology, Delhi Technological University, Delhi, India
e-mail: navneetab@dce.ac.in

M. P. Shah (ed.), *Microbial Technologies in Industrial Wastewater Treatment*,
https://doi.org/10.1007/978-981-99-2435-6_12

of more imminent species, for example, Uranium-238 (half-life of 4.5 billion years) decays to form Radium-226 with a half-life of 1600 years(Prakash et al. 2013a).

The waste coming from the nuclear reactors contains chelating agents which are used for decontaminating and cleaning-up reactors and equipment. These chelators, for example; EDTA, NTA, HEDTA, oxalic acid, tartaric acid, and so on form complexes with radionuclides and help in separating them. However, being prone to biodegradation, these chelates happen to precipitate ions as insoluble oxides and hydroxides. This hampers their flushing into the groundwater (Prakash et al. 2013a). One of the most efficient chelators ever considered is "citrate" due to its complexing with metal ions to form soluble metal citrates (Francis 1998).

2 Radionuclide Contamination—Sources and Hazards

The major contributor of radio waste is the "nuclear fuel cycle (NFCs)" which comprises the predominant operations, reprocessing, manufacture, decommissioning, recycling, and management of waste (Lloyd and Renshaw 2005). The NFCs associate with processes involved in electricity production through nuclear reactions. Even if nuclear testing is the foremost radio-pollution contributor, fuel reprocessing leads to higher contamination locally (Roh et al. 2015). Such areas are generally contaminated with radionuclides like Uranium-238, Neptunium-237, Uranium-235, Plutonium-239, Americium-241, Radium-226, Cesium-137, Technitium-99, and Strontium-90 (United States Nuclear Regulatory Commission 2021).

Based on the level of emission, the waste from the NFCs (shown in Fig. 1) is categorized into three levels; low-level wastes generated at all the stages of the cycle, medium-level wastes produced at reprocessing and operation of the cycle, and the longer half-life elements contributing to high-level of waste. The latter also comprises exceptionally reactive products that can undergo nuclear fission, and are generated during reprocessing (World Nuclear Association 2021).

The extent of exposure to these radiations is generally the deciding factor of the severity of the pathological conditions. Acute health effects like vomiting and nausea can worsen on prolonged exposure leading to hair loss, disorientation, low BP, mutations that might contribute to cancer, and maybe death. The in-utero developing embryo is extremely susceptible to radiations that might interfere with cellular developments (Bogutskaya et al. 2011). Different types of radionuclides and the associated health hazards are enlisted in Table 1.

3 Possible Remediation of Radionuclides

The traditional concept stated that radioactive waste since produced in smaller quantities needed not to be treated immediately and thus was stored, which ultimately reduced radiation due to degradation. Initially, these wastes were embedded 500 m

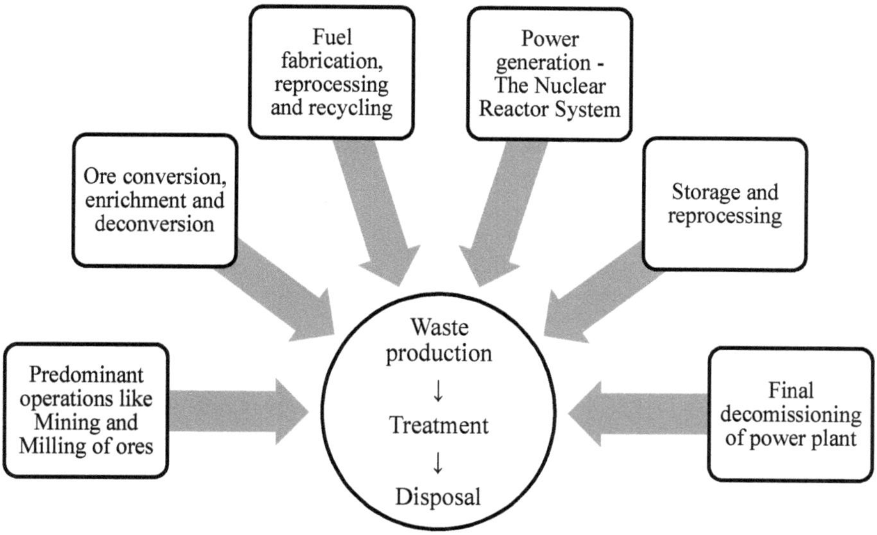

Fig. 1 Schematic representation of different phases of nuclear fuel cycle

Table 1 List of some radionuclides and the potential hazards they cause

Radionuclide	Characteristics & Hazards	Reference
Uranium (U)	α emitter Generally, does not penetrate the skin barrier, but upon ingestion causes kidney, liver, or bone cancer. Inhalation might cause lung cancer	(National Center for Environmental Health (NCEH), Emergency Management, Radiation 2018)
Neptunium (Np)	Rare but life-threatening exposures. Might cause hematopoietic complications and cancer depending upon exposure limits	(National Library of Medicine)
Plutonium (Pu)	e^-, β particles and γ emitter Not readily absorbed in the gastrointestinal (GI) tract. Can be inhaled and might travel to bones and liver through blood or lymph eventually causing cancer. Causes genetic alterations	(United States Nuclear Regulatory Commission 2020)
Technetium (Tc)	β particle emitter Has the capacity to get accumulated in thyroid glands and GI tract. Sometimes used in medical analysis, but is easily excreted from the body	(United States Environmental Protection Agency 2021a)
Americium (Am)	α particles and γ radiations If swallowed or inhaled, gets concentrated in muscles, bones, and liver and persists in the body for a very long time. Is carcinogenic	(United States Environmental Protection Agency 2021b)
Cesium (Cs)	β particles and γ rays Can cause "acute radiation sickness" and even death. Can burn soft and muscle tissues, but also leads to cancer	(United States Environmental Protection Agency 2021c)

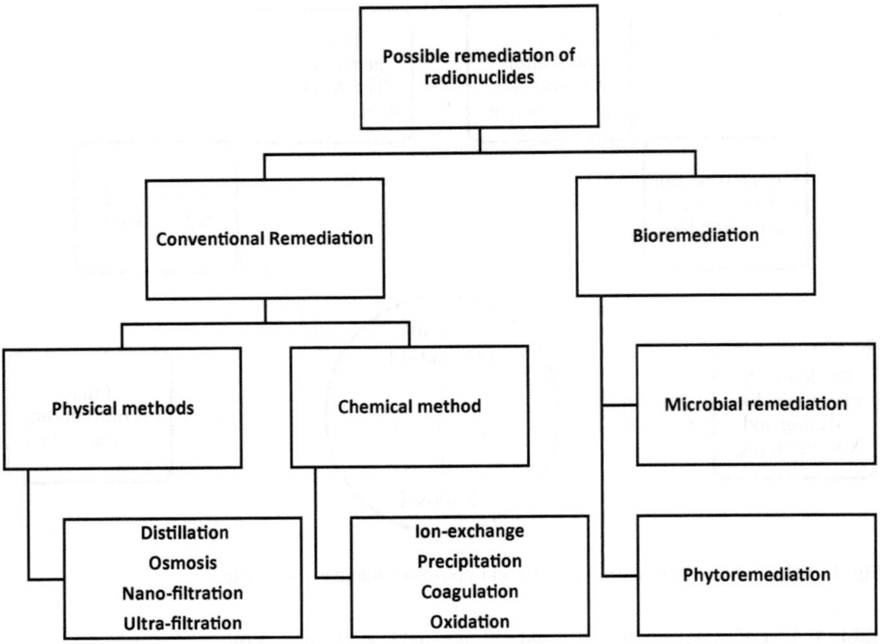

Fig. 2 Different remediation strategies adopted for the elimination of radionuclides

down into the land, for time-called recovery as these can also serve as a remarkable source of energy (World Nuclear Association 2021). Various methodologies adopted for the elimination of radionuclides from environment are depicted in Fig. 2.

4 Bioremediation of Radionuclides

Bioremediation is the superintended process of cleaning up contaminations from the environment through the employment of living organisms, for example, microbes and plants. There are different ways to do it and is considered a cost-effective way of treating pollutants, on a modest technology basis, as compared to other remediation strategies (Lloyd and Renshaw 2005). The exorbitant physio-chemical remediation approaches have made the interest in microbes and plants intense (Kumar et al. 2007).

4.1 Microbial Remediation

This technique engages the microbial metabolic actions to clean up the soil and water radionuclide contamination. For example, bacterial species *Shewanella*, *Geobacter*, *Kineococcus*- and *Deincoccus* effectively work on radioactive waste by affecting the solubility, vigor, and bioavailability of radionuclides (Chung et al. 2010; Shah MP. 2020). The conventional excavation of contaminated soil is now replaced with microbial remediation by *Desulfuromusa ferrireducens*(Green et al. 2012). The general mechanism of intramural microbes is either immobilizing the target radionuclide or enhancing its removal. The site-specific interactions of microorganisms with radionuclide involve redox reactions which biotransform the waste material making it soluble and thus help in flushing out easily (Amachi et al. 2010). The general mechanism adopted by microorganisms for remediation of radionuclides is represented in Fig. 3.

The biological alteration in the oxidation state of radionuclides is the preliminary step in bioremediation. The solubility and transportability of toxic mercury are altered upon methylation to $[Hg(CH_3)_2]$ which is also less toxic (Wang et al. 2012). There are several mechanisms through which microorganisms metabolize radionuclides to contribute to bioremediation (Hegazy and Emam 2011), depending on various factors including availability of nutrients, environmental parameters, electron donors, and electron acceptors. Some of these strategies are:

- Altering pH or Eh

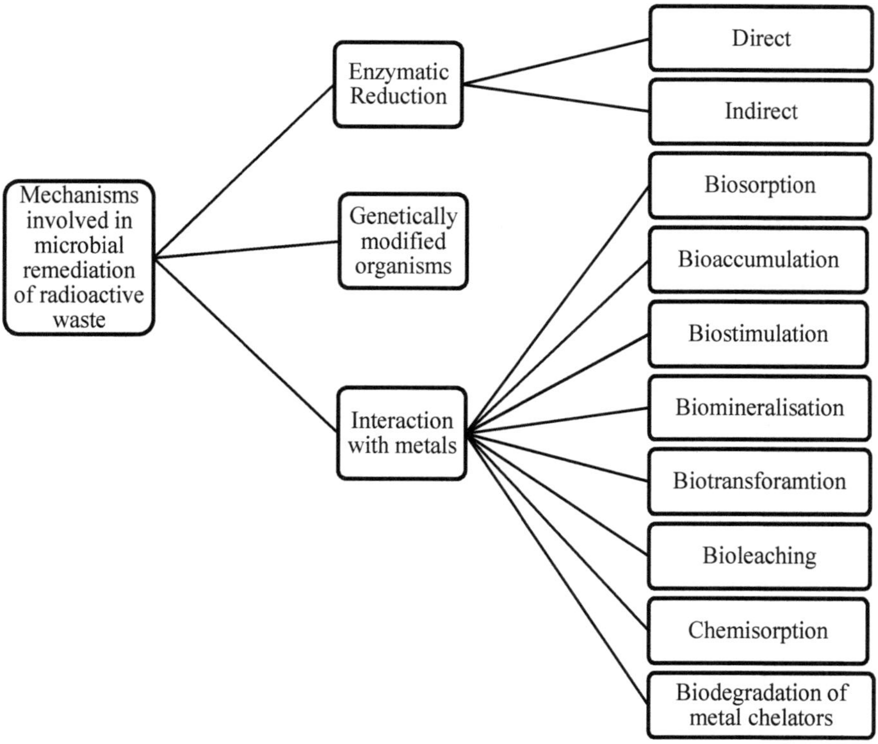

Fig. 3 The general mechanism involved in microbial remediation of waste along with the key interactions of microbes and their enzymes with metals and waste components, linking them to bioremediation

Table 2 Enlisted are some enzymes with their microbes that function in bioremediation of radio-waste

Microbial Species	Protein/Enzyme involved	Function	Reference
Shewanella putrefaciens	Cytochrome c in the periplasm of microbe	Reduces uranyl ion U(VI)	(Prakash et al. 2013a)
Desulfovibrio vulgaris	Cytochrome c + hydrogenase	Reduces U(VI) with hydrogenase as the e^- giver	(Lovley and Phillips 1994)
D. desulfuricans (G20)	Cytochrome c	Hydrogen-subjected reduction of U(VI)	(Prakash et al. 2013b)
Geobacter sulfurreducens	PpcA (a cytochrome homolog of periplasm)	Reduces Fe(III) and U(VI)	(Lloyd et al. 2003)
E. coli	Tc(VII) reductase and tri-component of FHL[1] (needs formate as hydrogen donor)	Reduction of Tc(VII) by electron transfer from dihydrogen	(Lloyd et al. 1997)
Desulfovibrio fructosovorans	Ni–Fe hydrogenase in periplasm	Reduces Tc(VII)	(De Luca et al. 2001)

- Redox reactions
- Surface accumulation or sorption
- Biodegradation of biomolecule-radionuclide complexes

4.1.1 Mechanisms for Microbial Remediation

Direct Enzymatic Modes

The reduced form of radionuclides is insoluble and hence found as precipitates. The radionuclides when converted to their oxidized states become soluble and hence get easily washed off. Some mechanisms of microbial remediation are mentioned below in Table 2.

The otherwise insoluble radionuclide Tc(VII) has a low capability of complexing with ligand and thus makes conventional remediation tedious. Thus, microbes like *S. putrefaciens*, *G. sulfurreducens*, and *G. metallireducens* reduce metals and precipitate the radionuclide, lowering its oxidative status to Tc(IV), the final insoluble state (Wildung et al. 2000). There are other microorganism with highly active FHL like *Desulfovibrio desulfuricans,* that can utilize electron from formate to reduce Tc(VII) (Prakash et al. 2013a). Several sulfur-reducing bacteria like *D. desulfuricans* when immobilized can be used to treat Tc(VII) available in low concentration against highly concentrated Nickel (Lloyd et al. 1999). In 2006, *Halomonas* (Tc-202), a

[1] FHL = formatehydrogenlyase.

novel marine strain was isolated, that could aerobically eliminate Tc(VII) from both solid and fluid material (Fujimoto and Morita 2006).

The other radionuclides present on-site include soft actinide metals namely; Thorium, Plutonium, Americium, and so on, which also have to be treated apart from Technetium and Uranium. These tetra-valent actinides, which can be easily immobilized in organic matrixes, have high ligand-binding capabilities and hence are obsequious to bioremediation (Prakash et al. 2013a). Some microorganisms that enzymatically reduce these radionuclides include *Geobacter sp.* and other iron-reducing microbes like *R. ferrireducens* (Kim et al. 2012; Shah MP., 2021). The one-step precipitation of actinides by iron-(III)-reducing microbes was otherwise contradicted by a scientific study in which cations are produced that can be bioprecipitated. For example, *S. putrefaciens* converts lead from (V) to (IV); which later is liberated by *Citrobacter sp.*, biomineralizes the cation by phosphorylation (Prakash et al. 2013a).

Indirect Enzymatic Modes

The metal- or sulfate-reducing microbes are capable of indirectly reducing the subterranean radionuclides. There is a coupled oxidation–reduction ongoing of hydrogen or organic compounds and iron (III ➔ II) or sulfur (IV ➔ II) respectively. The sulfur might form hydrogen sulfide by sulfur-reducing bacteria. These products are capable of undergoing subsequent chemical reductions that contribute to the production of other insoluble compounds (Van Hullebusch et al. 2005) which can be precipitated in the form of oxides or hydroxides.

Some microbes like *Microbacterium flavescens* can grow in the presence of radionuclides like uranium and americium, producing unidentified organic compounds and extracellular metabolites that might help with dissolving, transporting, and mobilizing the radionuclides from the soil into the cell (John et al. 2001).

Studies have identified enzymes produced by microbes extracellularly, in the presence of metals like Plutonium and Thorium. These enzymes in turn have increased the elutes of Pu and Th, forming complexes, from the soil column as compared to controls due to enhanced mobility (Panak and Nitsche 2001). The iron-chelating siderophores are produced ubiquitously by microbes from elements like Pu, U, Th, and Fe. These elements have thus been found to share biochemical similarities of iron-sequestering, which not only increase the solubility of radionuclides but also its bioavailability. *P. aeruginosa* produces extracellular chelators capable of bioaccumulating U (Premuzic et al. 1985). This elicited recognition of other chelators that could bioaccumulate metals like Th and so on, which contribute to radionuclide contamination. Some of the other efficient actinide solubilizing siderophores are enterobactin, catecholate, and desferrioxamine ligands which were tested on plutonium oxides (Brainard et al. 1992).

Biosorption and Bioaccumulation

Metal sorption is the ultimate microbe-metal interaction. The bacterial slime and capsules have negatively charged membranes that secrete polysaccharides that are capable of sequestering the cationic metal ions through various unrecognized mechanisms. These metal cations bind to the lipopolysaccharides' O-side chains assisted by carboxyls. The gram-negative bacteria have side chains that are negatively charged which affects the hydrophobicity of cells (Langley and Beveridge 1999). *C. indica*, a marine brown alga upon Ca-pre-treatment effectively adsorbs radionuclide Uranium (Khani et al. 2006). Other radionuclide biosorbants are *Firmicutes* and *C. freudii*(Xie et al. 2008). There is an electrostatic interaction between the phosphate group of *Citrobacter* LPS and uranyl phosphate that leads to the formation of nucleation site and ultimately its precipitation (Macaskie et al. 2000). The resting bacterial cells changed the oxidation state of plutonium (VI ➔ V) through endogenous respiration and led to its sorption which was also found to be in positive correlation with temperature (Panak et al. 2002). Biosorption has been found to be faster and efficient than bioreduction. However, alone on-site biosorption is not possible because the radionuclide-contaminated sites have lower ground biomass. RDT and stimulated microbial culture against radionuclide can be an effective way of enhancing biosorption (Prakash et al. 2013a).

Biostimulation

Biostimulation engages specific microbial communities intending to reduce the bioavailability of the radionuclide contaminant on-site. For example, iron- and sulfate-reducing microbes could effectively decrease the spread of Uranium from soil to groundwater (Vrionis et al. 2005). The metal-reducing bacteria are energetically supported by the nitrate, the e^- acceptors, where nitric acid is present as the co-contaminant. This is an example of biostimulation (DiChristina 1992). However, nitrate as a co-contaminant in sediment limits the microbial reduction of U(VI) due to the presence of heavy metals (Finneran et al. 2002). As a way out, the nitrate and heavy metal had to be treated off-site, followed by in-situ reduction of U(VI) through biostimulation was performed(Wu et al. 2006). Incorporation of a competent carbon source to the co-contaminated sites has enhanced bioreducing effects on radionuclides. For example, adding ethanol as a reducing agent of nitrate at co-contaminated sites was effective and led to successfully immobilized U(VI) (Nyman et al. 2006). This can be simply understood as "bioaugmentation."

A large number of efficient radionuclides-reducing microbes like *Geobacter sp.* and *Shewanella sp.* be limited by the presence of non-reducible co-contaminating heavy metals. They have resistant effects on microbial-reducing activity thus limiting the in-situ bioremediation process. The endogenous microbes at the Oak Ridge FRC were found to contain a segment of the heavy-metal resistant gene (Martinez et al. 2006).

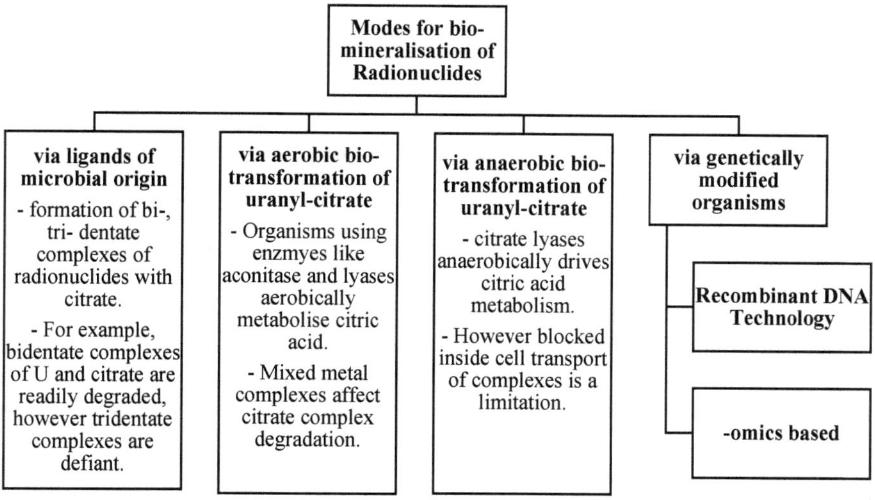

Fig. 4 Different modes of biomineralization of radionuclides

The other method of bioremediating sites contaminated with Uranium having a concentration over 1000 μm is at lowered pH and increased concentration of nitric acid. This ex-situ method involved elevation of pH with the addition of an external carbon source that stimulated the proliferation of radionuclide-reducing bacteria and denitrifying bacteria. This complex loop system was found to be effective in reducing contaminant concentrations (Wu et al. 2006). Another method introduced for the removal of high Uranium concentrations involved the ethanol-biofilm-associated-biofilm approach (Marsili et al. 2007).

Radionuclide Biomineralization

Microorganisms have varied mechanisms of interacting with metal ions, either by transforming them through general immobilization or by producing biofilms that bind to metallic ions. This precipitates the insoluble minerals. The same microorganism would interact with different metals differently, for example; *Citrobacter sp.* It enzymatically produces poly-crystallized metal deposits as phosphates inside and surrounding the cell wall. The influx-outflux of UO_4 and PO_4 respectively creates a gradient, which removes Uranium from the solution. The major enzymes involved in this mechanism are the acid-phosphatase of the outer membrane, the gene of which was also cloned in *P. aeruginosa* for similar results (Keasling et al. 2000). Most of the biomineralization mechanisms are discussed in Fig. 4.

The existence of multi-metal complex in the environment is a limitation to such transformation technologies and thus demands research (Dodge et al. 2002). Another limitation is the restricted transportation of metal complexes like uranyl-citrate, inside the cell which is assumed by labeled analyses. Further, when a cell-free isolate of the test organisms was taken to degrade these un-transported complexes, they were degraded rapidly hence showing efficacy (Francis et al. 2002).

4.1.2 Genetically Modified Microbes and Their Bioremedial Action

The genetically engineered microbe-mediated environmental remediation engages biotransformation of waste, with enhanced solubility and mobility of elements as compared to the native microbial activity. Recombinant DNA technology has been adopted as a way to introduce desired characters into microbes for enhanced surface assimilation of waste components, especially metals. Genetically engineered polypeptide constructs have been designed for accoutred metal fusion with bacterial membranes (Prakash et al. 2013a). Some of the performed microbial alterations are discussed in Table 3.

It is the microbial protein (or enzyme) interaction with waste that has to be studied for a better analysis of potential genes that have to be engineered for desired outcomes. Advanced genomics and proteomic studies can be carried out for identification of those genes and proteins to metabolize the targeted radioactive waste (Nagaraj and Singh 2010). The metabolic pathways can be selected through genome-wide analysis of transcriptome (Prakash et al. 2013a). However, there is a certain limitation to the use of GMOs in bioremediation which is further discussed.

4.1.3 Factors Affecting Microbial Remediation

Factors governing the uptake of radionuclides by microorganisms are as follows.

1. *pH and temperature for activity*—Microbial enzymes are functional at a particular pH and temperature range, deviation from which would affect the efficiency of the process.
2. *Nutrient availability*—Microbes need various micro-and macro-nutrients as essential sources for their energy and growth.
3. *Moisture content*—Water is required for maintaining an optimal osmotic pressure that supports the growth and functionality of microbial cells.
4. *Electron acceptors*—Depending on the mode of bioremediation, the need for e^- acceptors for microbes arises.

Other factors like microbial species involved, inter-and intra-species interaction as well as the environmental components also govern the efficacy of the remediation process. Sometimes, microbes require add-ons which enhance the permeability and thus facilitate the uptake of radionuclides by the cell for further metabolism(Tekere 2019).

Table 3 Some of the genetically engineered strains that have contributed toward microbial remediation of radioactive waste are described along with mode of engineering and the targeted outcome

Strain produced	Modification introduced	Outcome	Reference
Recombinant *E. coli*	Alteration in transporter gene • *merTP* of *S. marcescens* • *nixA* of *H. pylori*	5X sorption of Uranium	(Beckwith et al. 2001; Prakash et al. 2013b)
Engineered *Deinococcus radiodurans*	Cloning of *E. coli* transporter *merA* gene	Catabolizing radio waste including Hg and toluene as a C- and energy source	(Brim et al. 2000)
Engineered *D. geothermalis* (a close relative of *D. radiodurans*)	Transformation by introducing the same plasmid designed for *Deinococcus radiodurans*	The initially thermophilic, radiation-resistant bacteria with new in-situ remediation properties. It can also reduce Mercury(II) along with other contaminants	(Brim et al. 2003)
Engineered *Geobacteriaceae metallireducens*	A fumarate transporter encoding gene *dcuB* of *G. sulfurreducens* introduced	Made it possible for *G. metallireducens* to grow on fumarate. Allowed terminal electron transportation. The activity of succinate dehydrogenase and fumarate reductase was introduced	(Butler et al. 2006)

4.1.4 Advantages of Microbial Remediation Strategy

The presence of diverse microorganism communities in abundance at varied sites provides an insight into bioremediation. Application of -omics technology is easy and viable in microbial communities, with frequent quality checks. It is found that the microbes are more efficiently able to degrade pollutants when acting in the consortium, attributed to their individual metabolic activities being clubbed up, rather than as isolates (Azubuike et al. 2016). Figure 5 depicts various advantages conferred by the use of microorganisms for remediation.

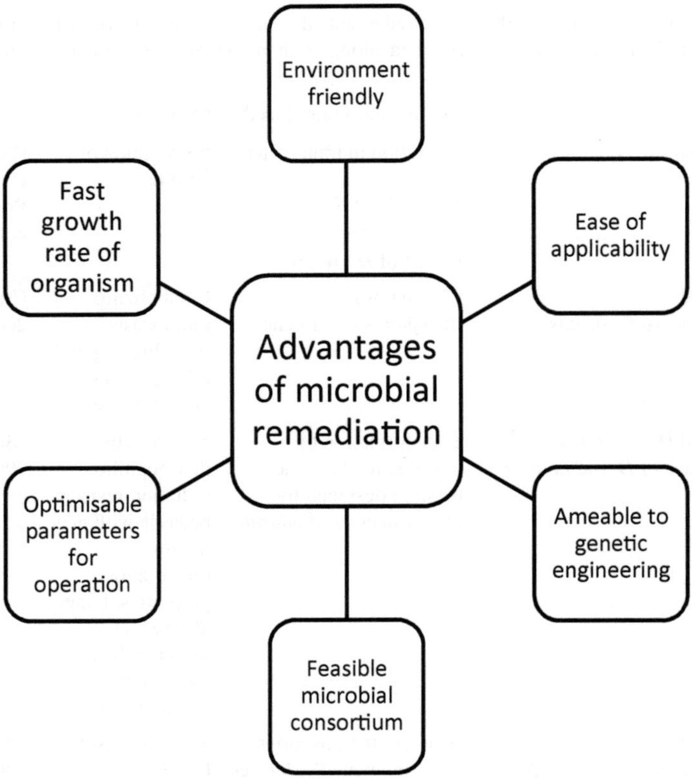

Fig. 5 Advantages of opting for microbial remediation

4.1.5 Limitations of Microbial Remediation Strategy

The unavailability of pollutant-degrading microbes or the necessity of a typical environment for the viability of such species is a limitation. However, limited levels of nutrients, bio-availability of pollutants along with co-contaminants are also deciding factors for the efficiency of such treatments. There are grounds for competitive action among the indigenous and introduced species in case a consortium is required to degrade pollutants. The other equipped techniques like immobilization are a draining process and are not always feasible. There is a need to add chemicals to the aged pollutants in order to increase its bioavailability to the remediating microbe (Azubuike et al. 2016).

Most of the GM microbes have been efficient in catabolizing radio wastes; however, strategies for in-situ remediation of soil and water still need to be developed. The companionship of toxic metals, anions, chelators, and other competing substances on-site limits bioremediation. The reversibility of reduction leading to remobilization of radionuclides through biotic and abiotic factors is another concern. Achieving an optimal condition for effective bioremediation is also important.

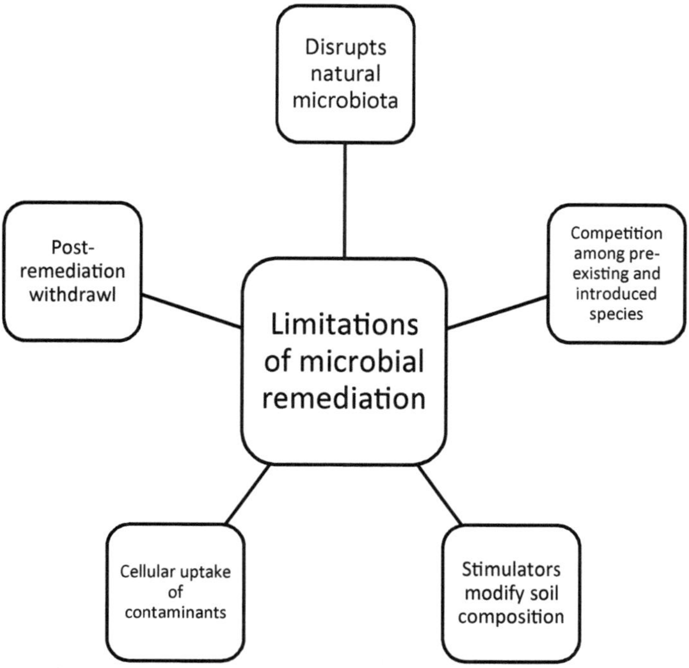

Fig. 6 Limitations of opting for Microbial remediation

Bioaugmentation and biostimulation deal with the introduction of foreign species of microbes or additives, that might hamper the natural microflora. Maintaining an optimal physicochemical environment for microbial growth along with target enhanced in-situ remedial activity is still difficult to achieve. We must precisely analyze the heterogeneously co-contaminated sites for developing standard treatment strategies (Prakash et al. 2013a). Figure 6 enlists various disadvantages faced while using microorganisms for remediation.

4.2 *Phytoremediation Strategies*

Phytoremediation is the use of plants (wild variety or genetically modified) or plant parts or plant extracts to eliminate or remove environmental pollutants (Peuke and Rennenberg 2005; Ali et al. 2013). Phytoremediation is a technique that can effectively provide solution to the problem such as contaminated soil with radioactive compounds and other pollutants (McGrath et al. 2002). It helps in the effective removal of the bio-available portion of the pollutants. Phytoremediation involves various techniques such as oxidation by phytocompounds, volatilization, and sometimes coupling with microbial remediation to increase the overall efficiency of the process (Yadav and Kumar 2020).

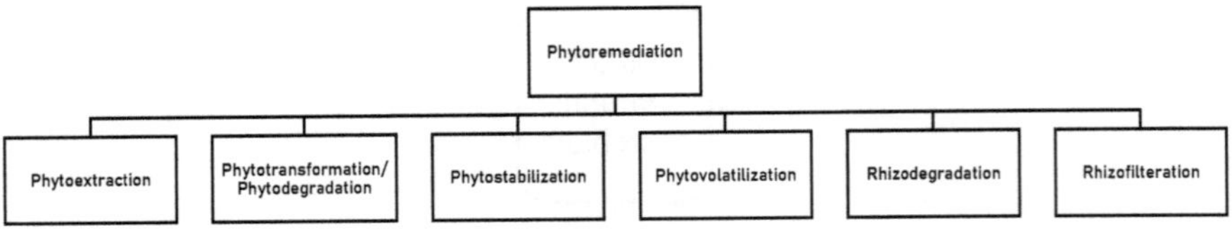

Fig. 7 Various mechanisms covered under the umbrella of phytoremediation

Phytoremediation strategies inexpensively and efficiently remove the radioactive contaminants present in the soil, sediments, and water. Various features of plants that help them in the removal of contaminants are translocation capability, bioaccumulation, and the ability to degrade various contaminants. These properties of plants help them to accumulate radioactive waste in different plant parts such as leaves, stems, and roots and thus eventually cleaning up the contaminants from the environment (Pavel and Gavrilescu 2008). The various mechanism involved in phytoremediation is phytoextraction which involves the removal of pollutants by accumulating the pollutants in the plant biomass, phytotransformation/ phytodegradation refers to the breakdown of pollutants and their uptake from the environment, phytostabilization aims at reducing the bioavailability either by binging or by immobilizing the pollutants so that they are not available in the environment, and phytovolatilization is the process of converting pollutants in the growth medium to a volatile compound which gets released in the atmosphere (Suman et al. 2018; Yan et al. 2020). Figure 7 overlays the various mechanisms that can be employed by plants for the remediation of radionuclides.

4.2.1 Mechanism Employed by the Plants for Phytoremediation

Phytoextraction

Radionuclides are eliminated from the soil without causing any changes in the soil structure and soil fertility. Vascular plants take up chemical substances from the roots and then transport these chemicals to harvestable parts of the plant through the vascular system. These parts containing the radionuclide wastes are then harvested and concentrated and eventually disposed of in procedural ways as radioactive wastes are disposed off. The efficiency of decontamination depends upon factors including the amount of bioavailable radionuclides present in the system, uptake rate of contaminants by the plant roots, and the transport regulation of the radionuclide via the vascular transport system. This technique is useful in large areas with minimal contamination (Dushenkov 2003; Yadav and Kumar 2020). The efficiency of phytoextraction can be expressed as bioaccumulation coefficient/ transfer coefficient/ concentration ratio/ soil–plant transfer factor represented in Eq. (1).

$$\text{Bioaccumulation Coefficient} = \frac{\text{Particular radionuclied concentration in plant shoots}}{\text{Radionuclied concentration in soil}} \quad (1)$$

Phytodegradation

The contaminants present in the soil are broken down by the plants either internally through various metabolic pathways of the plants or externally by producing contaminant-degrading enzymes and releasing them into the soil. These degraded pollutants can be used by the plants as nutrients. The radionuclide waste cannot be degraded by any external agents nor the rate of radioactive decay can be altered using any process (Ibeanusi and Grab 2004; Yadav and Kumar 2020).

Phytostabilization

This method holds or binds the contaminants such as radionuclide at a particular place and holds it. The mobility of the pollutants through wind is restricted, thus preventing secondary contamination and the bioavailability of the contaminants is also reduced simultaneously. Phytostabilization can be achieved by either immobilizing by absorption or precipitating the contaminants (Ibeanusi and Grab 2004; Yan et al. 2021). This method finds its utility as an *in-situ* remediation strategy for low levels of radionuclide-contaminated areas or vast areas where other in situ methods cannot be feasibly applied. This method lowers the associated risks in the case of a radionuclide with smaller half-lives. The root system prevents soil erosion from the contaminated areas and also minimizes the rate of water percolation and thus ultimately preventing radionuclide from leaching. This method is beneficial for open-pit uranium mines and for controlling tailings from strips. Uranium toxicity in plants is a major challenge faced by this method. Plants with high transpiration rates and with fast-developing deep root systems are preferred for phytostabilization. This method possibly reduces the risk associated with the radionuclide on humans and the environment. Phytostabilization does not eliminate the source of the radionuclide from the site to stop further release (Dushenkov 2003; Yan et al. 2021).

Phytovolatilization

This process makes use of the plant's capability to transpire large volumes of water. Along with a large amount of water, radionuclide also reaches the air and thus removing the contaminant from water (McGrath et al. 2002). It is primarily employed for the removal of ^{3}H (tritium), a radioactive isotope of hydrogen. Tritium is usually present as tritiated water in environmental settings. This leads to the incorporation of ^{3}H into the hydro cycle. Being a weak β-emitter, its effect is easily shielded in the air and thus causes no radiation exposures whereas if incorporated in water or any organic substance then it poses serious health hazards if absorbed by the body. Phytovolatilization is operationally cheaper than any other method conventionally used to treat ^{3}H-contaminated water. This method reduces the risk posed by ^{3}H water by diverting the path of exposure to the public rather than removal or isolation (Dushenkov 2003; Vangronsveld et al. 2009).

Rhizofilteration

Plant roots accumulate, precipitate, and concentrate contaminants from the effluent or soil water (Prasad and De Oliveira Freitas 2003). The properties of plant suitable for rhizofilteration includes the ability to produce large root biomass with the large surface area when cultivated in hydroponics, high uptake efficiency, ability to take up the contaminants, and ability to tolerate high contaminant concentrations (McGrath et al. 2002). The pioneering study to employ rhizofilteration can be traced back to the 1950s in Russia by Timofeeva-Ressovskia et al. The study was deployed in a freshwater pond contaminated with several radionuclides. The results showed that the water plants were able to significantly accumulate the radionuclide with high BC levels. This method can be used for a significant reduction in the concentration of radionuclide employing a pond system with a slow water flow rate so that the plant roots get enough time to accumulate the contaminants (Dushenkov 2003). This method is feasible for the removal of radionuclide from an aqueous medium (Hossain 2020).

Rhizodegradation:

Rhizodegradation involves enhancing the rate and extent of the natural process of biodegradation of soil contaminants via., the process of mineralization and transformation by the roots of the plant and the microorganisms associated with the rhizosphere(Qixing et al. 2011).

Table 4 contains examples of various plants used for the remediation of different radionuclide species from contaminated areas.

4.2.2 Genetically Modified Plants for Phytoremediation

With the advancements in genetic engineering, the biochemical pathways and the associated genes which help in hyperaccumulation of different compounds including heavy metals and that of a radionuclide can be easily identified and characterized for understanding the process better. A better understanding of the uptake, translocation, and hyperaccumulation, and other associated regulatory processes enables the use of molecular tools to develop a plant which is economical and highly efficient for phytoremediation (Fulekar et al. 2009). The plant suitable for genetic modifications should have high biomass production, should have remediation properties

Table 4 Examples of phytoremediation of the different radionuclides

Radionuclide Species	Plant species for remediation	Method of Phytoremediation	References
^{134}Cs and ^{238}U	*Triticum aestivum* and *Helianthus giganteus*	Phytoextraction	(Zargan and Fakharyfar 2016)
^{134}Cs and ^{60}Co	*Helianthus annuus*	Phytoextraction	(Achmad and Hadiyanto 2018)
^{137}Cs	*Catharanthus roseus*	Phytoextraction	(Fulekar et al. 2010)
^{137}Cs	*Salix viminalis* and *S. alba*	Phytoextraction	(Rodzkin et al. 2019)
^{137}Cs and ^{90}Sr	*Panicum virginatum*	Phytoextraction	(Entry and Watrud 1998)
^{226}Ra and ^{40}K	*Phragmites australis*	Phytoextraction	(Yan and Luo 2016)
^{232}Th	*Rumex acetosa*	Phytoextraction	(Yan and Luo 2016)
^{238}U	*Cibotium barometz*	Phytoextraction	(Yan and Luo 2016)
^{238}U, ^{232}Th and ^{40}K	*Olea europaea*	Phytoextraction	(Bany Fawaz and Abu-El-Sha'r 2021)
^{239}Pu	*Vetiveria zizanoides*	Phytoextraction	(Singh et al. 2016)
^{137}Cs	*Helianthus* sp., *Sorghum bicolor*, *Amaranthus* sp. and *Fagopyrum esculentum*	Phytostabilization	(Suzuki et al. 2012)
^{137}Cs and ^{90}Sr	*Paspalum notatum, Sorghum halpense* and *Panicum virginatum*	Phytostabilization	(Entry et al. 2001)
^{238}U	*Helianthus* sp., *Chrysopogon zizanioides* and *Megathyrsus maximus*	Phytostabilization	(Roongtanakiat et al. 2010)
^{40}K, ^{232}Th, ^{206}Ra, ^{137}Cs and ^{90}Sr	*Stipa capillata, Festuca valesiaca, Agropyron cristatum*	Phytostabilization	(Prasad 2007)
^{3}H	*Pinus*sp	Phytovolatilization	(Lewis and Pelt 2002)
^{137}Cs and ^{60}Co	*Eichhorina crassipes*	Rhizofilteration	(Saleh 2012)
^{137}Cs and ^{90}Sr	*Cladophora glomerta*and *Elodea canadensis*	Rhizofilteration	(Dushenkov 2003)
^{238}U and ^{137}Cs	*Phaseolus vulgaris*	Rhizofilteration	(Dinis and Fiúza 2021)

by nature, and ease of genetic modulations. For genetic modulation, several target gene families must be identified and targeted that play a key role in hyperaccumulation and transport of radionuclide to enhance the phytoremediation capability of the plant (Doty 2008). Glutamylcysteinesyntlietase has been overexpressed using genetic engineering in *Populus angustifolia* and *Silene cucubalis* to increase their hyperaccumulation capability as compared to their wild counterparts (Fulekar et al.

2009). Plants including *Brassica juncea*, *Liriodendron tulipifera,* and *Helianthus annuus* have also been engineered to enhance their hyperaccumulation properties (Eapen and D'Souza 2005). Three common approaches used for genetic engineering of plants for phytoremediation include: 1. increasing the frequency of uptake sites, 2. expanding sites for intracellular binding, and 3. altering uptake specificity to minimize competition from unwanted ions. Metabolic pathways can be incorporated into plants that help them to hyperaccumulate the radionuclides or help in the process of phytovolatilization (Mello- Farias et al. 2011).

4.2.3 Factor Affecting Phytoremediation

Factors governing the uptake of radionuclide waste by plants are as follows:

1. *Plant Species*: Plants with advanced capability to accumulate radionuclide are selected for the success of phytoremediation.
2. *Properties of Medium*: Biotic and abiotic factors such as moisture content, presence, and concentration of organic matter, temperature, pH influences the rate of uptake of radionuclide by the plants.
3. *Root Zone*: The root zone determines the extent of adsorption and accumulation or metabolization of the contaminants inside the plant tissue.
4. *Presence of Chelators*: The rate of uptake of the radionuclide can be increased by enhancing the bioavailability of radionuclide by adding biodegradable physiochemical agents such as chelators and micronutrients (Malhotra et al. 2014).

The capability of plants in adapting to the climatic conditions and soil of polluted sites also affects the rate of phytoremediation. The rate and efficiency of phytoremediation are also determined by the soil type, nature of contaminants, and the bioavailability of the compounds (Yadav and Kumar 2020).

Baker and Brooks suggested that "metal/ radionuclide hyperaccumulating plants should abide by a standard that the concentration of metal/ radionuclide stored inside the plant can be more than that of the surrounding soil." Transfer factor or TF, based on their study can be represented as Eq. (2).

$$\text{Transfer factor (TF)} = \frac{\text{target element concentration in the plant}}{\text{target element concentration in the tailings/soil}} \quad (2)$$

TF acts as an index that governs the growth of an element and its transfer from soil to plants (Baker and Brooks 1989).

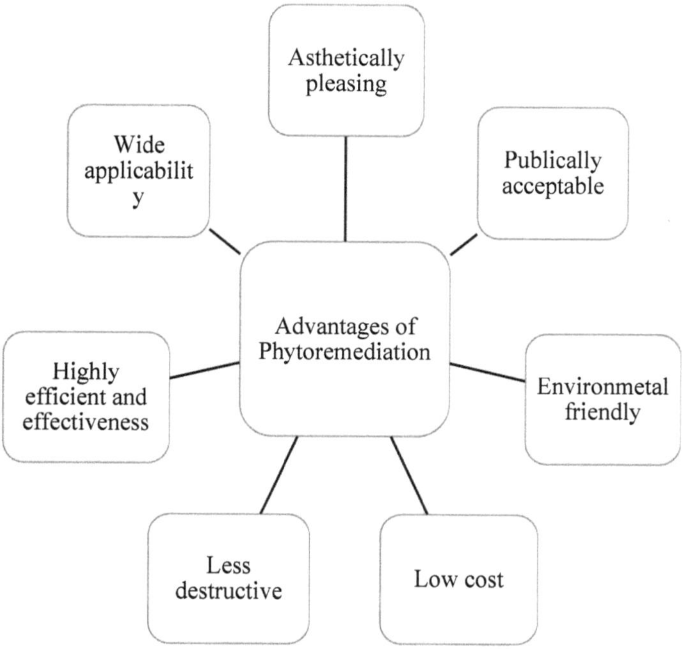

Fig. 8 Advantages of opting for Phytoremediation

4.2.4 Advantages of Opting for Phytoremediation

The phytoremediation method is more publicly acceptable and is comparably eco-friendly. Phytoremediation strategies are highly effective in minimizing the contaminant load and thus allow the reusability of the land for other purposes. The cost of investment along with low running cost and can be used over a vast environmental setting. These strategies are best suited for low levels of contaminants spread over larger areas. This method contains risk to a greater extent and also leads to the generation of plant residues containing radionuclide metals (Eapen et al. 2007; Malhotra et al. 2014). Figure 8 depicts various advantages conferred by the use of plants for remediation.

4.2.5 Limitations of Opting for Phytoremediation

Despite being heavily advantageous this method also has some limitations such as this process is a very time taking and labor-intensive process. Enormous amounts of biomass are produced, the aging process of plants, and the impact of contaminants on plant growth limits the large-scale usability of the process. Plant growth rate also limits the process and time required for remediation much more as compared to the traditional methods used. Other factors include the excavation and disposal of the

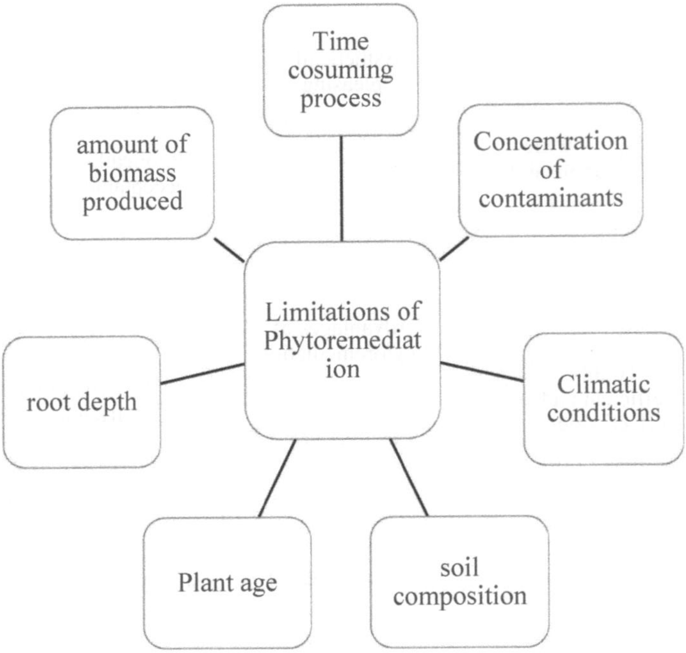

Fig. 9 Limitations of opting for Phytoremediation

plant biomass, is a long-driven process, time taken for the growth, climatic conditions, root system development, quality of soil, and the number of contaminants present in the soil (McGrath et al. 2002; Malhotra et al. 2014). Figure 9 depicts various disadvantages faced by the use of plants for remediation of radionuclides.

5 Challenges and Proposed Ways

5.1 Micro-Remediation Challenges

Remediation of radionuclide employing microorganisms has made a lot of progress in the recent past and is the most feasible and effective way of eliminating these radionuclides from the environment. Despite all the advantages, some major challenges are accompanying this remediation procedure which includes optimization of the operational conditions and procedure for an effective and sustainable process in the presence of inhibitors such as toxic metals, chelators, and competing anions. There are also possibilities of reoxidation and remobilization of the radionuclides by the metabolic processes of microorganisms and other abiotic factors and thus contaminating the environment again. To combat these challenges, care must be taken before choosing additives to increase bioaugmentation and biostimulation so that these additives do not disrupt the natural microbiome of the area (Prakash et al. 2013a).

5.2 *Phytoremediation Challenges*

Most of the plants capable of phytoremediation have a slow growth rate and have small biomass. This slows down the rate of the entire process of remediation and thus to make phytoremediation more fast and effective plants with high growth rate and hyperaccumulation properties need to be discovered naturally or can be developed by using recombinant DNA technologies. High concentrations of contaminants reduce the growth rate of plants and thus slowing down the entire process. By employing genetic engineering, high metal tolerance can be induced so that plants can easily grow under highly contaminated areas (McGrath et al. 2002). It is a general notion that engaging edible plants in phytoremediation would be dangerous to the population consuming it. Hence making way for these hazards into the food chain is strictly not recommended (Sarra et al. 2014). Fiber crops are considered best for phytoremediation, for they are not consumed by organisms and hence pose no serious threat to health.

The application of GMOs is limited due to regulatory and ethical issues. The exotic species introduced taking over the typical microbiota creates uneasiness. Addressing the compatibility of these recombinants with other organisms and the environment is an essential task. Imagine on-site remediation using genetically modified microorganisms. Removal of these species after remediation would be a tedious task. The enzymes might be immobilized to achieve results, but the process is economically draining and requires a lot of optimizations as per site.

6 Conclusion

With the advancement in the scientific world, there has also been an associated risk accompanying those advancements. Likewise, with the advancement in the field of atomic research and exploitation of atomic energy in instances of war and other associated activities, there has been an increase in the concentration and variety of radionuclide waste. These radionuclides are also generated by anthropological instances. These radionuclides have harmful impacts on the human as well as on the environment. The radiations emitted by these radionuclides may induce cancer and other disorders and can cause birth defects or even death in the case of infants or prenatal. Thus, removal or containment of these radionuclides becomes necessary to prevent the exposure of radiations. The major problem with the remediation of radionuclides is that these compounds cannot be degraded by external activities and will only degrade on their own. Thus, removal from the environment is the only way to contain the radiations emitted by these radionuclides. Various strategies have been devised to remove radionuclides including the physical methods (membrane and osmosis technologies, membrane distillation, nano-filtration, ultra-filtration), chemical methods (sorption, ion-exchange, precipitation, coagulation and flocculation, oxidation process), and biological methods (microorganisms based and plant based) (Hossain 2020). Among all the available methods, biological methods are widely used because of technically no harmful impact and minimal input of cost and energy. Microbial remediation involves the use of ubiquitously present simple yet metabolically active microorganisms such as bacteria, algae, fungi, and several protozoans. These organisms can be used effectively on-site with minimal amenities. The bioavailability of contaminants triggers the activation of enzymatic or other processes which help in clearing off the radionuclide from the environment by accumulating them or by rendering them unavailable by stabilizing them. An omics-based study can be a good choice for exploiting microbial species for bioremediation. The cell-free approach of bioremediation involving microbial metabolites can also be a new area of research. Phytoremediation strategies offer plant-based methods to remove radionuclides from the environment by either binding to the radionuclide and thus making it unavailable for the system or by accumulating radionuclides to various parts of the plants. Phytoremediation also involves a strategy to volatilize the radionuclide and thus change the way of exposure to the radiation. Phytoremediation can also be used in hydroponic settings where contaminated water can be fed and plant roots take up the radionuclides. Genetic engineering increases the efficiency of both plants and microbes for radionuclide remediation. Genetic engineering enhances the uptake efficiency, and increases the tolerance against the radionuclides.

7 Future Perspectives and Research Opportunities

Effective in-situ remediation strategies can be designed through virtual screening of genomes of microorganisms, selecting and engineering strains, along stimulations. Element-definite classification of strain is a better scheme for radio-waste-bioremediation. Genetically modified plants with higher remediation capacity should be used for in-situ remediation projects to decrease the time required for remediation.

References

Achmad CAA, Hadiyanto (2018) Phy. E3S Web Conf 73:3–6. https://doi.org/10.1051/e3sconf/20187305027

Ali H, Khan E, Sajad MA (2013) Phytoremediation of heavy metals-concepts and applications. Chemosphere 91:869–881. https://doi.org/10.1016/j.chemosphere.2013.01.075

Amachi S, Minami K, Miyasaka I, Fukunaga S (2010) Ability of anaerobic microorganisms to associate with iodine: 125I tracer experiments using laboratory strains and enriched microbial communities from subsurface formation water. Chemosphere 79:349–354. https://doi.org/10.1016/j.chemosphere.2010.02.028

Azubuike CC, Chikere CB, Okpokwasili GC (2016) Bioremediation techniques–classification based on site of application: principles, advantages, limitations and prospects. World J Microbiol Biotechnol 32. https://doi.org/10.1007/s11274-016-2137-x

Baker AJ, Brooks RR (1989) Terrestrial higher plants which hyper- accumulate metallic elements—a review of their distribution, ecology and phytochemistry. Biorecovery 1:81–126

Bany Fawaz EH, Abu-El-Sha'r WY, (2021) Uptake and distribution of natural ^{238}U, ^{232}Th, and ^{40}K Radionuclides in Olive Trees Grown in Hausha Area, Jordan. J Hazard, Toxic, Radioact Waste 25:04021023. https://doi.org/10.1061/(asce)hz.2153-5515.0000621

Beckwith CS, McGee DJ, Mobley HLT, Riley LK (2001) Cloning, expression, and catalytic activity of Helicobacter hepaticus urease. Infect Immun 69:5914–5920. https://doi.org/10.1128/IAI.69.9.5914-5920.2001

Bogutskaya NG, Zuykov MA, Naseka AM, Anderson EB (2011) Normal axial skeleton structure in common roach Rutilus rutilus (Actinopterygii: Cyprinidae) and malformations due to radiation contamination in the area of the Mayak (Chelyabinsk Province, Russia) nuclear plant. J Fish Biol 79:991–1016. https://doi.org/10.1111/j.1095-8649.2011.03078.x

Brainard JR, Strietelmeier BA, Smith PH et al (1992) Actinide binding and Solubilization by microbial Siderophores. Radiochim Acta 58–59:357–364. https://doi.org/10.1524/ract.1992.5859.2.357

Brim H, McFarlan SC, Fredrickson JK et al (2000) Engineering Deinococcus radiodurans for metal remediation in radioactive mixed waste environments. Nat Biotechnol 18:85–90. https://doi.org/10.1038/71986

Brim H, Venkateswaran A, Kostandarithes HM et al (2003) Engineering Deinococcus geothermalis for bioremediation of high-temperature radioactive waste environments. Appl Environ Microbiol 69:4575–4582. https://doi.org/10.1128/AEM.69.8.4575-4582.2003

Butler JE, Glaven RH, Esteve-Núñez A et al (2006) Genetic characterization of a single bifunctional enzyme for fumarate reduction and succinate oxidation in Geobacter sulfurreducens and engineering of fumarate reduction in Geobacter metallireducens. J Bacteriol 188:450–455. https://doi.org/10.1128/JB.188.2.450-455.2006

Chung AP, Lopes A, Nobre MF, Morais PV (2010) Hymenobacter perfusus sp. nov., Hymenobacter flocculans sp. nov. and Hymenobacter metalli sp. nov. three new species isolated from an uranium mine waste water treatment system. Syst Appl Microbiol 33:436–443. https://doi.org/10.1016/J.SYAPM.2010.09.002

De Luca G, De Philip P, Dermoun Z et al (2001) Reduction of Technetium(VII) by Desulfovibrio fructosovorans is mediated by the nickel-iron hydrogenase. Appl Environ Microbiol 67:4583–4587. https://doi.org/10.1128/AEM.67.10.4583-4587.2001

DiChristina TJ (1992) Effects of nitrate and nitrite on dissimilatory iron reduction by Shewanella putrefaciens 200. J Bacteriol 174:1891–1896. https://doi.org/10.1128/jb.174.6.1891-1896.1992

de Dinis ML, Fiúza A (2021) Mitigation of uranium mining impacts—a review on groundwater remediation technologies. Geosci 11. https://doi.org/10.3390/geosciences11060250

Dodge CJ, Francis AJ, Gillow JB et al (2002) Association of uranium with iron oxides typically formed on corroding steel surfaces. Environ Sci Technol 36:3504–3511. https://doi.org/10.1021/es011450+

Doty SL (2008) Enhancing phytoremediation through the use of transgenics and endophytes. New Phytol 179:318–333

Dushenkov S (2003) Trends in phytoremediation of radionuclides. Plant Soil 249:167–175. https://doi.org/10.1023/A:1022527207359

Eapen S, D'Souza SF (2005) Prospects of genetic engineering of plants for phytoremediation of toxic metals. Biotechnol Adv 23:97–114. https://doi.org/10.1016/j.biotechadv.2004.10.001

Eapen S, Singh S, D'Souza SF (2007) Phytoremediation of metals and radionuclides. Environ Bioremediation Technol 75:189–209. https://doi.org/10.1007/978-3-540-34793-4_8

Entry JA, Watrud LS (1998) Potential remediation of ^{137}Cs and ^{90}Sr contaminated soil by accumulation in alamo switchgrass. Water Air Soil Pollut 104:339–352. https://doi.org/10.1023/A:1004994123880

Entry JA, Watrud LS, Reeves M (2001) Influence of organic amendments on the accumulation of ^{137}Cs and ^{90}Sr from contaminated soil by three grass species. Water Air Soil Pollut 126:385–398. https://doi.org/10.1023/A:1005201220596

Finneran KT, Housewright ME, Lovley DR (2002) Multiple influences of nitrate on uranium solubility during bioremediation of uranium-contaminated subsurface sediments. Environ Microbiol 4:510–516. https://doi.org/10.1046/j.1462-2920.2002.00317.x

Francis AJ (1998) Biotransformation of uranium and other actinides in radioactive wastes. J Alloys Compd 271–273:78–84. https://doi.org/10.1016/S0925-8388(98)00028-0

Francis AJ, Joshi-Tope GA, Dodge CJ, Gillow JB (2002) Biotransformation of uranium and transition metal citrate complexes by clostridia. J Nucl Sci Technol 39:935–938. https://doi.org/10.1080/00223131.2002.10875622

Fujimoto K, Morita T (2006) Aerobic removal of technetium by a marine Halomonas strain. Appl Environ Microbiol 72:7922–7924. https://doi.org/10.1128/AEM.00819-06

Fulekar MH, Singh A, Bhaduri AM (2009) Genetic engineering strategies for enhancing phytoremediation of heavy metals. African J Biotechnol 8:529–535. https://doi.org/10.4314/ajb.v8i4.59858

Fulekar MH, Singh A, Thorat V et al (2010) Phytoremediation of ^{137}Cs from low level nuclear waste using Catharanthus roseus. Indian J Pure Appl Phys 48:516–519

Green SJ, Prakash O, Jasrotia P et al (2012) Denitrifying bacteria from the genus Rhodanobacter dominate bacterial communities in the highly contaminated subsurface of a nuclear legacy waste site. Appl Environ Microbiol 78:1039–1047. https://doi.org/10.1128/AEM.06435-11

Hegazy AK, Emam MH (2011) Accumulation and soil-to-plant transfer of radionuclides in the Nile delta coastal black sand habitats. Int J Phytoremediation 13:140–155. https://doi.org/10.1080/15226511003753961

Hossain F (2020) Natural and anthropogenic radionuclides in water and wastewater: sources, treatments and recoveries. J Environ Radioact 225:106423. https://doi.org/10.1016/j.jenvrad.2020.106423

Ibeanusi VM, Grab DA (2004) EPA radionuclide biological remediation resource guide, pp 1–68

John SG, Ruggiero CE, Hersman LE et al (2001) Siderophore mediated plutonium accumulation by Microbacterium flavescens (JG-9). Environ Sci Technol 35:2942–2948. https://doi.org/10.1021/es010590g

Keasling JD, Van Dien SJ, Trelstad P et al (2000) Application of polyphosphate metabolism to environmental and biotechnological problems. Biochem 65:324–331

Khani MH, Keshtkar AR, Meysami B et al (2006) Biosorption of uranium from aqueous solutions bynonliving biomass of marinealgae Cystoseira indica. Electron J Biotechnol 9:100–106. https://doi.org/10.2225/vol9-issue2-fulltext-8

Kim SJ, Koh DC, Park SJ et al (2012) Molecular analysis of spatial variation of iron-reducing bacteria in riverine alluvial aquifers of the Mankyeong River. J Microbiol 50:207–217. https://doi.org/10.1007/s12275-012-1342-z

Kumar R, Singh S, Singh OV (2007) Bioremediation of radionuclides: emerging technologies. Omi A J Integr Biol 11:295–304. https://doi.org/10.1089/omi.2007.0013

Langley S, Beveridge TJ (1999) Effect of O-side-chain-lipopolysaccharide chemistry on metal binding. Appl Environ Microbiol 65:489–498. https://doi.org/10.1128/aem.65.2.489-498.1999

Lewis CM, Pelt R Van (2002) Natural remediation at Savannah River Site Catherine M. Lewis and Robert Van Pelt Bechtel Savannah River Inc., Savannah River Site P.O. Box 616, Aiken, SC 29801. In: WM'02 Conference, pp 1–12

Lloyd JR, Cole JA, Macaskie LE (1997) Reduction and removal of heptavalent technetium from solution by Escherichia coli. J Bacteriol 179:2014–2021. https://doi.org/10.1128/jb.179.6.2014-2021.1997

Lloyd JR, Leang C, Hodges Myerson AL et al (2003) Biochemical and genetic characterization of PpcA, a periplasmic c-type cytochrome in Geobacter sulfurreducens. Biochem J 369:153–161. https://doi.org/10.1042/BJ20020597

Lloyd JR, Renshaw JC (2005) Bioremediation of radioactive waste: radionuclide–microbe interactions in laboratory and field-scale studies. Curr Opin Biotechnol 16:254–260. https://doi.org/10.1016/J.COPBIO.2005.04.012

Lloyd JR, Ridley J, Khizniak T et al (1999) Reduction of technetium by Desulfovibrio desulfuricans: biocatalyst characterization and use in a flowthrough bioreactor. Appl Environ Microbiol 65:2691–2696. https://doi.org/10.1128/aem.65.6.2691-2696.1999

Lovley DR, Phillips EJ (1994) Reduction of chromate by *Desulfovibrio vulgaris* and its c_3 cytochrome. Appl Environ Microbiol 60:726–728

Macaskie LE, Bonthrone KM, Yong P, Goddard DT (2000) Enzymically mediated bioprecipitation of uranium by a Citrobacter sp.: a concerted role for exocellular lipopolysaccharide and associated phosphatase in biomineral formation. Microbiology 146:1855–1867. https://doi.org/10.1099/00221287-146-8-1855

Malhotra R, Aggarwal S, Agarwal R, Gauba P (2014) Phytoremediation of radioactive metals. Indo Glob J Pharm Sci 04:75–79. https://doi.org/10.35652/igjps.2014.40

Marsili E, Beyenal H, Di Palma L et al (2007) Uranium immobilization by sulfate-reducing biofilms grown on hematite, dolomite, and calcite. Environ Sci Technol 41:8349–8354. https://doi.org/10.1021/es071335k

Martinez RJ, Wang Y, Raimondo MA et al (2006) Horizontal gene transfer of PIB-type ATPases among bacteria isolated from radionuclide- and metal-contaminated subsurface soils. Appl Environ Microbiol 72:3111–3118. https://doi.org/10.1128/AEM.72.5.3111-3118.2006

McGrath SP, Zhao J, Lombi E (2002) Phytoremediation of metals, metalloids, and radionuclides. Adv Agron 75:1–56. https://doi.org/10.1016/s0065-2113(02)75002-5

Mello- Farias de PC, Soares Chaves AL, Leoneti C (2011) Transgenic plants for enhanced phytoremediation—physiological studies. Genet Transform. https://doi.org/10.5772/24355

Nagaraj NS, Singh OV (2010) Using genomics to develop novel antibacterial therapeutics. Crit Rev Microbiol 36:340–348. https://doi.org/10.3109/1040841X.2010.495941

National Center for Environmental Health (NCEH), Emergency Management, Radiation and CB (2018) Radiation emergencies. In: Centers Dis. Control Prev. https://www.cdc.gov/nceh/radiation/emergencies/isotopes/uranium.htm

National Library of Medicine Neptunium, Radioactive. In: WebWISER. https://webwiser.nlm.nih.gov/substance?substanceId=417&identifier=Neptunium, Radioacti

Nyman JL, Marsh TL, Ginder-Vogel MA et al (2006) Heterogeneous response to biostimulation for U(VI) reduction in replicated sediment microcosms. Biodegradation 17:303–316. https://doi.org/10.1007/s10532-005-9000-3

Panak PJ, Booth CH, Caulder DL et al (2002) X-ray absorption fine structure spectroscopy of plutonium complexes with Bacillus sphaericus. Radiochim Acta 90:315–321. https://doi.org/10.1524/ract.2002.90.6.315

Panak PJ, Nitsche H (2001) Interaction of aerobic soil bacteria with plutonium(VI). Radiochim Acta 89:499–504. https://doi.org/10.1524/ract.2001.89.8.499

Pavel LV, Gavrilescu M (2008) Overview of ex situ decontamination techniques for soil cleanup. Environ Eng Manag J 7:815–834. https://doi.org/10.30638/eemj.2008.109

Peuke AD, Rennenberg H (2005) Phytoremediation. EMBO Rep 6:497–501

Prakash D, Gabani P, Chandel AK et al (2013a) Bioremediation: a genuine technology to remediate radionuclides from the environment. Microb Biotechnol 6:349–360. https://doi.org/10.1111/1751-7915.12059

Prakash D, Gabani P, Chandel AK et al (2013b) Bioremediation: a genuine technology to remediate radionuclides from the environment. https://doi.org/10.1111/1751-7915.12059

Prasad MNV, De Oliveira Freitas HM (2003) Metal hyperaccumulation in plants—biodiversity prospecting forphytoremediation technology. Electron J Biotechnol 6:110–146. https://doi.org/10.2225/vol6-issue3-fulltext-6

Prasad MNV (2007) Grasses tolerant to radionuclides growing in Kazakhstan nuclear test sites exhibit structural and ultrastructural changes—implications for phytoremediation and involved risks. Terr Aquat Environ Toxicol

Premuzic ET, Francis AJ, Lin M, Schubert J (1985) Induced formation of chelating agents by Pseudomonas aeruginosa grown in presence of thorium and uranium. Arch Environ Contam Toxicol 14:759–768. https://doi.org/10.1007/BF01055783

Shah Maulin P (2020) Microbial bioremediation & biodegradation. Springer

Qixing Z, Zhang C, Zhineng Z, Weitao L (2011) Ecological remediation of hydrocarbon contaminated soils with weed plant. BioOne 2:97–105. https://doi.org/10.3969/j.issn.1674-764x.2011.02.001

Rodzkin A, Khroustalev B, Kundas S et al (2019) Potential of energy willow plantations for biological reclamation of soils polluted by ^{137}cs and heavy metals, and for control of nutrients leaking into water systems. Environ Clim Technol 23:43–56. https://doi.org/10.2478/rtuect-2019-0078

Roh C, Kang CK, Lloyd JR (2015) Microbial bioremediation processes for radioactive waste. Korean J Chem Eng 32:1720–1726. https://doi.org/10.1007/s11814-015-0128-5

Shah Maulin P (2021) Removal of refractory pollutants from wastewater treatment plants. CRC Press

Roongtanakiat N, Sudsawad P, Ngernvijit N (2010) Uranium absorption ability of sunflower, vetiver and purple guinea grass. Kasetsart J, Nat Sci 44:182–190

Saleh HM (2012) Water hyacinth for phytoremediation of radioactive waste simulate contaminated with cesium and cobalt radionuclides. Nucl Eng Des 242:425–432. https://doi.org/10.1016/j.nucengdes.2011.10.023

Sarra A, Bruno C, Salah R et al (2014) Bioaccumulation and photosynthetic activity response of Kenaf (Hibicus cannabinus L.) to Cadmium and Zinc. Greener J Agric Sci 4:091–100. https://doi.org/10.15580/gjas.2014.3.1216131031

Shehata SM, Badawy RK, Aboulsoud YIE (2019) Phytoremediation of some heavy metals in contaminated soil. Bull Natl Res Cent 43. https://doi.org/10.1186/s42269-019-0214-7

Singh S, Fulzele DP, Kaushik CP (2016) Potential of Vetiveria zizanoides L. Nash for phytoremediation of plutonium (239Pu): Chelate assisted uptake and translocation. Ecotoxicol Environ Saf 132:140–144. https://doi.org/10.1016/j.ecoenv.2016.05.006

Suman J, Uhlik O, Viktorova J, Macek T (2018) Phytoextraction of heavy metals: a promising tool for clean-up of polluted environment? Front Plant Sci 871:1–15. https://doi.org/10.3389/fpls.2018.01476

Suzuki Y, Saito T, Tsukada H (2012) Phytoremediation of radiocesium in different soils using cultivated plants. Proc Int … 1–4

Tekere M (2019) Microbial bioremediation and different bioreactors designs applied. In: Biotechnology and Bioengineering. IntechOpen, p 13

United States Environmental Protection Agency (2021a) Radionuclide Basics: Technetium-99. https://www.epa.gov/radiation/radionuclide-basics-technetium-99

United States Environmental Protection Agency (2021b) Radionuclide Basics: Americium-241. https://www.epa.gov/radiation/radionuclide-basics-americium-241

United States Environmental Protection Agency (2021c) Radionuclide Basics: Cesium-137. https://www.epa.gov/radiation/radionuclide-basics-cesium-137

United States Nuclear Regulatory Commission (2021) Radionuclides. https://www.nrc.gov/reading-rm/doc-collections/cfr/part020/appb/index.html

United States Nuclear Regulatory Commission (2020) Backgrounder on Plutonium. https://www.nrc.gov/reading-rm/doc-collections/fact-sheets/plutonium.html

Van Hullebusch ED, Peerbolte A, Zandvoort MH, Lens PNL (2005) Sorption of cobalt and nickel on anaerobic granular sludges: isotherms and sequential extraction. Chemosphere 58:493–505. https://doi.org/10.1016/j.chemosphere.2004.09.017

Vangronsveld J, Herzig R, Weyens N et al (2009) Phytoremediation of contaminated soils and groundwater: lessons from the field. Environ Sci Pollut Res 16:765–794. https://doi.org/10.1007/s11356-009-0213-6

Vrionis HA, Anderson RT, Ortiz-Bernad I et al (2005) Microbiological and geochemical heterogeneity in an in situ uranium bioremediation field site. Appl Environ Microbiol 71:6308–6318. https://doi.org/10.1128/AEM.71.10.6308-6318.2005

Wang J, Feng X, Anderson CWN et al (2012) Remediation of mercury contaminated sites—a review. J Hazard Mater 221–222:1–18. https://doi.org/10.1016/j.jhazmat.2012.04.035

Wildung RE, Gorby YA, Krupka KM et al (2000) Effect of electron donor and solution chemistry on products of dissimilatory reduction of technetium by Shewanella putrefaciens. Appl Environ Microbiol 66:2451–2460. https://doi.org/10.1128/AEM.66.6.2451-2460.2000

World Nuclear Association (2021) Nuclear Fuel Cycle Overview. https://world-nuclear.org/information-library/nuclear-fuel-cycle/introduction/nuclear-fuel-cycle-overview.aspx

Wu WM, Carley J, Fienen M et al (2006) Pilot-scale in situ bioremediation of uranium in a highly contaminated aquifer. 1. Conditioning of a treatment zone. Environ Sci Technol 40:3978–3985. https://doi.org/10.1021/es051954y

Xie S, Yang J, Chen C et al (2008) Study on biosorption kinetics and thermodynamics of uranium by Citrobacter freudii. J Environ Radioact 99:126–133. https://doi.org/10.1016/j.jenvrad.2007.07.003

Yadav D, Kumar P (2020) Phytoremediation of hazardous radioactive wastes. In: Assessment and management of radioactive and electronic wastes. IntechOpen, p 13

Yan A, Wang Y, Tan SN et al (2020) Phytoremediation: a promising approach for revegetation of heavy metal-polluted land. Front Plant Sci 11:1–15. https://doi.org/10.3389/fpls.2020.00359

Yan L, Van LQ, Sonne C et al (2021) Phytoremediation of radionuclides in soil, sediments and water. J Hazard Mater 407:124771. https://doi.org/10.1016/j.jhazmat.2020.124771

Yan X, Luo XG (2016) Uptake of uranium, thorium, radium and potassium by four kinds of dominant plants grown in uranium mill tailing soils from the southern part of China. Radioprotection 51:141–144. https://doi.org/10.1051/radiopro/2015031

Zargan J, Fakharyfar M (2016) Phytoremediation of cesium and uranium contaminated soils by plants. J Appl Environ Biol Sci 6:173–182

Bioremediation of Petroleum Sludge

Asmita Kumari, Nidhi Solanki, and Navneeta Bharadvaja

1 Introduction

Petroleum hydrocarbons are a vital source of raw materials and energy production for many industries. The demand for petroleum hydrocarbons is steadily increasing because of the rising population and industrialization (Varjani and Upasani 2017). The US Environmental Protection Agency has listed them as refractory compounds, and they are considered major environmental contaminants (Yuniati 2018). Petroleum is a dark, thick, sticky liquid that primarily consists of substantial quantities of hydrogen and carbon. It gets its name from a Latin term that means "rock oil." During the production of crude oil, petroleum refineries generate significant amounts of sludge. Hazardous metals, aromatic and polyaromatic hydrocarbons, as well as water and oil, are present in the sludge, posing a threat to humans and plants (Ahmad 2017). Many practices like the offshore petroleum and onshore petroleum industries, municipal and industrial runoff, and effluent release, contribute to petroleum hydrocarbon contamination (Varjani and Upasani 2017). They generate deliberate or accidental oil spills during the transportation, production, or refining of petroleum oil, polluting the soil, groundwater, and oceans (Al-Hawash et al. 2018). Accidental or unintended oil spills as a result of human or industrial operations pollute the environment adversely (Yuniati 2018). Biomagnification of PHs causes mutations or death of indigenous microbes and plants (Das and Chandran 2011). Sludge is treated by recovering the oil using demulsifiers, and subsequently, hard particles of the sludge are disposed off using a variety of disposal methods (Islam 2015). Landfill, chemical oxidation, solvent extraction, natural attenuation, and other physicochemical methods for the degradation of hydrocarbons are all known. However, these procedures are expensive, time-consuming, and inefficient (Cai et al. 2021). The limits of

A. Kumari · N. Solanki · N. Bharadvaja (✉)
Plant Biotechnology Laboratory, Department of Biotechnology,
Delhi Technological University, Delhi, India
e-mail: navneetab@dce.ac.in

M. P. Shah (ed.), *Microbial Technologies in Industrial Wastewater Treatment*,
https://doi.org/10.1007/978-981-99-2435-6_13

physical and chemical sludge disposal technologies have prompted researchers to look into other feasible options. Bioremediation is a carbon–neutral, environmentally beneficial, cost-effective, and noninvasive method of soil remediation (Yuniati 2018). Biodegradation allows for complete mineralization and production of very few secondary metabolites and by-products, therefore it is preferred over physicochemical treatment of phenol-containing wastewater (Banerjee 2021). During the bioremediation, process microbes catalyze the breakdown of toxic compounds into innocuous substances such as H2O, CO2 in aerobic settings, and CH4 in anaerobic conditions (Haritash and Kaushik 2009; Thamer et al. 2013). During the bioremediation process, microbes degrade hydrocarbons and absorb them as carbon and energy sources for their development (Fritsche and Hofrichter 2001) Various methods of bioremediation involve bioslurry, biosparging, bioventing, phytoremediation, biopiles, phycoremediation, biocomposting, landfarming, etc. Many microorganisms, including yeast, bacteria, fungi, and algae, have been found to be capable of degrading hydrocarbons (Haritash and Kaushik 2009). Eight bacteria, 21 fungi, and four yeasts were isolated from petroleum-polluted soils and cyanobacterial mats in Indonesia (Chaillan et al. 2004). Isolated bacterial species belonged to the genera *Gordonia, Brevibacterium, Aeromicrobium, Dietzia, Burkholderia*, and *Mycobacterium. Aspergillus, Penicillium, Fusarium, Amorphoteca, Neosartorya, Paecilomyces, Talaromyces*, and *Graphium* were among the fungi found. Yeasts were *Candida, Yarrowia,* and *Pichia* (Chaillan et al. 2004). *Rhodotorula rubra, Candida famata, Saccharomyces cerevisiae, Candida lipolytica,* and *Protothecazopfii* are the other algae and fungal species used in bioremediation (Das and Chandran 2011). Simple unbranched aliphatic hydrocarbons are easily degraded as compared to complex aromatic hydrocarbons (Al-Hawash et al. 2018; Shah 2020). pH, temperature, salinity, pressure, concentration, and type of hydrocarbons, as well as genetic and metabolic properties of bacteria, all have an impact on the microbial breakdown of petroleum hydrocarbons (Thamer et al. 2013). Microbes produce biosurfactants, which are surface-active compounds produced by microbes that improve bioavailability and ramp up the bioremediation process. These chemicals aid in the cracking, solubilization, and emulsification of hydrocarbons, allowing bacteria to more easily degrade them (Varjani and Upasani 2017). This chapter delves into the mechanisms of petroleum sludge bioremediation, as well as the strategies and factors that influence bioremediation.

2 Bioremediation

Bioremediation is the biodegradation of petroleum hydrocarbons and transforming pollutants into less hazardous compounds. Through the process of bioremediation, alteration of mobility of contaminants is done thus preventing the spread of toxicity from the pollutants (Yuniati 2018). Hazardous pollutants are converted to innocuous chemicals such as carbon dioxide, methane, and biomass. To detoxify these pollutants, the technique of bioremediation utilizes microorganisms. In the process of bioremediation microorganisms meet their metabolic needs by consuming toxic xenobiotics present in the contaminant. Long-chain hydrocarbons present in the sludge can be degraded aerobically or anaerobically by microorganisms like bacteria as well as eukaryotes such as algae and fungi (Banerjee 2021).

Other techniques used for the remediation of petroleum sludge include physiochemical technologies. Solidification, incineration, soil vapor extraction, chemical treatment, and solvent extraction processes are carried out in this technique (Sarkar et al. 2020). Biodegradation allows for complete mineralization and production of very few secondary metabolites and by-products, therefore it is preferred over physicochemical treatment of phenol-containing wastewater (Banerjee 2021). Further, this chemical treatment approach is cost-intensive that also lead to economic check. Froman efficiency point of view, physiochemical treatment methods lead to the production of more persistent secondary metabolites due to incomplete conversion of contaminants (Sarkar et al. 2020; Varjani and Upasani 2016). Dispersion, sorption, volatilization, and use of chemicals like hydrogen peroxide are some physicochemical methods used to treat petroleum contaminants. These procedures also account for limitations of the high cost of excavation and chemicals used (Varjani 2016; Chandra et al. 2013). Therefore to overcome such issues and to adopt more sustainable and natural remedies bioremediation is preferred. Furthermore, the benefits include the flexibility and adaptability of the process to the environment. Here the microorganisms that can break down novel synthetic chemical compounds emerge over time (Ward 2004). The procedures are considered to be environmentally benign and can be employed on-site, and in many cases, in situ, with dilute or extensively disseminated contaminants (Ward 2004; Iwamoto and Nasu 2001). Natural attenuation processes can take anywhere from 5 to 25 years for the degradation process to complete, in-situ subsurface processes take about 0.5–3 years, while ex-situ subsurface processes take < 1 year for the degradation process (Ward and Singh 2004).

3 Microorganisms Involved in Bioremediation

Several bacteria, fungi, and yeast species are capable of decomposing petroleum hydrocarbons. These microorganisms are found in their native environments. Methane to C40 molecules is among the biodegradable hydrocarbon compounds (Jafarinejad 2016; Baniasadi and Mousavi 2019) (Fig. 1).

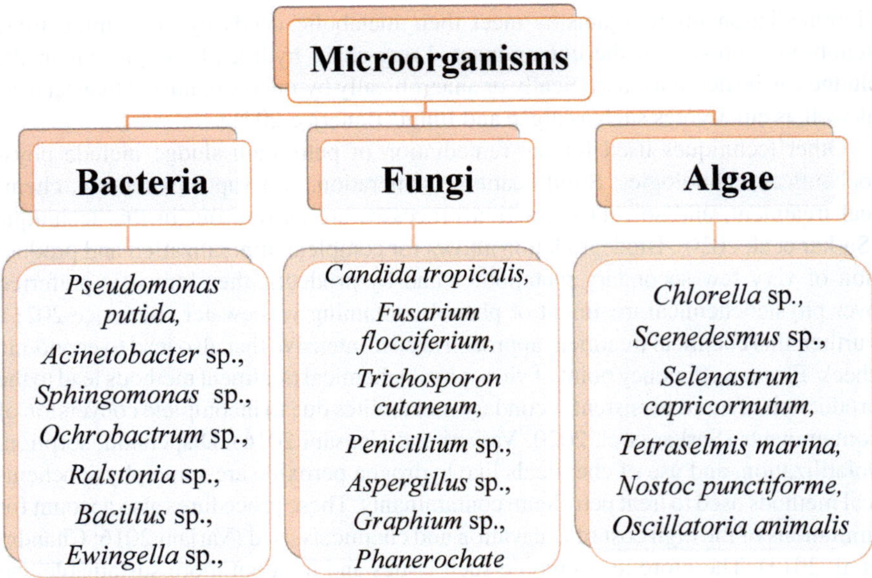

Fig. 1 Microorganism involved in bioremediation

4 Techniques for Accelerated Remediation

4.1 Addition of Surfactant

Since bioremediation uses local or exogenous microorganisms to degrade petroleum hydrocarbons and is a widely used technique, studies have been done to devise strategies that enhance this process of remediation and make it fast and cost-effective (Suganthi et al. 2018). Surfactants being the one, these are amphipathic, surface-active chemicals produced by microorganisms that lower the surface tension between two fluids. It can form a stable emulsion with water by forming micelles. Such emulsifiers improve the surface area of hydrophobic molecules and make them available for microorganisms to act upon (Zhang et al. 2010). Surfactants also decompose heavy metals present in the petroleum sludge such as chromium, arsenic, cadmium, lead, and nickel (Suganthi et al. 2018). These heavy metals accumulate in soil and can have a significant impact on health. The two common biosurfactants include rhamnolipids and Tween 80 (Naeem and Qazi 2020). Other examples of surfactants include Triton X-100, and Afonic 1412-7 (Johnson and Affam 2019). Since chemical surfactants might have concerns due to their resistance to biodegradation therefore biosurfactants are preferred over these since they are more environment friendly (Johnson and Affam 2019). Genera *Pseudomonas* is the most well-known bacteria that can synthesize biosurfactants naturally. *Pseudomonas aeruginosa* among other *Pseudomonas* species is used extensively for the synthesis of glycolipid-type biosurfactants (Das and Chandran 2011).

4.2 *Addition of Algal–bacterial Consortium*

Microalgae, which account for over half of worldwide photosynthetic activity, provide oxygen to aerobic bacteria, which use it to remove organic pollutants and absorb CO_2 (Tang et al. 2010). Microalgae can boost bacterial activity by assisting with co-metabolic breakdown or production of biosurfactants and extracellular materials. Furthermore, bacteria can promote microalgae development by producing growth-promoting compounds (Chaillan et al. 2006). Interaction of phototrophic algal species and heterotrophic bacteria can maximize the breakdown of crude oil. Microalgal proliferation is associated with bacteria in the natural environment. A natural algal–bacterial consortium is built when there is a symbiotic association between microalgae and bacteria associated with those microalgae. Microalgae have even been considered a partial function of the accompanying bacteria (Tang et al. 2010). In a study by Naeem and Qazi (2020), Shah MP (2020) for the elimination of phenanthrene, they used an algal–bacterial consortium composed of *Chlorella sorokiniana* and *Pseudomonas migulae*. The algal–bacterial consortium could detoxify 200–500 mg/L of phenanthrene under photosynthetic conditions along with organic solvents. The degradation rate was determined to be 24.2 $gm^{-3}\ h^{-1}$under ideal conditions, and microalgal species further aided the process by producing biosurfactants.

5 Mechanism of Microbial Degradation

Most of the pollutants degrade at the fastest rate under aerobic conditions. Aerobic degradation is the most efficient mechanism for the breakdown of petroleum hydrocarbons which begins with an oxidative process of intracellular attack of organic contaminants (Olajire 2014). Activation and incorporation of oxygen are done with the help of enzyme oxygenase and peroxidase. This is followed by the transformation of contaminants into intermediates of the metabolic pathway, the tricarboxylic acid cycle. These intermediates form major precursor metabolites for cell biomass synthesis (Olajire 2014; Kour et al. 2021). Microbiological degradation occurs in a specific order, depending on the molecules being degraded. For example, the breakdown of aliphatic hydrocarbons is easy and fast followed by aromatic hydrocarbons and cycloalkanes. Branched and polynucleated compounds are the most difficult ones. In general with an increase in the degree of halogenation, the efficiency of biodegradability decreases (Claro et al. 2019).

Monoterminal oxidation is a common way for alkanes to break. The initial stage in this procedure is to introduce molecular oxygen into a hydrocarbon and oxidize the methyl group at the terminal end to produce primary alcohol and is subsequently oxidized, yielding aldehyde and fatty acid (Varjani 2016). The next step is β-oxidation, which eliminates acetyl coenzyme-A, this results in the shortening of fatty acid to a two-carbon molecule (Varjani 2016; Abbasian et al. 2015).

Aromatic hydrocarbons present in the sludge are broken down into dicarboxylic acid through ring cleavage reaction (Varjani 2016; Hendrickx et al. 2006). Protocatechuates and catechols are produced as a result of ortho- or meta-cleavage pathways for benzene rings. These are then transformed into tricarboxylic acid cycle intermediates (Varjani 2016).

An oxidase system converts cyclic alkanes to cyclic alcohols, which are then dehydrogenated to ketones. A monooxygenase system is preceded by a lactone ring, which is cleaved by the enzyme lactone hydrolase (Abbasian et al. 2015). It is difficult to obtain pure cultures capable of degrading cycloalkanes since two oxygenase systems are rarely present in the same bacteria (Varjani 2016).

6 Methods of Bioremediation

The viability of a biological treatment approach is highly dependent on the limiting variables and also the location of the pollutant. It even relies on as to whether the contaminated soil is to be left in place or scraped and transported to an offsite for remediation. Remediation of pollutants can be done on-site as well as off-site (Ossai et al. 2020; Hamzah et al. 2013). The type of ex-situ remediation method to be used is decided based on the type and amount of pollutant present and is then carried out in compliance with government laws and is commonly used because it has a safe impact on the environment. In contrast to ex-situ, in-situ is less expensive as no excavation is required and also it does not damage the soil structure. It also enhances microbial breakdown in toxic polluted soil by bringing out the interactions between contaminants and biomass (Kour et al. 2021).

6.1 Bioaugmentation

When indigenous microbial communities are insufficient for degrading soil contaminants such as petroleum hydrocarbons, bioaugmentation is applied. When a specific hydrocarbon like polynuclear aromatic hydrocarbons has to be degraded which cannot be done by native microbes, bioaugmentation is used (Jafarinejad 2016; Das and Dash 2014). To reinforce the naturally occurring microbes, microorganisms with higher photocatalytic degradation capacity are supplied to the polluted area and this can be done using a variety of methods. Nonindigenous microorganisms from other contaminated settings are frequently employed to populate the site of action. An alternative method is autochthonous bioaugmentation in which microbes are isolated from the polluted site. Then these microbes are grown in bioreactors under lab conditions (Lim et al. 2016). These mass-produced microorganisms are used as inoculums at the site of the contaminant. Replanted microorganisms to the target area can shorten the time period it takes to initiate bioremediation. The problem of adaptation by the microorganisms is also solved when the inoculums of improved native microorganisms are added. Microbes selected for this purpose are based on the criteria of their physiology and metabolic activity. Preliminary laboratory studies for microbial screening improve the chances of effective bioremediation. The most effective bioaugmentation consortium studied include "*Pseudomonas fluorescens* and *Pseudomonas putida* mixed with *Aeromonas hydrophila* and *Alcaligenes xylosoxidans*, in addition to *Xanthomonas* sp., *Gordonia* sp., *Stenotrophomonas maltophilia*, and *Rhodococcusequi*" (Baniasadi and Mousavi 2019; El Fantroussi and Agathos 2005; Szulc et al. 2014).

6.2 Biostimulation

It is the technique of increasing the native biodegradation rate of organic contaminants by supplying restricted nutrients and co-substrates to enhance the activity of microorganisms (Soleimani et al. 2013; Simpanen et al. 2016). Under certain water and soil conditions, microbial activity and growth might be restricted. The poor degradation of oil sludge is a result of an imbalanced equilibrium of nutritional additives (Singh et al. 2012; Crivelaro et al. 2010). Because oil sludge has a limited quantity of nutrients, biostimulation is one of the options for dealing with this problem. This could be because some nutrients are unavailable in oil sludge. After all, they are elements of complex molecules inaccessible to degrading microbes (Kour et al. 2021). Changing the parameters such as temperature, and oxygenation can also be performed. All these actions are carried out with the goal of accelerating the activity of oil-degrading bacteria and can also be referred to as fertilization (Jafarinejad 2016). Bioremediating microbes include *Alcaligene, Pseudomonas, Acinetobacter,* and *Rhodococcus* that mineralize the pollutants and convert compounds in oil sludge to CO_2 and H_2O (Kour et al. 2021).

6.3 Phytoremediation

Phytoremediation uses green plants to eliminate toxins from the soil by forming a symbiotic relationship with microbes present in the soil (Adam 2001). The procedure utilizes green plants' natural processes to clean up surrounding contaminated soil with organic and inorganic compounds. Phytoextraction, phytotransformation, rhizodegradation, and phytovolatilization are some of the mechanisms by which plants hydrolyze (Hussain et al. 2018; Ossai et al. 2020). Plants have built-in enzymatic and absorption mechanisms that remove, sequester, and transport pollutants.

They provide nutrition and habitat to aerobic and anaerobic microbes in a symbiotic relationship where microbes degrade the contaminants (Ecobiol 2006). Cost-effectiveness and long-term application are advantages of employing phytoremediation. It further eliminates land disturbance, transit, and legal costs involved with offsite remediation (Schnoor et al. 1995; Das and Chandran 2011).

6.4 Phycoremediation

Microalgae are used in phycoremediation to eliminate toxins from the soil such as heavy metals, and petroleum hydrocarbon. Numerous algal species are employed for this like *Anabaena, Cylindretheca, Chlamydomonas, Dunaliella, Nostoc, Oscillatoria, and Ulva* (Naeem and Qazi 2020; Olguín and Sánchez-Galván 2011). Algae are well-known for their adaptability, as they can develop in a variety of environments, including autotrophic, mixotrophic, and heterotrophic settings. Mixotrophic algae are effective bioremediation and carbon sequestration agents (Subashchandrabose et al. 2013). Since algae can perform photosynthesis they are capable of releasing oxygen as well as increase BOD levels and eliminate excess nutrients from polluted water (Fathi et al. 2013). After the completion of the breakdown of contaminants, mineral uptake by the microalgae occurs by physical adsorption on the cell surface and then intracellular uptake and absorption (Ossai et al. 2020).

6.5 Landfarming

In landfarming, contaminated soil is extracted and physically separated by sieve and then dispersed over the soil of thickness 0.4 m and is replanted regularly till the contaminants are degraded. This activity boosts the action of native microorganisms and makes it easier for them to degrade pollutants aerobically (Kumar et al. 2020; Wiliiams 2002). This technique of bioremediation is best suited for pollutants with volatile organic compounds and low molecular weight compounds as a pollutant (Guarino et al. 2017). A higher rate of degradation by the microorganisms is recorded in the top plow layer (10–15 cm). To improve hydrocarbon breakdown, additives such as inocula, fertilizer, and biosurfactants can also be added in the process. Long processing time is required since landfarming settings do not provide ideal and controlled parameters for microbial activity (Ward et al. 2003). But to overcome this issue pulverized limestone is employed to manage the pH of the soil (Kumar et al. 2020).

6.6 Biopile

Biopiles include the application of two techniques, landfarming and composting in a designed cell that is well aerated with blowers. The cell is richly supplied with nutrients and a leachate collection system x (Benyahia and Embaby 2016; Ossai et al. 2020). This method entails piling of extracted sludge followed by biostimulation and aeration to boost microbial action for degradation of pollutants. This method can treat a huge amount of contaminated soil even in harsh conditions in a given space. Oxygen, water, micronutrients, organic materials, and leachate management system are the fundamental elements of this procedure (Dias et al. 2015). Biopiles can be formed in two ways, static piles where the shape of the pile is a pyramid with elevation ranging between 0.8 and 2 m. The height of biopiles formed depends on the type of aeration system employed. While the other is dynamic biopile where the piles are continuously plowed and flipped in order to maintain and maximize the aeration and availability of nutrients to the native microbes (Koning et al. 2001; Kumar et al. 2020).

6.7 Bioventing

It's a type of bioremediation that employs a forced oxygen supply to expedite the remediation process (Dupont 1993; Lee et al. 2006). Air is considered as the most efficient carrier of oxygen than water (Brown et al. 1996). Bioventing uses the forced supply of oxygen to the unsaturated contaminated zones of petroleum hydrocarbon (Kuppusamy et al. 2020). Traditional bioventing methods used an electric-powered blower to forcefully push air into polluted zones. This procedure, however, was both costly and inefficient. Currently, an alternate form of bioventing uses renewable wind and solar energy to inject air into a contaminated site in a reliable and efficient manner (Dominguez et al. 2012). The soil must be porous enough to allow air to travel through it for bioventing to work (Vogel 1995). The rate of bioremediation depends upon the microbial activity and metabolism that remediate the petroleum hydrocarbons (Mosco and Zytner 2017). It is typically used in the low or moderate contamination zones of pHs, which decompose more readily under aerobic than anaerobic circumstances (Lee and Swindoll 1993). Benzene, toluene, ethylbenzene, and xylene are light and simple hydrocarbons that decompose more easily than complex aromatic hydrocarbons (Brown et al. 1996; Hoeppel et al. 1991). The branched and aromatic hydrocarbons take approximately eight to ten years to degrade through bioventing. Soil warming is done to enhance the efficiency of bioventing. The method's disadvantages include its sluggish rate of deterioration and the fact that it cannot be used for extremely volatile chemicals (Brown et al. 1996).

6.8 Biosparging

It's a bioremediation technology during which pressurized air is pumped into aquifers to speed up biodegradation (Kao et al. 2008). Accidental oil leaks during petroleum refining contaminate groundwater, resulting in water pollution (Kim et al. 2014). Biosparging is a method in which microbial activity is enhanced by supplying oxygen at low rates under high pressure in unsaturated contaminated zones (Macaulay and Rees 2014). It increases the interaction between soil and groundwater by increasing mixing in the saturated zone (Vidali 2001). The exterior oxygen supply increases the groundwater oxygen level which stimulates the rate of biodegradation (Johnson et al. 2001). The goal of low-rate oxygen delivery is to promote hydrocarbon biodegradation rather than volatilization (Sperry et al. 2001). This approach is excellent at removing low molecular weight volatile hydrocarbons like benzene, toluene, ethylbenzene, and xylene from the environment (Kumar et al. 2018). Biosparging and bioventing have the same basic biodegradation rate constraints (Brown et al. 1996).

6.9 Bioslurry

It's an ex-situ bioremediation method that entails using bioreactors to digest soil, liquid, or sediments (Kuppusamy et al. 2016). The system can be used in a variety of feed modes, including continuous, semicontinuous, and batch, and in aerobic, anoxic, and anaerobic settings (Concetta Tomei and Daugulis 2013). The polluted soil or liquid is fed into the bioreactor as a slurry and allowed to remediate under the bioreactor's control conditions. As an inoculum, soil-native microorganisms or bacteria with specific metabolic activity against a specific hydrocarbon are utilized (Kuppusamy et al. 2016). To improve the efficacy of bioremediation, conditions such as pH, temperature, nutrient input, oxygen, and mixing are managed and tuned in the bioreactor (Lumia et al. 2020). It improves reaction kinetics and emissions management (Ossai et al. 2020). Additional specialized nutrients are supplied to enhance microbial growth and metabolism in order to speed up bioremediation rates (Ossai et al. 2020). A variety of petroleum hydrocarbons are degraded by this method of bioremediation (Kuppusamy et al. 2016). Bioslurry has been shown to have higher rates of remediation than other in situ methods available (Prasanna et al. 2008). This approach provides for greater control parameter optimization and improved mass transfer rates (Lumia et al. 2020). The high cost, prolonged treatment time, soil extraction, and transfer to the treatment region are the biggest downsides of this strategy (Concetta Tomei and Daugulis 2013; Ossai et al. 2020) (Table 1).

Table 1 Techniques and microorganisms used in bioremediation

Microorganisms	Substrate	Method of Bioremediation	%TPH Reduction	References
Methylococcus	Alkanes and cycloalkanes	Biostimulation	96	(Chandra et al. 2013)
Aspergillus sp., *Penicillium* sp.	Polychlorinated biphenyls (PCBs)	Bioslurry	13	(Naeem and Qazi 2020; Baniasadi and Mousavi 2019)
Rhodococcus	cycloalkanes	Natural attenuation by creating 2 biopiles	81	(Naeem and Qazi 2020)
Acinetobacter	C10-C30 alkanes	Biopile + phytoremediation	80	(Chandra et al. 2013; Naeem and Qazi 2020)
Mycobacterium	C5-C16 alkanes	Biostimulation	> 90	(Naeem and Qazi 2020)
Alcanivorax sp., *Cycloclasticus* sp.,	Aliphatic and aromatic compounds	Isolated bacterial consortium	82.3	(Baniasadi and Mousavi 2019; Naeem and Qazi 2020)
Chlorella vulgaris, *Arthrospira platensis*	Inorganic compounds	Addition of biosurfactants, biostimulation	80–90	(Olguín and Sánchez-Galván 2011)
Candida tropicalis, *Candida maltosa*	C10-C16 alkanes, fatty acids	Biopile	60–70	(Chandra et al. 2013; Naeem and Qazi 2020)
Pseudomonas sp.	C5-C16 alkanes, alkyl benzenes	Bacterial consortium, biopile	> 80	(Chandra et al. 2013; Naeem and Qazi 2020)
Methylocystis	C1-C5 (halogenated) alkanes	Bioaugmentation	70–80	(Chandra et al. 2013)

7 Factors Affecting the Efficiency of Bioremediation

Bioremediation is the process of transforming hazardous compounds into less dangerous substances such as CO2 and H2O employing indigenous bacteria. As a result, bioremediation efficacy is reliant on the ability and variables impacting the activity of the bacteria that degrade the contaminants. [1] Environmental elements such as pH, temperature, nutrition, oxygen, salinity, and water activity are all key influences on microbial activity, [2] molecular structure, and the nature of hydrocarbons [3] functional activity and characteristics of the microbes (Al-Hawash et al. 2018; Sarkar et al. 2020; Varjani and Upasani 2017).

7.1 Temperature

It is one of the key determinants of bioremediation rate. It alters microbial development and metabolism, gas solubility, viscosity, volatilization, and diffusion of PHs, as well as the physicochemical characteristics of hydrocarbons (Atlas 1975; Sarkar et al. 2020). Because the enzymatic activity is limited at low temperatures, the rate of breakdown is slow (Al-Hawash et al. 2018). Moreover, low temperature increases the viscosity and the rate of volatilization (Foght et al. 1996). On the other hand, high temperatures reduce viscosity and increase the rate of diffusion, solubility, and volatilization of the hydrocarbon (Aislabie et al. 2006; Okoh 2006; Varjani and Upasani 2017). In the soil environment, the rate of hydrocarbon deterioration is best when the temperature is around 30 and 40 degrees Celsius. Bioremediation occurs at a different range of temperatures. For example, in the marine environment, the maximum decomposition rate occurs at temperatures between 15 and 20 degrees Celsius, while in the freshwater environment, the maximum rate of degradation occurs at temperatures between 20 and 30 degrees Celsius (Al-Hawash et al. 2018; Varjani and Upasani 2017; Yuniati 2018). In an experiment performed by Thamer et al. (2013), they used *Bacillus thuringiensis* bacteria to remediate crude oil in order to access its bioremediation capacity. *Bacillus thuringiensis* was incubated at 30 °C for 27 days in a bottle containing 2 ml of crude oil. The bacteria was shown to be capable of remediating 80 percent of the crude oil by creating 2.3 g/l of emulsion, which improved the accessibility of the hydrocarbon components (Thamer et al. 2013) (Fig. 2).

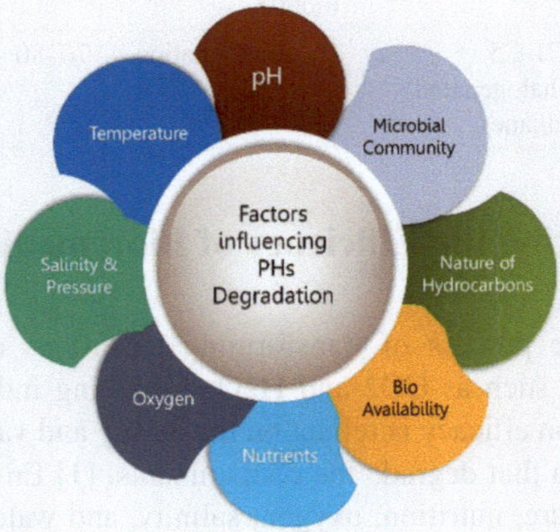

Fig. 2 Factors affecting petroleum hydrocarbon bioremediation

7.2 pH

pH is another physiological factor that influences the efficacy of bioremediation (Aislabie et al. 2006; Yuniati 2018). pH is a highly variable factor that affects the functional activity of the enzymes and membrane transport. Different enzymes have different optimum pH at which they show their best activity. Changes in pH in a variety of habitats affect the functional and catalytic activity of the enzymes. Different species engaged in the bioremediation process tend to grow in different pH ranges. Heterotrophic bacteria and fungi, for example, demand a neutral or alkaline pH to grow, while fungi may withstand an acidic pH. Several researchers have proven that when the pH is neutral or alkaline, the rate of bioremediation rises (Yuniati 2018). The influence of pH, reduction potential, and oxygen on the rates of degradation of two hydrocarbons, octadecane, and naphthalene, was studied by Hambrick and his co-workers. He discovered that mineralization rates are controlled by pH and reduction potential, with the highest rate of mineralization occurring at pH 8 with a high redox potential (Iii et al. 1980). Thavasi et al. investigate the impact of several variables such as pH, temperature, and salinity on the rate of *Pseudomonas aeruginosa* bioremediation of crude oil. The *Pseudomonas aeruginosa* was grown at a different range of pH, temperature, salinity (NaCl), and substrate (crude oil) conditions. The best bioremediation rates were found at 38 °C, pH 8.0, and 35 and 2.0 percent crude oil, respectively (Thavasi et al. 2007).

7.3 Salinity and Pressure

The major parameters that influence the rate of bioremediation are salinity and pressure. In the extreme hypersaline environments, salt lakes or deep seas, salinity, and pressure are the major determinants of the rate of bioremediation of hydrocarbons. Increased salt concentrations decrease the rate of bioremediation by affecting microbial growth, metabolism, and catalytic efficiency of the key enzymes involved in the process of remediation (Al-Hawash et al. 2018; Waikhom et al. 2020). In many bacterial species, high salt concentration causes osmotic shock and plasmolysis. Wardl and Brock (1978) conducted an experiment to study the effect of salinity on the bioremediation of glutamate and hexadecane from the samples collected from the Great Salt Lake. He found that the rate of hexadecane bioremediation was minimal, whereas the rate of glutamate bioremediation was very low in high salt concentrations. In addition to the excessive salinity, high pressure slows bioremediation, Wardl and Brock (1978), Schwarz et al. (1975) observed that mixed cultures at 4C at 1 atm degraded n hexadecane completely in 4 weeks. When the cultures were subjected to 500 atm pressure at 4C, the same microbial population took 32 weeks to bioremediate the n hexadecane. The rate of metabolism of bacteria is slowed by high pressure. As a result, the bioremediation rate is hindered. Hydrocarbon breakdown rates in the deep benthic zones of oceans are extremely slow due to the refractory fraction of hydrocarbons that survive for long periods of time (Varjani and Upasani 2017).

7.4 Nutrient Availability

Nutrient availability is critical for the bioremediation of petroleum hydrocarbons by microbes (Al-Hawash et al. 2018). Nutrients such as nitrogen, carbon, phosphorous, and potassium are limiting factors for the growth and metabolism of microbes (Aislabie et al. 2006). Biostimulation, also known as nutrient-enhanced bioremediation, is one of the most viable "new approaches" for oil-contaminated coastline remediation (Nikolopoulou and Kalogerakis 2008). However unlimited supply of nutrients above the optimum concentration retards the growth and biodegradation efficiency of microbes (Waikhom et al. 2020). Because microbes use nitrogen and phosphorus to generate biomass, nutrient availability at the bioremediation site is vital (Waikhom et al. 2020). Microbes get carbon from the hydrocarbon degradation and oxygen and hydrogen are provided by water (Al-Hawash et al. 2018). Marine ecosystems polluted by oil spills have low nitrogen and phosphorus concentrations but high carbon concentrations (Atlas 1981). However, due to the high demand for nitrogen and phosphorous by the plants that grow there, wetlands are unable to deliver appropriate amounts of N/P (Atlas 1981). To alter the N/P/K ratios, N/P/K is supplied from an external source in the form of N-P-K fertilizers, oleophilic fertilizers, ammonium salts, and phosphorous. Microbes thrive best in a 100:10:1 ratio of carbon, nitrogen, and phosphorus. (Kim et al. (2005) conducted a study to evaluate how adding nutrients, surfactants, and microbial communities affected bioremediation efficacy. He stated that using SRIF (slow-release inorganic fertilizer) with a C: N:P ratio of 100:10:3 increased the CO2 evolution rate by tenfold (Kim et al. 2005).

7.5 Oxygen

Microbes break down petroleum hydrocarbons by an oxidative process in which oxygenases and peroxidases oxidize the substrate, with molecular oxygen acting as an electron sink (Leahy and Colwell 1990; Okoh 2006). Degradation of petroleum hydrocarbons occurs more efficiently in aerobic conditions. The amount of oxygen available in the soil is determined by its texture, type, and microbial oxygen consumption rates (Al-Hawash et al. 2018). Nitrate and sulfate act as external electron donors under anaerobic settings, assisting in the degradation of hydrocarbons (Atlas 1981). Anaerobic decomposition occurs at a significantly slower pace than aerobic decomposition. Grishchenkov et al. (2000) investigated the biodegradation efficiency of *Pseudomonas sp.BS2201, BS2203*, and *Brevibacillus sp.BS2202* strains under aerobic and anaerobic conditions. Under aerobic and anaerobic conditions, microorganisms degraded 20–25% and 15–18% of total extractable material, respectively

(Grishchenkov et al. 2000). As a result, fast hydrocarbon breakdown does not occur in the oxygen deficit conditions. Anoxic sediments contaminated by hydrocarbons remain there as pollutants for many years (Atlas 1981).

7.6 Nature of Hydrocarbon

Degradability of petroleum hydrocarbons depends upon the concentration, and physical and chemical properties of the hydrocarbons (Al-Hawash et al. 2018). Simple hydrocarbons with a low molecular weight are more sensitive to microbial breakdown than complex hydrocarbons with a high molecular weight (Leahy and Colwell 1990). Biodegradability of hydrocarbons can be ranked as linear alkanes > branched alkanes > low-molecular weight alkyl aromatics > monoaromatics > cyclic alkanes > polyaromatics asphaltenes (Yuniati 2018). Aromatic- or saturates-containing hydrocarbons are more biodegradable than heavy asphaltic-naphthenic crude oils (Varjani and Upasani 2017). The concentration of petroleum hydrocarbons has an impact on bioremediation. The lag phase decreases as hydrocarbon concentration rises, and the level of mineralization and degradation rises. Extremely high hydrocarbon concentrations are harmful to microbial organisms, while extremely low hydrocarbon concentrations restrict carbon supply, slowing breakdown (Waikhom et al. 2020). Because of the bioavailability of the compound, the same hydrocarbon compound can be destroyed to varying degrees by the same bacterium (Varjani and Upasani 2017).

7.7 Bioavailability and Biosurfactants

The extent of bioremediation depends upon the bioavailability of petroleum hydrocarbons (Al-Hawash et al. 2018). The amount of a material that is physically and chemically available for bacteria to take up and use is referred to as bioavailability. Hydrophobic petroleum hydrocarbons are hydrocarbons with poor bioavailability and low water solubility, making them resistant to degradation (Al-Hawash et al. 2018). The bioavailability of a pollutant is governed by its chemical properties, such as hydrophobicity, solubility, biological activity, and environmental circumstances (Waikhom et al. 2020). Hydrocarbon bioremediation is dependent on interactions between hydrocarbon droplets and microbial cells, as well as diffusion into microbial cells through cell membranes (Varjani and Upasani 2017). Biosurfactants are bacteria-produced surface-active compounds that help in the emulsification of hydrocarbons, allowing microbes to metabolize them (Banat et al. 2010). They're amphipathic compounds synthesized as secondary metabolites during microbial growth in the stationary phase (Banat et al. 2010). They can either be released extracellularly or remain linked to microbial cells (Varjani and Upasani 2017). They are nontoxic and biodegradable, and they aid in lowering the oil's interfacial tensions and viscosity, making it more susceptible to biodegradation (Dias et al. 2012; Kavitha et al. 2014).

8 Conclusion

This chapter presented an outline of the restoration of petroleum hydrocarbons contaminating soil and water. There is a multitude of known remediation techniques, but no single approach is best suited for all kinds of pollutants. The choice of remediation is based on a thorough understanding of characteristics of pollutants and the environment and most importantly morphology and degrading capability of the microbes. Bioremediation is a more efficient method than existing physicochemical procedures which include toxic chemicals. It involves the degradation of harmful and toxic contaminants by microbes into nontoxic and simpler compounds. The efficacy of bioremediation depends upon various factors including types of microbes, pH, temperature, salinity, pressure, concentration, and nature of hydrocarbons. The metabolic and functional activities of microbes rely on these physical and chemical factors. These variables can be tweaked to speed up the bioremediation process. The extraordinary significance of extremophiles in bioremediation emphasizes the necessity for more research so that new species can be discovered and the processes they utilize to survive in such harsh settings can be investigated. Multi-omics research is currently insufficient, and more research is needed to better understand the ecology, epigenetics, and metabolism of bacteria involved in bioremediation. Microbes that have been genetically modified to improve their ability to degrade contaminants will have a great impact in improving bioremediation efficiency.

References

Abbasian F, Lockington R, Mallavarapu M, Naidu R (2015) A comprehensive review of aliphatic hydrocarbon biodegradation by bacteria. Appl Biochem Biotechnol 176:670–699. https://doi.org/10.1007/s12010-015-1603-5

Adam G (2001) A study into the potential of phytoremediation for diesel fuel contaminated soil, 1–427

Ahmad J (2017) Bioremediation of petroleum sludge using Effective Microorganism (EM) technology. Pet Sci Technol 35:1515–1522. https://doi.org/10.1080/10916466.2017.1356850

Aislabie J, Saul DJ, Foght JM (2006) Bioremediation of hydrocarbon-contaminated polar soils. Extremophiles 10:171–179

Al-Hawash AB, Dragh MA, Li S et al (2018) Principles of microbial degradation of petroleum hydrocarbons in the environment. Egypt J Aquat Res 44:71–76

Atlas RM (1975) Effects of temperature and crude oil composition on petroleum biodegradation

Atlas RM (1981) Microbial degradation of petroleum hydrocarbons: an environmental perspective

Banat IM, Franzetti A, Gandolfi I et al (2010) Microbial biosurfactants production, applications and future potential. Appl Microbiol Biotechnol 87:427–444

Banerjee A (2021) Toxic effect and bioremediation of oil contamination in algal perspective. Elsevier

Baniasadi M, Mousavi SM (2019) A comprehensive review on the bioremediation of oil spills

Benyahia F, Embaby AS (2016) Bioremediation of crude oil contaminated desert soil: effect of biostimulation, bioaugmentation and bioavailability in biopile treatment systems. Int J Environ Res Public Health 13. https://doi.org/10.3390/ijerph13020219

Brown RA, Hinchee RE, Norras RD, Wilson JT (1996) Biorernediation of petroleum hydrocarbons: a flexible, variable speed technology

Cai Y, Wang R, Rao P et al (2021) Bioremediation of petroleum hydrocarbons using acinetobacter sp. SCYY-5 isolated from contaminated oil sludge: strategy and effectiveness study. Int J Environ Res Public Health 18:1–14. https://doi.org/10.3390/ijerph18020819

Chaillan F, Gugger M, Saliot A et al (2006) Role of cyanobacteria in the biodegradation of crude oil by a tropical cyanobacterial mat. Chemosphere 62:1574–1582. https://doi.org/10.1016/j.chemosphere.2005.06.050

Chaillan F, le Flèche A, Bury E et al (2004) Identification and biodegradation potential of tropical aerobic hydrocarbon-degrading microorganisms. Res Microbiol 155:587–595. https://doi.org/10.1016/j.resmic.2004.04.006

Chandra S, Sharma R, Singh K, Sharma A (2013) Application of bioremediation technology in the environment contaminated with petroleum hydrocarbon. Ann Microbiol 63:417–431. https://doi.org/10.1007/s13213-012-0543-3

Claro EMT, Cruz JM, Montagnolli RN et al (2019) Microbial degradation of petroleum hydrocarbons: technology and mechanism. Microb Action Hydrocarb 125–141. https://doi.org/10.1007/978-981-13-1840-5_6

Concetta Tomei M, Daugulis AJ (2013) Ex situ bioremediation of contaminated soils: an overview of conventional and innovative technologies. CritAl Rev Environ Sci Technol 43:2107–2139

Crivelaro SHR, Mariano AP, Furlan LT et al (2010) Evaluation of the use of vinasse as a biostimulation agent for the biodegradation of oily sludge in soil. Braz Arch Biol Technol 53:1217–1224. https://doi.org/10.1590/S1516-89132010000500015

Das N, Chandran P (2011) Microbial degradation of petroleum hydrocarbon contaminants: an overview. Biotechnol Res Int 2011:1–13. https://doi.org/10.4061/2011/941810

Das S, Dash HR (2014) Microbial bioremediation: a potential tool for restoration of contaminated areas. Elsevier Inc.

Dias RL, Ruberto L, Calabró A et al (2015) Hydrocarbon removal and bacterial community structure in on-site biostimulatedbiopile systems designed for bioremediation of diesel-contaminated Antarctic soil. Polar Biol 38:677–687. https://doi.org/10.1007/s00300-014-1630-7

Dias RL, Ruberto L, Hernández E et al (2012) Bioremediation of an aged diesel oil-contaminated Antarctic soil: evaluation of the “on site” biostimulation strategy using different nutrient sources. Int Biodeterior Biodegrad 75:96–103. https://doi.org/10.1016/j.ibiod.2012.07.020

Dominguez RC, Leu J, Bettahar M (2012) Sustainable wind-driven bioventing at a petroleum hydrocarbon-impacted site. Remediation 22:65–78. https://doi.org/10.1002/rem.21321

Dupont R (1993) Fundamentals of bioventing applied to fuel contaminated sites. Environ Prog 12(1):45–53

Ecobiol J (2006) Phvtoremediation: processes and mechanisms. J Ecobiol 18:33–38

El Fantroussi S, Agathos SN (2005) Is bioaugmentation a feasible strategy for pollutant removal and site remediation? Curr Opin Microbiol 8:268–275. https://doi.org/10.1016/j.mib.2005.04.011

Fathi AA, Azooz MM, Al-Fredan MA (2013) Phycoremediation and the potential of sustainable algal biofuel production using wastewater. Am J Appl Sci 10:189–194. https://doi.org/10.3844/ajassp.2013.189.194

Foght JM, Westlake DW, Johnson WM, Ridgway HF (1996) Environmental gasoline-utilizing isolates and clinical isolates of Pseudomonas aeruginosa are taxonomically indistinguishable by chernotaxonornic and molecular techniques

Fritsche W, Hofrichter M (2001, May) Aerobic degradation by microorganisms. Microbiol Set 144–167

Grishchenkov VG, Townsend RT, Mcdonald TJ et al (2000) Degradation of petroleum hydrocarbons by facultative anaerobic bacteria under aerobic and anaerobic conditions

Guarino C, Spada V, Sciarrillo R (2017) Assessment of three approaches of bioremediation (Natural Attenuation, Landfarming and Bioagumentation – Assistited Landfarming) for a petroleum hydrocarbons contaminated soil. Chemosphere 170:10–16. https://doi.org/10.1016/j.chemosphere.2016.11.165

Hamzah A, Phan CW, Abu Bakar NF, Wong KK (2013) Biodegradation of crude oil by constructed bacterial consortia and the constituent single bacteria isolated from Malaysia. Bioremediation J 17:1–10. https://doi.org/10.1080/10889868.2012.731447
Haritash AK, Kaushik CP (2009) Biodegradation aspects of Polycyclic Aromatic Hydrocarbons (PAHs): a review. J Hazard Mater 169:1–15
Hendrickx B, Junca H, Vosahlova J et al (2006) Alternative primer sets for PCR detection of genotypes involved in bacterial aerobic BTEX degradation: distribution of the genes in BTEX degrading isolates and in subsurface soils of a BTEX contaminated industrial site. J Microbiol Methods 64:250–265. https://doi.org/10.1016/j.mimet.2005.04.018
Hoeppel RE, Hinchee RE, Arthur MF (1991) Bioventing soils contaminated with petroleum hydrocarbons
Hussain I, Puschenreiter M, Gerhard S et al (2018) Rhizoremediation of petroleum hydrocarbon-contaminated soils: improvement opportunities and field applications. Environ Exp Bot 147:202–219. https://doi.org/10.1016/j.envexpbot.2017.12.016
Iii GAH, Delaune RD, Patrick WH (1980) Effect of estuarine sediment pH and oxidation-reduction potential on microbial hydrocarbon degradation
Islam B (2015) Petroleum sludge, its treatment and disposal: a review
Iwamoto T, Nasu M (2001) Current bioremediation practice and perspective. J Biosci Bioeng 92:1–8. https://doi.org/10.1263/jbb.92.1
Jafarinejad S (2016) Petroleum waste treatment and pollution control, 1–362
Johnson OA, Affam AC (2019) Petroleum sludge treatment and disposal: a review. Environ Eng Res 24:191–201. https://doi.org/10.4491/EER.2018.134
Johnson PC, Johnson RL, Bruce CL, Leeson A (2001) Advances in in situ air sparging/biosparging. Bioremediation J 5:251–266. https://doi.org/10.1080/20018891079311
Kao CM, Chen CY, Chen SC et al (2008) Application of in situ biosparging to remediate a petroleum-hydrocarbon spill site: field and microbial evaluation. Chemosphere 70:1492–1499. https://doi.org/10.1016/j.chemosphere.2007.08.029
Kavitha V, Mandal AB, Gnanamani A (2014) Microbial biosurfactant mediated removal and/or solubilization of crude oil contamination from soil and aqueous phase: an approach with Bacillus licheniformis MTCC 5514. Int Biodeterior Biodegrad 94:24–30. https://doi.org/10.1016/j.ibiod.2014.04.028
Kim S, Krajmalnik-Brown R, Kim JO, Chung J (2014) Remediation of petroleum hydrocarbon-contaminated sites by DNA diagnosis-based bioslurping technology. Sci Total Environ 497–498:250–259. https://doi.org/10.1016/j.scitotenv.2014.08.002
Kim SJ, Choi DH, Sim DS, Oh YS (2005) Evaluation of bioremediation effectiveness on crude oil-contaminated sand. Chemosphere 59:845–852. https://doi.org/10.1016/j.chemosphere.2004.10.058
Koning M, Hupe K, Stegmann R (2001) Thermal processes, scrubbing/extraction, bioremediation and disposal. Biotechnol Set, 304–317. Wiley Online Library
Kour D, Kaur T, Devi R et al (2021) Beneficial microbiomes for bioremediation of diverse contaminated environments for environmental sustainability: present status and future challenges. Environ Sci Pollut Res 28:24917–24939. https://doi.org/10.1007/s11356-021-13252-7
Kumar B, Manoj K, Solanki K et al (2020) Waste to energy: prospects and applications
Kumar V, Shahi SK, Singh S (2018) Bioremediation: an eco-sustainable approach for restoration of contaminated sites. Microbial bioprospecting for sustainable development. Springer, Singapore, pp 115–136
Kuppusamy S, Maddela NR, Megharaj M, Venkateswarlu K (2020) Approaches for remediation of sites contaminated with total petroleum hydrocarbons. In: Total petroleum hydrocarbons. Springer, pp 167–205
Kuppusamy S, Palanisami T, Megharaj M et al (2016) Ex-situ remediation technologies for environmental pollutants: a critical perspective. Reviews of environmental contamination and toxicology. Springer, New York LLC, pp 117–192
Leahy JG, Colwell RR (1990) Microbial degradation of hydrocarbons in the environment

Lee MD, Swindoll CM (1993) Bioventing for in situ remediation. Hydrol Sci J 38(4):273–282
Lee TH, Byun IG, Kim YO et al (2006) Monitoring biodegradation of diesel fuel in bioventing processes using in situ respiration rate. Water Sci Technol 53:263–272. https://doi.org/10.2166/wst.2006.131
Lim MW, Von LE, Poh PE (2016) A comprehensive guide of remediation technologies for oil contaminated soil—present works and future directions. Mar Pollut Bull 109:14–45. https://doi.org/10.1016/j.marpolbul.2016.04.023
Lumia L, Rabbeni G, Giustra MG et al (2020) Treatment of contaminated sediments by bio-slurry reactors: study on the effect of erythromycin antibiotic. Chem Eng Trans 79:391–396. https://doi.org/10.3303/CET2079066
Macaulay B Rees D (2014) Bioremediation of oil spills: a review of challenges for research advancement
Mosco MJ, Zytner RG (2017) Large-scale bioventing degradation rates of petroleum hydrocarbons and determination of scale-up factors. Bioremediation J 21:149–162. https://doi.org/10.1080/10889868.2017.1312265
Naeem U, Qazi MA (2020) Leading edges in bioremediation technologies for removal of petroleum hydrocarbons. Environ Sci Pollut Res 27:27370–27382. https://doi.org/10.1007/s11356-019-06124-8
Nikolopoulou M, Kalogerakis N (2008) Enhanced bioremediation of crude oil utilizing lipophilic fertilizers combined with biosurfactants and molasses. Mar Pollut Bull 56:1855–1861. https://doi.org/10.1016/j.marpolbul.2008.07.021
Okoh AI (2006) Biodegradation alternative in the cleanup of petroleum hydrocarbon pollutants. Biotechnol Mol Biol Rev 1:38–50
Olajire AA, Essien JP (2014) Aerobic degradation of petroleum components by microbial consortia. J Pet Environ Biotechnol 5. https://doi.org/10.4172/2157-7463.1000195
Olguín EJ, Sánchez-Galván G (2011) Phycoremediation
Ossai IC, Ahmed A, Hassan A, Hamid FS (2020) Remediation of soil and water contaminated with petroleum hydrocarbon: a review. Environ Technol Innov 17. https://doi.org/10.1016/j.eti.2019.100526
Prasanna D, Venkata Mohan S, Purushotham Reddy B, Sarma PN (2008) Bioremediation of anthracene contaminated soil in bio-slurry phase reactor operated in periodic discontinuous batch mode. J Hazard Mater 153:244–251. https://doi.org/10.1016/j.jhazmat.2007.08.063
Sarkar J, Roy A, Sar P, Kazy SK (2020) Accelerated bioremediation of petroleum refinery sludge through biostimulation and bioaugmentation of native microbiome. In: Emerging technologies in environmental bioremediation. Elsevier, pp 23–65
Schnoor JL, Licht LA, McCutcheon SC et al (1995) Phytoremediation of organic and nutrient contaminants: pilot and full-scale studies are demonstrating the promise and limitations of using vegetation for remediating hazardous wastes in soils and sediments. Environ Sci Technol 29:318. https://doi.org/10.1021/es00007a747
Schwarz JR, Walker JD, Colwell DRR, Colwell RR (1975) Deep-sea bacteria: growth and utilization of n-hexadecane at in situ temperature and pressure
Shah Maulin P (2020) Microbial bioremediation & biodegradation. Springer
Simpanen S, Dahl M, Gerlach M et al (2016) Biostimulation proved to be the most efficient method in the comparison of in situ soil remediation treatments after a simulated oil spill accident. Environ Sci Pollut Res 23:25024–25038. https://doi.org/10.1007/s11356-016-7606-0
Singh B, Bhattacharya A, Channashettar VA et al (2012) Biodegradation of oil spill by petroleum refineries using consortia of novel bacterial strains. Bull Environ Contam Toxicol 89:257–262. https://doi.org/10.1007/s00128-012-0668-x
Soleimani M, Farhoudi M, Christensen JH (2013) Chemometric assessment of enhanced bioremediation of oil contaminated soils. J Hazard Mater 254–255:372–381. https://doi.org/10.1016/j.jhazmat.2013.03.004
Sperry KL, Stanley C, Kay J (2001) Field trial of biosparging with oxygen for bioremediation of volatile organic compounds. Remediat j: J Environ Cleanup Costs, Technol Tech 11(4):47–62

Subashchandrabose SR, Ramakrishnan B, Megharaj M et al (2013) Mixotrophic cyanobacteria and microalgae as distinctive biological agents for organic pollutant degradation. Environ Int 51:59–72. https://doi.org/10.1016/j.envint.2012.10.007

Suganthi SH, Murshid S, Sriram S, Ramani K (2018) Enhanced biodegradation of hydrocarbons in petroleum tank bottom oil sludge and characterization of biocatalysts and biosurfactants. J Environ Manag 220:87–95. https://doi.org/10.1016/j.jenvman.2018.04.120

Szulc A, Ambrozewicz D, Sydow M et al (2014) The influence of bioaugmentation and biosurfactant addition on bioremediation efficiency of diesel-oil contaminated soil: feasibility during field studies. J Environ Manag 132:121–128. https://doi.org/10.1016/j.jenvman.2013.11.006

Tang X, He LY, Tao XQ et al (2010) Construction of an artificial microalgal-bacterial consortium that efficiently degrades crude oil. J Hazard Mater 181:1158–1162. https://doi.org/10.1016/j.jhazmat.2010.05.033

Thamer M, Al-Kubaisi AR, Zahraw Z et al (2013) Biodegradation of Kirkuk light crude oil by Bacillus thuringiensis, Northern of Iraq. Nat Sci 5:865–873. https://doi.org/10.4236/ns.2013.57104

Thavasi R, Jayalakshmi S, Balasubramanian T, Banat IM (2007) Effect of salinity, temperature, pH and crude oil concentration on biodegradation of crude oil by Pseudomonas aeruginosa

Varjani SJ (2016) Microbial degradation of petroleum hydrocarbons. Biores Technol 223:277–286. https://doi.org/10.1016/j.biortech.2016.10.037

Varjani SJ, Upasani VN (2016) Biodegradation of petroleum hydrocarbons by oleophilic strain of Pseudomonas aeruginosa NCIM 5514. Bioresour Technol 222:195–201. https://doi.org/10.1016/j.biortech.2016.10.006

Varjani SJ, Upasani VN (2017) A new look on factors affecting microbial degradation of petroleum hydrocarbon pollutants. Int Biodeterior Biodegrad 120:71–83

Vidali M (2001) Bioremediation. an overview. Pure Appl Chem 73(7):1163–1172

Vogel CM (1995) Overview of bioventing technology for the remediation of petroleum hydrocarbon contamination. Clean-up of former soviet military installations: identification and selection of environmental technologies for use in Central and Eastern Europe. Springer, Berlin Heidelberg, pp 109–111

Waikhom D, Ngasotter S, Devi LS et al (2020) Role of microbes in petroleum hydrocarbon degradation in the aquatic environment: a review. Int J Curr Microbiol Appl Sci 9:2990–2903. https://doi.org/10.20546/ijcmas.2020.905.342

Ward DM, Brock TD (1978) Hydrocarbon biodegradation in hypersaline environments. Appl Environ Microbiol 35(2):353–359

Ward O, Singh A, Van Hamme J (2003) Accelerated biodegradation of petroleum hydrocarbon waste. J Ind Microbiol Biotechnol 30:260–270. https://doi.org/10.1007/s10295-003-0042-4

Ward OP (2004) The industrial sustainability of bioremediation processes. J Ind Microbiol Biotechnol 31:1–4. https://doi.org/10.1007/s10295-004-0109-x

Ward OP, Singh A (2004) Evaluation of current soil bioremediation technologies. Appl Bioremediation Phytoremediation 1:187–214. https://doi.org/10.1007/978-3-662-05794-0_9

Wiliiams J (2002) Bioremediation of contaminated soils: a comparison of in situ and ex situ techniques

Yuniati MD (2018) Bioremediation of petroleum-contaminated soil: a Review. In: IOP conference series: earth and environmental science. Institute of Physics Publishing

Zhang X, Li J, Thring R, Huang Y (2010) Surfactant enhanced biodegradation of petroleum hydrocarbons in oil refinery tank bottom sludge. J Can Pet Technol 49:34–39. https://doi.org/10.2118/137211-PA

Development and Implementation of the Integrated Technology for Biological Detoxication of Ionic Mercury in Industrial Wastewater

Pawel Gluszcz

1 Introduction

The issue of process integration in industrial technologies is becoming increasingly popular, both in scientific and practical terms (Klemes et al. 2013). Thanks to the synergistic effect of the interaction of several processes taking place in the same apparatus and at the same time, it is possible to obtain much better quality and economic outcomes than in "traditionally," i.e., sequentially implemented technology even leading to the same goal. For this reason, striving to integrate processes while creating new or modifying known technologies fits well into the strategy of sustainable development—the only way to protect the natural environment from global catastrophe, which may be caused by the destructive activity of a rapidly developing industry (Kalitvenzeff et al. 2001; Dunn and El-Halwagi 2003).

In this paper the results of research focused on one of them—the integrated technology of biological detoxification of toxic ionic mercury present in industrial wastewater. In this case, increased efficiency and improved parameters of the wastewater treatment technology were obtained by combining a typical sorption of impurities on activated carbon with a simultaneous biological process—enzymatic bioreduction of Hg (II) ionic mercury to its elementary form. Elementary mercury is relatively harmless to living organisms in the aquatic environment, due to very low solubility of metallic mercury in water, and may be easily removed from the bioreactor in this form. In this paper a brief description of the way to achieve positive results of such an integrated process is presented. The development of the technology started from the research of the basics of the process, i.e., balance and kinetics data of sorption of ionic and metallic mercury from aqueous solutions onto the adsorbent bed (Gluszcz et al. 2004), and ending with the construction of the installation, which was applied

P. Gluszcz (✉)
Lodz University of Technology, Lodz, Poland
e-mail: pgluszcz@p.lodz.pl

M. P. Shah (ed.), *Microbial Technologies in Industrial Wastewater Treatment*,
https://doi.org/10.1007/978-981-99-2435-6_14

in a full industrial scale at the chemical plant in Tarnów-Mościce, Poland (Gluszcz et al. 2009).

2 Toxicity and Disposal Potential of Mercury

Mercury is one of the most dangerous heavy metals. Unlike many other compounds, which are essential for life in small quantities and toxic only at high concentrations, mercury plays no positive role in metabolic processes and is harmful to living organisms even in the smallest doses (Boening 2000). Its toxic effects are related to the affinity to functional thiol, carboxylic and amino acid groups and it is based on blocking the biochemical functions of these compounds. The harmful effects of mercury on living cells are irreversible since the metal in any form binds permanently to the sulfhydryl groups of enzymes and proteins, leading to their deactivation and, as a result, to disturbances in the metabolic processes of living cells. The most sensitive to the destructive effects of this element is the central nervous system of animals and humans (Hyman 2004). Mercury therefore has an extremely adverse, toxic effect on all living organisms and is disproportionately slowly removed from the environment, compared to the rate and intensity of its anthropogenic, industrial emissions. An additional risk is that the metal accumulates in the tissues of living organisms, which leads to the accumulation of toxic effects at subsequent trophic levels in organisms forming the food chain (Gochfeld 2003).

Environmental pollution by mercury, due to its specific physical and chemical properties, is global in nature and no region of the world is currently free from mercury compounds pollution (Shahid et al. 2020). Restrictions on the industrial use of mercury were introduced in EU countries many years ago, but in economically less developed countries, both continental and global, mercury continues to be widely used and released into the environment in large quantities, freely spreading in the atmosphere and accumulating in the aquatic ecosystems of the world (Sysalova et al. 2017; Rashid et al. 2022).

Mercury has been used in many applications since ancient times, but nowadays most of it is used and consumed in the chemical industry, mainly for the simultaneous production of chlorine and sodium hydroxide in the so-called chlor-alkali technology (Gluszcz et al. 2013). This method involves electrolysis of the NaCl solution in apparatus where thin layer of the liquid mercury forms the cathode. The advantage of amalgam technology is that it is possible to obtain Cl_2 and NaOH of very high quality and in higher concentrations than in other available methods; the obvious disadvantage is the emission of toxic mercury into the environment. In properly designed installations it should be kept in a closed loop, but in practice always a certain amount of mercury is released through different streams: through air, sewage, solid waste, and the products themselves, and the resulting waste must be treated using costly technologies. These include precipitation in the form of mercury sulfide (HgS), chemical or electrochemical reduction of ionic mercury, adsorption on various types of sorbents/biosorbents, and ion exchange (Hua et al. 2020). These solutions are not always fully effective and require significant investment and maintenance costs, so not all of them have been widely used on an industrial scale. The main disadvantage of using activated carbons for the sorption of mercury, especially in ionic form, is the large amount of hazardous solid waste generated, which has to be landfilled in hazardous waste dumps and requires chemical stabilization. Partial regeneration of coal by roasting (retorting) is only possible in the case of metallic mercury sorption.

The group of relatively new sorption methods used to remove heavy metals includes biosorption, i.e., a physicochemical process of removing contamination by biological material (biomass) (Abbas et al. 2014; Shah 2020). There are two types of biosorption: passive, independent of the biosorbent's life activity, and active—dependent on the metabolism and the rate of transport of metal ions to the cell interior. The removal of heavy metals using biosorbents is of great interest because of its natural, biological origin, ease and low acquisition costs, and reproducibility. The problem, however, remains, as in the case of active coals, as there is a need to utilize used biomass containing large quantities of toxic compounds.

Active biosorption, which consists of active (safe for the microorganism) transport of metal ions to the cell interior and enzymatic intracellular biotransformation, is closely related to the metabolism of the living organism (Rajerdran et al. 2003). Biotransformation of heavy metal ions is a natural defensive mechanism of microorganisms and this ability occurs in many microorganisms present in the environment. As regards toxic mercury in ionic form, metabolic conversion results in an enzymatic reduction of mercury existing in water-soluble compounds to an elementary (metallic) form that is much less harmful to living organisms (Tekere 2020). The application of such a technology allows for fast and effective treatment of large amounts of wastewater, as well as the recovery of metallic mercury and its return to the production process (Wagner-Doebler et al. 2000).

For many years, the hydrazine method has been used to neutralize mercury in sewage in Polish chemical plants operating chlor-alkali installations (Zakłady Azotowe in Tarnów-Mościce, Zakłady Chemiczne Oświęcim and Zakłady "Rokita" in Brzeg Dolny). Due to its high reduction potential, hydrazine can effectively remove up to 97% of ionic mercury from the aqueous phase, with an initial maximum concentration of up to 17 mg/dm^3. After reduction, metallic mercury is removed from the suspension on vibration filters with a layer of powdery active carbon. In this way, the mercury content of wastewater can be reduced to a maximum of around 500 μg/dm^3, which unfortunately exceeds the limit values for industrial wastewater released to environment by a factor of five. An additional significant disadvantage of this technology is the high toxicity and carcinogenic effect of hydrazine, which, in the case of less careful dosing in the reduction process, can be released to the aquatic environment together with the treated wastewater.

As can be easily seen, the methods of mercury neutralization in wastewater used so far in industry sometimes require the use of additional toxic substances (such as hydrazine precipitation and other chemical methods), are energy-intensive (e.g., electrochemical reduction) and do not allow to completely solve the problem of mercury emission to the environment, as they also generate large amounts of sediments (precipitation method, coagulation) or sewage (ion exchange) with a high content of mercury compounds. Moreover, known chemical and physicochemical methods, even those considered to be effective and economically viable, in fact only lead to a postponement of the problem of their permanent removal from the environment, so new technologies for removing mercury from wastewater, especially cheap and environmentally friendly ones, are still being sought. The method meeting such conditions seems to be the detoxification of mercury in industrial wastewater by bioreduction using specific strains of living bacteria.

The first research works on the use of microorganisms for the remediation of mercury from sewage were undertaken at the German National Research Centre for Biotechnology (GBF) in Braunschweig (Germany), under the direction of prof. W-D. Deckwer, in cooperation with the Department of Bioprocess Engineering at the Technical University of Łódź, Poland, already in the years 1993–1999 (Ledakowicz and Deckwer 1993; Shah 2021). Research conducted at GBF allowed to design a pilot installation for bioremediation of industrial wastewater from chlor-alkali plants (Wagner-Doebler 2003). The main element of this installation was a bioreactor filled with pumice stones on which *Pseudomonas* bacteria, selected earlier as a result of microbiological tests, were immobilized in the form of a biological membrane. Despite the relatively high effectiveness of the bioreduction process, the concentration of mercury at the outlet from the bioreactor did not reach the assumed low-enough values and an additional stage of "polishing" of the wastewater by traditional sorption on activated carbon was necessary.

The analysis of the operation of the aforementioned installation gave rise to the idea of its modification through the use of active carbon fixed-bed instead of pumice stones to immobilize bacteria in the bioreactor. According to the thesis, the use of active carbon, which, unlike pumice, has high mercury sorption properties, would allow to integrate the process of adsorption and biotransformation in one apparatus and, as a result, would lead to increased efficiency of the method, acceleration of the process, increased resistance of the system to mercury concentration fluctuations and simplification of the installation itself, thus reducing the costs of its construction and operation. In addition, it would be easier to recover metallic mercury and regenerate bioreactor fillings than with pumice material. This thesis became the starting point and the basis for the research presented below, and then for the application of the integrated bioremediation method in the industrial scale at Zakłady Azotowe in Tarnów-Mościce, ZAT, Poland.

3 Purpose and Scope of Experimental Studies

The main objective of the research was to demonstrate greater effectiveness and other advantages of the integrated method of mercury bioremediation in industrial wastewater and to develop and apply the technology in an industrial scale in one of the Polish chlor-alkali plants. This required extensive laboratory-scale research to determine the basic operational parameters of the planned integrated process.

The scope of research work included:

- determination of optimal values of parameters of the adsorption process of various forms of mercury (Hg^{2+}, Hg^0) on active carbon of different types and under different conditions (it should be stressed that there were no literature data concerning the sorption of metallic mercury from aqueous solutions);
- determination of equilibrium and kinetic data for sorption of mercury on active carbon and verification of these data in the adsorption column of a laboratory scale (flow system, with distributed parameters);
- determining the conditions of the mercury biotransformation process in industrial wastewater in an integrated bioreactor with an activated carbon fixed-bed and comparing the efficiency of the integrated process with the technique used previously on a laboratory scale;
- development and verification of a mathematical model of the process of sorption of mercury on activated carbon in a batch system and in the flow-through column, taking into account the sorption of Hg(II) and Hg(0) and the possible bioreduction reaction of mercury by microorganisms;
- adaptation of the integrated bioremediation technology to industrial conditions—change of the scale from laboratory to the industrial one;
- design, construction, and commissioning of an industrial wastewater bioremediation installation at Zakłady Azotowe in Tarnów, ZAT, and demonstration of the effectiveness of the integrated technology in real conditions over a long period of time.

4 Discussion of the Results

In order to design and subsequently scale-up the technology for bioremediation of mercury from wastewater, it was necessary to determine the parameters for the mass transfer in the liquid phase in relation to mercury. This was particularly relevant for metallic mercury, Hg(0), which is released in the bioreactor in this process and for which such data were generally not available in literature. The research plan provided for both forms of mercury the determination of equilibrium parameters of the sorption process on active carbon using the batch method and the verification of these data in the activated carbon bed placed in the absorption column, i.e., in the flow-through system.

The aim of the first stage of the study was to select the most suitable type of active carbon for mercury sorption, to determine the isotherms of mercury in various conditions on the selected type of active carbon and to initially determine the optimal conditions of the sorption process. The results of this stage of the study were discussed in the paper (Głuszcz et al. 2004).

Taking into account only parameters of a mercury sorption process, activated carbon should have a high sorption capacity of the different forms of mercury and give a high process rate (high diffusion rates of mercury in grains and/or a short diffusion path). This would imply the choice of a carbon with possible fine grains (e.g., powdery coal). On the other hand, due to the expected function as a carrier for microorganisms, carbon grains should be of such a structure and size that they are easily and "willingly" inhabited by bacteria. At the same time, the technical condition—the necessity to ensure high intensity of liquid flow through the bioreactor (maximizing the efficiency of the process), entails the necessity to minimize the resistance to flow through the fixed-bed, and thus to maintain a high porosity of the bed throughout the exploitation of the installation; this means in practice that the grains of a carbon should be relatively large and resistant to mechanical stresses. The simultaneous fulfillment of these opposing conditions was not possible, therefore the selection of a carbon with optimal properties was not an easy matter and required detailed research.

The literature containing data on the properties of specific active carbons and their use in environmental engineering is very extensive (Graydon et al. 2009; Zhao et al. 2010; Diamantopoluou et al. 2010). The sorption capacity of activated carbon depends on many factors: the material of which it is made and the way it was processed, the way in which it was activated, its granulation, porosity, and of course, the conditions of a sorption process: mainly the pH of the solution, the temperature and the contact time of the phases. Additional improvement of carbon adsorption properties can be achieved through targeted changes in the chemical structure of the material surface. The so-called carbon impregnation, i.e., the incorporation of some elements into the surface layer or the application of appropriate chemical substances to the active surface, to which the removed contamination has a high affinity, gives particularly good results. In the case of mercury, impregnation with, e.g., iodine, silver, and most often sulfur compounds is used, which results from the high affinity of mercury to sulfur.

Taking into account the above conditions, 8 types of active carbon were analyzed in detail, differing in the raw material of which the given activated carbon was made, the activation method, granulation degree, and content of admixtures (sulfur or iodine). The selection of specific types of activated carbon for research was based on, among others, guidelines from the literature on the subject, professional experience of representatives of the chlor-alkali industry, availability and price of the material, as well as the possibility of using it for immobilization of microorganisms. In particular, it seemed promising to use sulfurous coal, i.e., coal with sulfur admixture, which should be particularly useful for sorption of mercury due to its high affinity to sulfur. The basic properties of different types of carbon used in the study were collected in Table 1. The isotherms of sorption of mercury were determined by contacting the same amount of active carbon with solutions of mercury chloride of different initial concentration within the range of 10–650 mg Hg/dm^3, selected so as to obtain the widest possible and uniform coverage of the line of isotherm with experimental points. The range of solution concentration, temperature, and pH corresponded approximately to the amplitude of variability of the actual parameters of industrial wastewater.

The analysis of the experimental data obtained allowed to conclude that it is best described by the "classical" Langmuir isothermal equation (Fig. 1).

Table 1 Basic properties of different types of activated carbon used in the study

Name	POOL (1–3)	POOL (1–3) + S	POOL (0.5–1.6)	DG (1–3) + S	Merck (dusty)	DGC (0.1–0.5)	Norit 3	DG (1–3) + J
bulk density [kg/m^3]	400 ± 50	400 ± 50	510 ± 30	460 ± 50	~ 300	430 ± 30	~ 600	460 ± 50
grain d_{av} [mm]	2,39	2,41	0,91	2,21	0,08	0,38	4,31	2,27
Impregnation content [% wt.]	–	Sulfur ~ 5	–	Sulfur > 5	–	–	Sulfur ~ 10	Iodine ~ 5
raw material	charcoal	charcoal	charcoal	coal	charcoal	coconut shells	charcoal pressed	coal

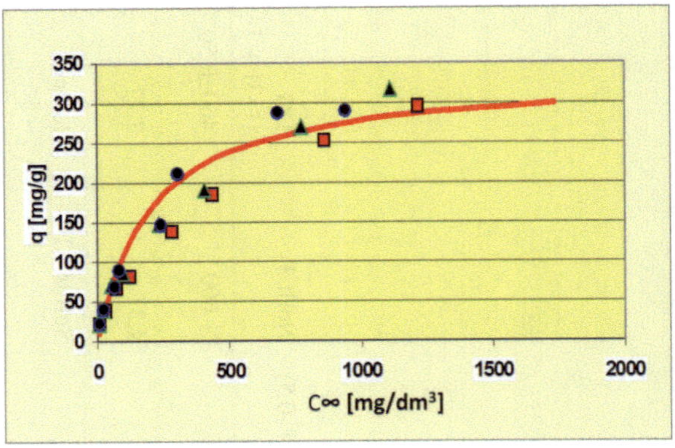

Fig. 1 Example of Hg^{2+} sorption isotherm for DG(1–3) + S carbon (temp. 20 °C, pH = 4) (points: experimental results from 3 experiments in the same conditions)

Using this equation it was possible to determine the Langmuir constants (q_m and b) and then to identify the diffusion coefficients of ionic mercury in the sorbent grains.

The kinetics of the sorption process was studied by measuring the changes of the mercury concentration in a given solution in time. For this purpose, the same amount of 100 cm^3 solution of mercuric chloride of the same initial concentration was added simultaneously to several flasks containing the same portion of adsorbent. The number of flasks with identical contents corresponded to the number of predicted samples taken during the process (the flask was not returned to the shaker after sampling); each flask therefore corresponded to the specified residence time. The sorption process was carried out at the constant temperature and pH; the values of these parameters were identical to those applied for the isotherms determination.

On the basis of the results analysis, it was found that the largest saturation sorption capacity was found in coal with the smallest granulation (Merck dusty, DGC 0.1–0.5). However, due to the correct operation of the designed continuous flow bioreactor (low flow resistance in the bed, possibility of immobilization of microorganisms), the choice of the sorbent with possibly large grains, i.e., DG(1–3) + S or POOL(1–3) + S carbon, was more advantageous, even at the cost of a certain reduction of coal sorption capacity. On the other hand, the comparison of the results for coal types POOL(1–3) and POOL(1–3) + S showed that sulfur impregnation caused a slight increase in the value of the q_m parameter, but at the same time a decrease in the value of the effective diffusion coefficient in grain, D_p. The increase in temperature caused a decrease in the maximum capacity of both sorbents, but this effect was much greater in the case of carbon impregnated with sulfur. This was evidence of a different mechanism of mercury binding by sulfuric carbon. It can be argued that as a result of high affinity of mercury to sulfur, the adsorption of Hg^{2+} on the surface of coal with sulfur admixture is clearly chemisorption in contrast to "pure" physical adsorption, predominant in the case of ordinary activated carbon; this may explain the stronger dependence of sorption capacity on temperature in the case of sulfur impregnated carbon. Also the increase in a pH value of the solution had a negative influence on the sorption. At higher pH values the process was slower than in acidic environment; coal sorption capacity was also slightly lower. This effect was observed both for sulfuric and nonimpregnated carbon.

Finally, the following conclusions were drawn on the basis of the studies carried out at this stage (Gluszcz et al. 2004):

- of the 8 tested types of activated carbon as a bioreactor filler for the removal of mercury from wastewater, DG(1–3) + S or POOL(1–3) + S carbons seem to be the best option. Taking into account the sorption capacity, diffusion coefficients, granulation, and mechanical strength of the grains, DG(1–3) + S persulfurized carbon was selected for further investigations;
- sulfur impregnation had no significant effect on the increase of sorption capacity of the sorbent in comparison with unsulfurized coal, but it caused a decrease in the diffusion coefficient in a grain;
- in solutions with pH = 4, the rate of mercury sorption was higher and the carbon saturation capacity was slightly higher than at higher pH values (7–9);
- the maximum sorption capacities of the examined coals of similar granulation were similar and exceeded 100 mg Hg/g carbon; the effective diffusion coefficients of Hg^{2+} ions in the grains of sorbent were about 10 times lower than the diffusion coefficient in water.

Continuing the research, which was to lead to the proper design of the integrated bioreactor, experiments were started in the flow-through system, i.e., in the column with a fixed-bed of selected active carbon, through which a solution of mercury chloride was passed (Fig. 2).

The glass column, equipped with a thermostatic jacket, had outlet ports at four levels and at the outlet; this allowed the solution to be sampled at different bed heights in order to gain a better understanding of the sorption process and to analyze the distribution of concentrations along the column height. The aim of the experiments was

Fig. 2 Laboratory-scale apparatus for mercury sorption parameters investigation in a flow-through system

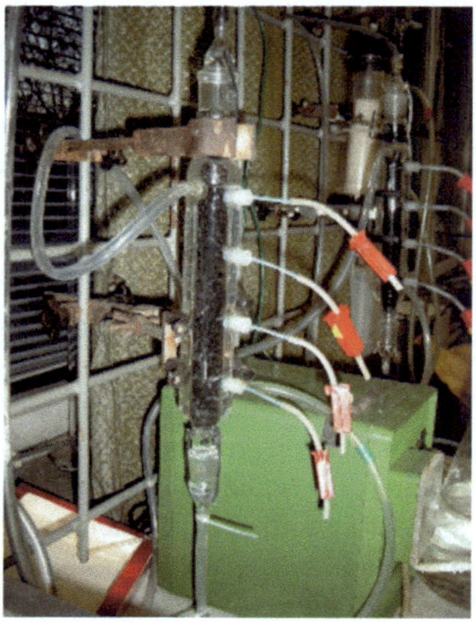

to obtain the curves of the bed breakthrough with respect to ionic mercury and on their basis to determine the effective sorption capacity of the coal deposit and the coefficients of longitudinal dispersion in the bed, to examine the dependence of these parameters on the flow rate of the solution through the column, its pH reaction and mercury concentration, and to verify the previously determined sorption parameters of the selected activated carbon. The coefficient of longitudinal dispersion, D_L, effective coefficient of mass transfer in the liquid phase on the surface of active carbon, k_f, and the coefficient of diffusion in a grain, D_p, were calculated using a mathematical model of the process, described briefly below. During the experimental works, it was also checked whether the performance of the fixed bed and its parameters do not change significantly when the direction of liquid flow changes (gravitationally downward or vertically upward), because for practical reasons the sewage in the target industrial installation should flow from the bottom of the bioreactor upward. At this stage of the study, DG(1–3) + S type of persulphated activated carbon, selected as the best at the earlier stage of the study, was used. Considering the use of this carbon for immobilization of microorganisms in the process of bioremediation, the influence of the components of culture medium for microorganisms on the ability of mercury adsorption on the examined carbon was also discussed.

Analyzing the breakthrough curves obtained when feeding the column with mercury chloride solution from the top and bottom, it was found that with a gravitational down-flow of liquid, the breakthrough of the column fixed-bed occurs much earlier. In the case of the top supply, the column cross-section may have an uneven bracing distribution and channel flow, even at low flow rates. In this case, a strong longitudinal dispersion takes place in the bed, which results in an earlier breakthrough of the deposit (appearance of mercury at the outlet), even if the sorbent is not yet fully saturated. From the point of view of the effective sorption capacity of the bed, the supply from a top is therefore less advantageous. When the column is fed from the bottom, the solution is filled evenly throughout the entire cross-section, and the fluid front moves approximately in a plug flow manner in the column. Under these conditions, there is almost no undesirable longitudinal dispersion, and the breakthrough of the bed occurs later, usually only after total saturation of the sorbent (of course, if the flow rate is sufficiently low in relation to the diffusion rate and sorption rate in the sorbent grains). From the point of view of the sorption process, the supply from the bottom is therefore more advantageous. It should be taken into account, however, that in an industrial practice such a supply method carries the risk of stratification of the deposit, especially at higher flow rates. This can lead to the formation of free spaces in the carbon bed, channeling and even rinsing of the sorbent grains from the column. In the experimental apparatus such undesirable phenomena were observed at the flow rate above 10 BV/h (Bed Volume per hour). Previous experiments with the bioremediation process have shown that, due to the rate of the mercury bioreduction reaction, the liquid flow rate should not exceed 2–3 BV/h. Therefore, it was decided that during further experiments in the columns, the flow direction of the liquid from the bottom to the top should be used. Then, the experimentally obtained bunch of curves of the carbon bed breakthrough at different flow rates of the liquid along the column and at different levels (i.e., at different total height of the deposit) allowed to determine the desired parameters of the bed performance. Their analysis leads to the conclusion that the coefficient of longitudinal dispersion and the coefficient of mass transfer increase significantly with the increase of the liquid flow rate. The diffusion coefficient in sorbent grains also increases, but relatively less than the other two parameters. The influence of temperature on sorption conditions, observed during the studies, was variable and depended on the flow rate of the liquid. At high flow rates, i.e., above 1 ml/min, the breakthrough took place at 20 °C at the earliest, and at 40 °C at the latest, i.e., the increase of the temperature resulted in an increase in the effective capacity of the bed. A reverse relationship was observed at low flow rates (less than 0.25 ml/min). Under such conditions, the increase in temperature led to faster penetration of the carbon deposit, thus reducing the effective capacity of the bed. This effect is probably related to the opposite influence of two phenomena: longitudinal dispersion in the bed and diffusion in the pores of the sorbent grains. As shown in Table 2, at low temperatures the diffusion coefficient has a lower value, i.e., when the flow velocity of the liquid through the deposit is high, the mercury ions are "washed out" from the column before they can overcome the diffusion path into the grain and saturate the sorbent. The effective capacity of the carbon bed is then determined by the diffusion rate, and not by the actual (equilibrium) sorption capacity of the bed. On the other hand, at a low flow rate, the residence time of the solution in the bed is long enough so that the mercury ions can transfer into the grains and be adsorbed there. The effective capacity of the deposit (and the associated breakthrough time) is then determined not by the diffusion rate, but by the actual capacity of the active carbon, decreasing, as we know, with the increase of temperature (in the case of the sulphated carbon used here). When designing an integrated bioreactor with an activated carbon bed in a larger scale, it is therefore necessary to carefully analyze and consider this simultaneous, opposing influence of the liquid flow rate and temperature on the course of the sorption process and the time of the deposit breakthrough, taking into account the specific industrial conditions.

The influence of pH on the process was much smaller. A lower value (e.g. pH = 4) allowed to obtain better conditions of mercury sorption, thanks to which the moment of breakthrough of the deposit was delayed and its effective sorption capacity increased. Such an effect was not desirable from the point of view of an integrated bioremediation process, as the microorganisms used to require the maintenance of the pH value at 7, and then the conditions of mercury sorption are less favorable. Fortunately, due to the relatively small influence of liquid acidity, the value of the solution pH should not have a significant impact on the industrial-scale installation performance.

Table 2 D_L, k_f, and D_p dependence on the temperature of the sorption process

Temperature [°C]	D_L*10^5 [m^2/s]	k_f*10^8 [m/s]	D_p*10^{10} [m^2/s]
20	2.31	0.76	2.13
30	2.78	1.52	3.41
40	2.64	1.04	7.42

The final conclusions reached on the basis of the experiments carried out in the continuous flow column with different deposit heights and under different process conditions were as follows (Gluszcz et al. 2006):

- there is practically no longitudinal dispersion in the liquid phase when feeding the column with the solution from a bottom and the conditions of the sorption process are then more favorable, but at higher flow rates (above 10 BV/h) it can lead to stratification of the bed or leaching of the sorbent grains from the reactor;
- there is a complex dependence of the effective capacity of the deposit on the temperature and flow rate of the solution: at a lower temperature the values of the mercury diffusion coefficient in the grains of the sorbent are smaller and at higher flow rates of the liquid through the column the breakthrough occurs earlier, although the carbon is not yet fully saturated with mercury; at higher temperatures and at lower flow rates the effective capacity of the bed increases, although the equilibrium sorption capacity of active coal decreases;
- the effective capacity of the deposit decreases slightly with increasing pH of the solution;
- an increase in the concentration of mercury in the solution at the inlet to the column caused an increase in the effective capacity of the deposit.

The results obtained in this part of the study enabled a detailed analysis of the interdependence of operational parameters of the mercury sorption process in the flow column, selection of optimal conditions for the process, and determination of its kinetic parameters.

Analysis of the results of experimental studies enabled to formulate a mathematical model of mercury adsorption process in pores of active carbon grains in a fixed-bed with continuous flow of a liquid along the column.

When formulating a mathematical model of such a process two basic steps of mass transfer should be considered: adsorption of an active liquid component from the solution onto a single grain of the sorbent and transport of the component along the bed height in the direction of the flow of the liquid phase. Each of these partial processes is described by different model equations and is characterized by different kinetic parameters.

5 Mathematical Model of Mass Exchange Between Solution and Grain of Sorbent

On the way of mercury ions (or atoms) between the main stream of the liquid phase and the solid surface within the pores of a sorbent grain, at least two steps can be distinguished: the transfer of mass from the liquid phase core into the outer surface of the grain through the boundary layer on the liquid side and the transport of the ion from the grain surface to its interior through a pore-filling solution. Due to the microporous structure of an activated carbon grain, it is usually assumed that the transport of mass inside the grain takes place only by diffusion, whereas in the surrounding solution it may take place both by diffusion and convection. The thickness of the laminar film on the grain surface depends on the mixing conditions of the liquid phase. If the process is carried out in a closed (batch) system (of the "perfect mixing tank" type), it is usually relatively easy to obtain/assume good mixing conditions in the liquid phase. This allows a constant concentration of the component to be assumed for the whole volume of the solution, or even to omit the resistance of the mass transfer to the solid surface (due to intensive mixing). In such conditions, the decisive influence on the whole process has the rate of ions diffusion inside the coal grain, characterized by the diffusion coefficient, D_p.

The mass balance equations for sorbent grain were formulated on the basis of the process scheme presented in Fig. 3.

The basic assumptions for the mathematical model were therefore as follows:

- the conditions of perfect mixing may be assumed in the solution surrounding the grain and the total mass transfer resistance is located in the boundary layer on the surface of the grain of the sorbent,
- there is a nonlinear gradient of an active component concentration in the boundary layer,
- the mass transport in pores of a grain is carried out exclusively by diffusion,
- the adsorption process onto the inner surface of the grain pores is immediate (takes place much faster than the preceding diffusion transport of mercury ions) and conditions of the interphase equilibrium prevail on the surface of the solid,
- the adsorption process is isothermal,
- the equilibrium of the sorption is described by the Langmuir's isotherm,

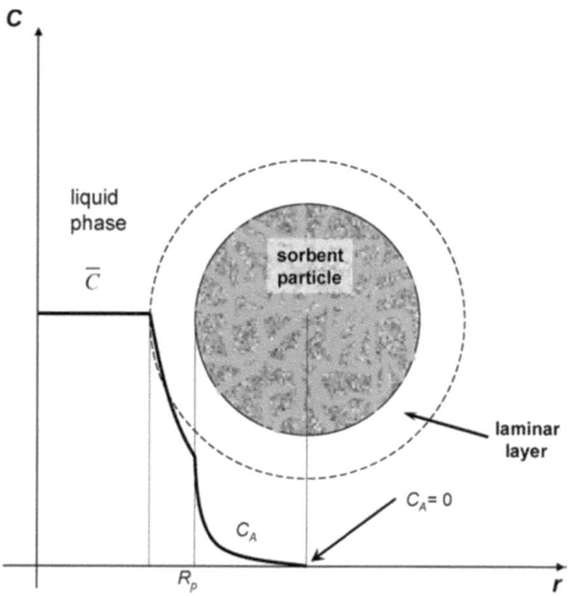

Fig. 3 Profile of Hg^{2+} concentration around and inside of a sorbent particle

- the value of the diffusion coefficient in pores is independent of concentration,
- all sorbent grains are spheres of equal diameter and density.

Under these conditions, the balance of the component in the solution surrounding the grain of sorbent is expressed by the equation:

$$-V\frac{d\overline{c}}{dt} = \int\limits_{A_p} k_f\left(c_0 - c|_{r=R_p}\right)dA, \tag{1}$$

and the equation for mass balance within the pores of a grain has the form:

$$\frac{\partial c_A}{\partial t} + \left(\frac{1-\varepsilon_p}{\varepsilon_p}\right)\rho_s\frac{\partial q_A}{\partial t} = D_{Ap}\frac{1}{r^2}\frac{\partial}{\partial r}\left(r^2\frac{\partial c_A}{\partial r}\right) \tag{2}$$

where q_A is the amount of adsorbate mass adsorbed per unit mass of adsorbent, as described by the Langmuir isothermal relationship:

$$\frac{q_A}{q_{Am}} = \frac{b_A\,c_A}{1+b_A\,c_A} \tag{3}$$

where b_A is the constant and q_{Am} is the maximum equilibrium capacity of the sorbent specific for component A.

Assuming a much higher rate of adsorption in comparison to diffusion in pores, it is justified to enter a relation:

$$\frac{\partial q_A}{\partial t} = \frac{\partial q_A}{\partial c_A}\frac{\partial c_A}{\partial t} = \frac{b_A\,q_{Am}}{(1+b_A\,c_A)^2}\,\frac{\partial c_A}{\partial t} \tag{4}$$

The substitution of this relation to Eq. (2) and the appropriate transformation leads to the expression:

$$\left[1+\left(\frac{1-\varepsilon_p}{\varepsilon_p}\right)\frac{b_A\,q_{Am}\rho_s}{(1+b_A c_A)^2}\right]\frac{\partial c_A}{\partial t} = D_{Ap}\,\frac{1}{r^2}\frac{\partial}{\partial r}\left(r^2\frac{\partial c_A}{\partial r}\right) \tag{5}$$

The following boundary conditions were assumed ($t \geq 0$):

$$\left.\frac{\partial c_A}{\partial r}\right|_{r=0} = 0 \tag{6a}$$

$$-\varepsilon_p D_{Ap}\left.\frac{\partial c_A}{\partial r}\right|_{r=R_p} = k_f\left(c_{A0} - c_A|_{r=R_p}\right), \tag{6b}$$

and the initial conditions ($t = 0$):

$$c_A = 0 \quad \text{for} \quad 0 \leq r \leq R_p \tag{7}$$

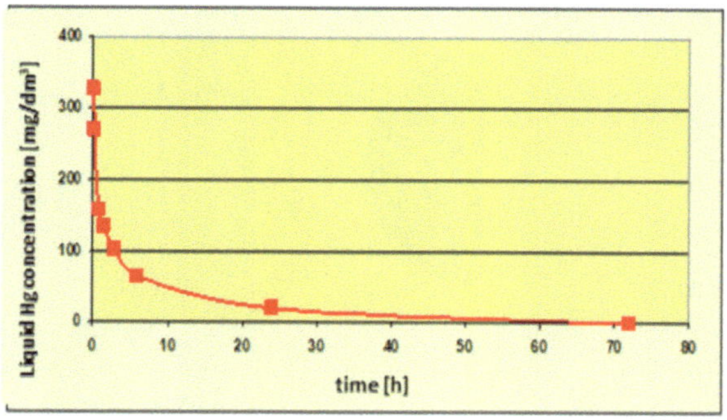

Fig. 4 Changes of mercury concentration in the solution during Hg^{2+} ions sorption onto the active carbon in a batch system (points–experimental values, line–values calculated from the model)

The model of mass exchange between solution and a sorbent grain, described by the above system of equations, was solved by finite element method in transient mode.

The experimental verification of the model was performed using the results of the kinetic studies of the ion exchange process carried out in a batch system with mixing, discussed in (Gluszcz et al. 2004, 2005). An example of a graph showing a comparison of experimental and model-based changes in the concentration of mercury ions in a solution under transient conditions in a closed system (in a flask) is shown in Fig. 4. A very good agreement of experimental data and the values of process operating parameters calculated over the whole applied range was obtained.

Model calculations for the process of gradually saturation of carbon grains with Hg^{2+} ions made it possible to observe changes in the distribution of mercury concentration in internal pores of a sorbent grain in subsequent periods of time. An example of the results of such calculations and their visualization is presented in the following Figs. 5a–5e.

These graphs illustrate a movement of the forehead of the ions' solution inside the grain and changes of the concentration gradient in the pores within the grain until it reaches equilibrium in the liquid phase.

The model presented was used to identify the diffusion coefficients of Hg^{2+} ions and elementary mercury atoms, Hg^0, in pores of active carbon grains, based on relevant experimental data. The values of diffusion coefficients obtained in this way were then used in the mathematical model of a flow-through column with an fixed activated carbon bed.

6 Mathematical Model of a Mass Exchange in an Activated Carbon Bed in a Column

When a mercury sorption process from the stream of a liquid is carried out in a continuous-flow column bioreactor, i.e., in a flow-through system, a constant concentration of the component cannot be assumed for the total volume of the liquid phase

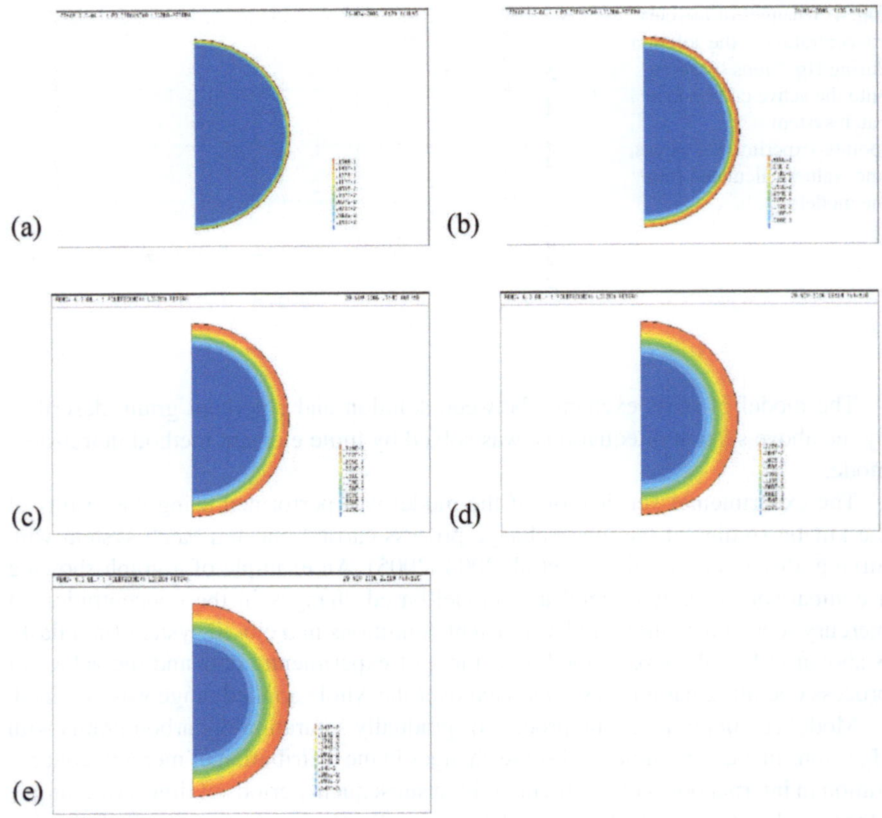

Fig. 5 Distribution of mercury concentration inside the sorbent grain after: (**a**) 45 min. (**b**) 180 min., (**c**) 48 h., (**d**) 96 h., (**e**) 172 h. from the start of the sorption process

at any time. Moreover, due to worse mixing conditions in the liquid stream, the influence of the liquid film mass transfer into the grain surface on the overall rate of Hg mass transfer is greater than in closed systems with good mixing. For the liquid stream flowing along the column usually a plug flow model may be assumed or, closer to reality, a plug flow model with superimposed longitudinal dispersion, which can be characterized by the so-called longitudinal (axial) dispersion coefficient, D_L. The equations of the mathematical model of the sorption process in a column with fixed carbon bed were formulated on the basis of the process diagram presented in Fig. 6.

The following simplifying assumptions have been made for the process:

a) the total liquid phase mass transfer resistance is located within the boundary layer existing on the surface of the sorbent grain; it is characterized by a diffusion coefficient for the diffusion direction perpendicular to the surface of the grain,
b) the mass transport in the pores of a grain is carried out exclusively by diffusion,

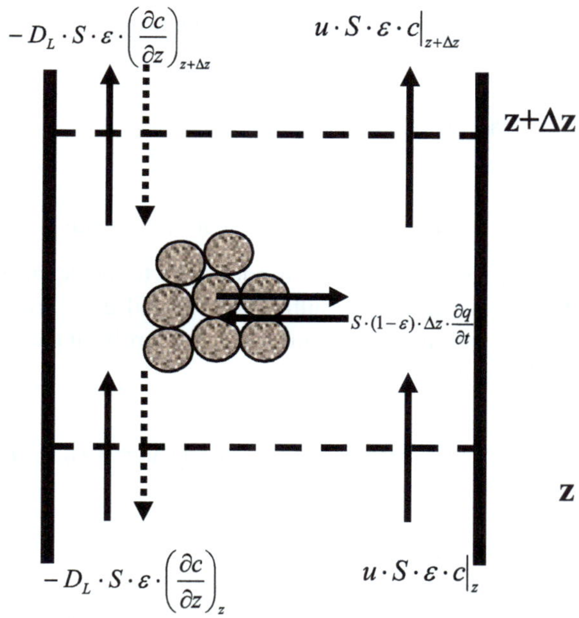

Fig. 6 Scheme of the mass exchange process in the column with activated carbon fixed bed

c) the adsorption on the inner surface of the grain pores takes place much faster than the preceding process of the component diffusion and the conditions of interphase equilibrium prevail on the solid surface,
d) the adsorption process is isothermal,
e) the equilibrium of the sorption process is described by the Langmuir isotherm,
f) the value of the diffusion coefficient in pores is independent of concentration,
g) sorbent grains are spheres of equal diameter and density,
h) the flow of the liquid along the column height is characterized by two-component rates: average flow rate and dispersion rate (not numerically solved), which can be taken into account using the longitudinal (axial) dispersion coefficient, D_L,
i) the concentration of the active component in the entire cross-section is the same (no flow in the radial direction and no radial concentration gradient in the deposit).

Inside the grain of a sorbent, of course, there is the same mass transfer mechanism as described earlier for the batch process, but from the side of the flowing liquid phase, the sorption process in the column is "seen" as the transport of the component to the solid phase through the boundary layer, at the rate determined here by the diffusion coefficient in the radial direction, D_{AR}. Taking this into account, the balance equation for component A in the liquid phase may be expressed in a form:

$$\frac{\partial c_A}{\partial t} + v_z \frac{\partial c_A}{\partial z} = D_{AR} \frac{1}{r} \frac{\partial}{\partial r}\left(r \frac{\partial c_A}{\partial r}\right) + D_{AL} \frac{\partial^2 c_A}{\partial z^2} \tag{8}$$

The boundary conditions take on a form: ($t \geq 0$),

- for the grain:

1. $$\left.\frac{\partial c_A}{\partial r}\right|_{r=0} = 0 \tag{9a}$$

2. c_A continuous in the calculations domain

The condition (9a)-(2) is equivalent to the equality of two streams at the solid–liquid interface: one resulting from the diffusion from the surface of the grain to its interior and the other from mass transport from the bulk of the liquid to the surface of the grain:

$$-D_{Ap}\left.\frac{\partial c_A}{\partial r}\right|_{r=R_p} = k_{fcol}\left(c_{A0} - c_A|_{r=R_p}\right) = -D_{AR}\left.\frac{\partial c_A}{\partial r}\right|_{r=R_p}, \tag{9b}$$

- for the column inlet:

$$-D_L\left.\frac{\partial c_A}{\partial z}\right|_{z=0} = v_z\left(c_{A0} - c_A|_{z=0}\right) \tag{10a}$$

- for the column outlet:

$$\left.\frac{\partial c_A}{\partial z}\right|_{z=L} = 0 \tag{10b}$$

Initial conditions ($t = 0$):

$$c_A = 0 \quad \text{for} \quad 0 \leq r \leq R_p \tag{11a}$$

$$c_A = 0 \quad \text{for} \quad 0 \leq z \leq L \tag{11b}$$

The above problem of mass exchange in a flow bioreactor was solved with the finite element method in a transient regime.

Further development of the flow-through bioreactor model required taking into account not only the process of ionic mercury sorption on activated carbon but also the activity of live biomass immobilized in the carbon bed, simultaneously reducing Hg^{2+} ions to Hg^0. For experimental verification of the latter case, it was necessary to determine the equilibrium and kinetic parameters of the sorption process of metallic mercury Hg^0 by DG(1–3) + S carbon, as well as the kinetic parameters of the mercury bioreduction reaction by living *Pseudomonas putida* bacteria.

Very few studies on the sorption of metallic mercury onto activated carbon are available in scientific literature. Occasionally one can find publications describing the research on the process of Hg^0 sorption from the gas phase (mainly from flue gases), but there are no reports dealing with aqueous solutions of metallic mercury. One reason for this is probably the experimental difficulties associated with the very low solubility of metallic mercury in water at moderate temperatures. However, even these few studies indicate that the sorption capacity of activated carbon for metallic mercury is much lower than for its ionic form. A series of experiments was carried out to determine the parameters of the Hg^0 sorption process on selected activated carbon as part of the research described herein.

Due to very low solubility of metallic mercury in water, it was not possible to obtain the sorption capacity of active carbon or to determine Hg^0 sorption isotherms in a "classical" way in thermostatic flasks with a fixed initial volume and concentration of solution. It was therefore necessary to develop a new technique for researching such a process. The research was carried on in two directions:

- an increase in the content of metallic mercury suspended in the liquid phase,
- sorption in a micro-flow column, in a circulating system, which would allow to use much larger amounts of solution in contact with the same amount of carbon sample than in the case of the "classical" method, in a closed system, in small capacity flasks.

The obtained results of these studies on the balance of sorption were collected in Table 3. The data contained therein indicate that with an increase in temperature from 5 to 50 °C the carbon saturation capacity in relation to Hg^0 decreases almost tenfold. Although such an effect was easy to predict qualitatively, the quantitative data obtained, which are not available in the literature and which are necessary for the proper understanding and description of the process and the design of mercury bioremediation facilities, were of great value. The sorption capacity of DG(1–3) + S carbon in the case of metallic mercury sorption was significantly lower (at least five times) than ionic mercury and amounted to 19.7 mg Hg^0/g carbon at 30 °C on average. Kinetic studies showed that the sorption rate of Hg^0 atoms was similar to that of Hg^{2+}.

Table 3 Parameters of the Langmuir isothermal equation for Hg^0 sorption onto activated carbon DG(1–3) + S

Temperature [°C]	5	15	25	35	50
Qm [mg/g]	71,79	35,36	21,19	17,42	6,21
b	8,21	4,55	11,15	24,41	26,53

7 Research on the Integrated Bioprocess in a Laboratory Scale

The integrated process of biological neutralization of ionic mercury present in aqueous solution by living bacteria immobilized on activated carbon in a laboratory flow-through bioreactor under various process conditions was investigated on a laboratory scale. The aim of the research of the process with active microorganisms was (Gluszcz et al. 2007):

- optimization of the composition of the nutrient solution for microorganisms colonizing the active carbon deposit in experimental column in order to obtain the best possible efficiency of mercury bioreduction;
- investigation of the influence of the nutrient components on the process of mercury sorption by activated carbon;
- determination of the influence of the NaCl salt present in a solution on the effectiveness of the bioprocess, as in practice industrial wastewater from chlor-alkali plants contains a certain amount of a brine;
- comparison of the effectiveness of the integrated bioreactor, containing an activated carbon as a sorbent for mercury and a carrier for the immobilization of microorganisms, with the effectiveness of the bioreactor filled only with porous pumice stone without mercury sorption properties;
- final verification, under laboratory conditions, of the validity of the concept of the integrated technology with sorption and biotransformation simultaneously taking place in the same apparatus, using in the research the real industrial wastewater from the Chemical Plant in Tarnów, Poland.

Enriched semisynthetic medium, the basic components of which are yeast extract and sucrose, is usually used as the primary medium in cultures of bacterial strains used for the bioremediation of mercury. Yeast extract, due to its complex composition, i.e., amino acids, peptides, water-soluble vitamins, pyrimidine and purine bases, mineral salts, and glucose, is a source of nitrogen compounds, organic carbon and B vitamins for microorganisms, while sucrose is the main source of carbon. Due to the properties of these substances, there was a concern that their presence in the solution flowing through the bioreactor may negatively influence the sorption properties of the carbon and the kinetics of mercury adsorption. From the point of view of the bioremediation technology, an important issue was the proper selection of the nutrient composition and the rate of its dosing during the process. According to the assumptions, active carbon, used as a bioreactor filling, was to serve not only as a carrier material for the immobilization of microorganisms, but also as a sorbent, thus increasing the effectiveness of mercury removal from aqueous solutions. The amount of nutrient solution and the concentration of components had to be selected in such a way as to avoid blockage and excessive inactivation of the deposit, on the one hand, and to ensure proper development and activity of microorganisms in a toxic environment, on the other. The studies were carried out in two ways: simultaneously different variants of the media were checked for high activity of bacteria and their resistance to fluctuations of mercury concentration in the solution, and the influence of these media on the properties of the sorbent filling the bioreactor was investigated.

Initially, the experiments were carried out using the above-mentioned typical enriched medium, changing the content of yeast extract, sucrose, and NaCl. It was found that these components, if present in quantities meeting the needs of microorganisms, have a significant adverse effect on the process: they significantly reduce the rate of adsorption and the sorption capacity of the deposit. This effect was clearly visible when comparing the mercury concentration at the outlet of two columns with "pure" adsorption (without the presence of microorganisms): the efficiency of the process of removing ionic mercury from the solution was much lower in the column fed with the solution of the nutrient medium than in the case of the inlet stream containing only mercury chloride in water. In view of this phenomenon, an attempt was made to use media with a different basic composition. An alternative to the enriched medium is synthetic media, prepared from precisely defined chemical compounds, with known qualitative and quantitative composition (as opposed to yeast extract). They are referred to as minimum nutrient media, because they always contain only the basic ingredients necessary for the life of the organism in the minimum necessary quantities. In this study, a mineral medium M9 was used, containing as a carbon source glucose or, for comparison or alternatively, sodium acetate. The efficiency of the mercury bioreduction process was higher than in the case of the enriched medium using both variants of the minimum medium. Mean values of Hg concentration observed at the outlet from bioreactors were below 0.1 mg/dm^3 (with the inlet concentration up to 16 mg Hg^{2+}/dm^3), which meant bioreduction efficiency of approx. 99%. In both cases, the bioreactors worked more steadily; the number of microorganisms remained stable throughout the experiment (about 600 h), regardless of the toxicity of the environment. In the case of bioreactors fed with glucose medium it was necessary to remove (rinse out) twice the excessively growing biomass, blocking the flow of the solution through the bed. After 500 h in one of the two parallel bioreactors, the breakthrough was observed and the process was finished. Bioreactors fed by acetate medium worked continuously without any complications until the end of the experiment, i.e., until about 600 h of the process.

The course of the process in bioreactors fed with mineral minimal medium was also compared with the reference process, in which the aforementioned optimized enriched medium was used. In the case of mineral nutrient medium the time for adaptation of microorganisms to new environmental conditions after inoculation of the column was longer than in the case of enriched medium; in the case of acetate medium the full bioreduction activity of the strain was achieved only after about 200 h from the beginning of the experiment. After this time, however, the efficiency of the bioremediation process was much higher than that of the enriched medium: the outlet mercury concentration dropped to approx. 50 $\mu g/dm^3$, even with the inlet concentration of up to 16 mg/dm^3, which gives a purification degree of the solution close to 100%. Similar results were obtained for the glucose and acetate substrate, but the advantage of the latter is the lack of excessive biomass growth and thus the possibility of longer and more stable work of the carbon bed.

No negative influence of NaCl concentration in wastewater on bioremediation process (up to 40 g NaCl/dm^3) was observed during the studies, which is important due to the fact that wastewater from chlor-alkali plants may contain significant amounts of brine. On the other hand, the flow rate of the liquid stream through the column had a significant influence on the effectiveness of the mercury bioreduction process; the highest effectiveness was achieved at the intensity of 1.6 BV/h.

The starting point for the whole series of experiments discussed here was the thesis that integration of the process of ionic mercury sorption, bioreduction, and reduced metallic mercury sorption in one bioreactor is possible and will allow to significantly increase the efficiency of mercury removal from aqueous solutions compared to the previously known technique, which is based only on Hg^{2+} bioreduction. The modification of the bioreactor was to replace the pumice deposit with granulated active carbon, which could serve both as a carrier for microorganisms and a sorbent for two forms of mercury. The final answer to the question about the sense and purpose of such a modification could only be given by comparison of the performance of two bioreactors, differing only in the type of filling, under the same process conditions. Such experiments have been carried out and the results of one of them is shown in Fig. 7.

As one can see, up to about 600 h the processes in both columns run very similar. During this time, the inlet concentration of mercury gradually increased from 1.5 mg/dm^3 to 8 mg/dm^3, obtaining in both bioreactors almost the same concentration at the exit from the column and a similar biomass content in the deposit. The input concentration of mercury was then increased to 18 mg/dm^3, simulating the shock loading of a bioreactor with Hg ions that could occur in industrial practice, e.g., as a result of

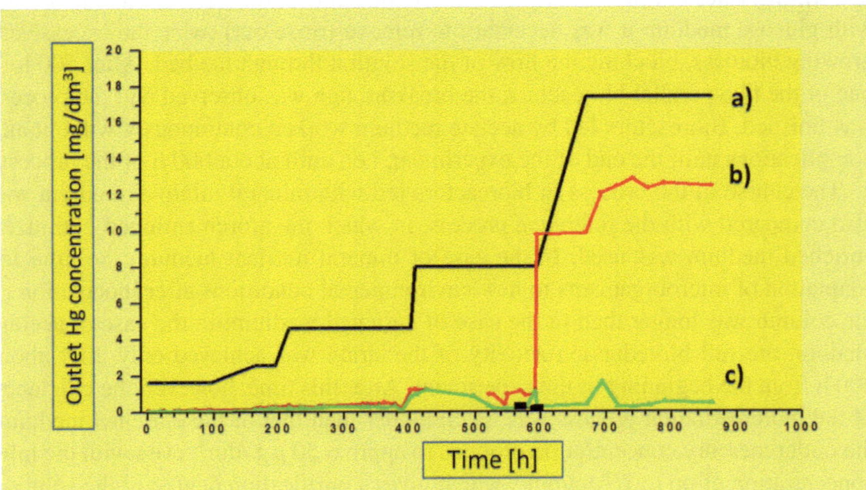

Fig. 7 Comparison of the process of ionic mercury bioremediation in two bioreactors: line a) - inlet Hg^{2+} concentration; line b)—pumice stone bed; line c)—activated carbon bed DG(1–3) + S (integrated bioreactor);

a failure. After this rapid change, in a bioreactor filled with pumice stones the breakthrough of the deposit occurred and the biomass was practically completely washed away; the bioremediation process was interrupted and the mercury concentration at the outlet reached 12–14 mg/dm^3. At the same time, the bioreactor with immobilized microorganisms on activated carbon continued to work steadily, retaining at least 97% of its capacity to remove mercury from wastewater. The temporary increase of the concentration at the outlet to about 500 $\mu g/dm^3$ was relatively small and short-lived. This experiment has shown that an integrated bioreactor can remove mercury from the solution at much higher concentrations than a pumice stone bioreactor and, in addition, is much more "flexible," resistant to sudden, abrupt changes in the mercury concentration in the inlet stream. At the moment of rapid increase in concentration, active carbon absorbs excess ionic mercury, acting as a buffer structure. Mercury accumulated on the surface of the sorbent can be gradually consumed by microorganisms when the toxicity of the flowing stream of liquid decreases. This is an additional advantage of the system because, according to previous studies, any rapid changes in mercury concentration, both up and down, are detrimental to microbial activity and reduce the effectiveness of the method. The presence of activated carbon reduces this danger and maintains high bioreactor efficiency even under changing conditions. The biofilm layer grows quickly on its surface and is much more resistant to washing out from the carbon than from pumice stones, even at high flow rates. All these observations and conclusions allowed us to conclude that the integration of the process of mercury bioreduction and sorption in the same bed leads to a significant increase in the effectiveness of the method, greater resistance of the system to inlet mercury concentration fluctuations, and the possibility of effective bioreactor operation at much higher mercury concentration values in wastewater.

In order to finally check the effectiveness of mercury removal from real industrial wastewater in the same research equipment, experiments were also conducted with the use of raw wastewater from two Polish chlor-alkali plants: one at Chemical Plant in Tarnów-Mościce and the other at Chemical Plant "Rokita" S.A. near Wrocław. Two columns, in which earlier studies of model wastewater bioremediation were carried out, were fed during 35 days of the experiment with real wastewater of "naturally" variable parameters; appropriate portions of the wastewater were delivered several times from the above-mentioned plants. The content of Hg^{2+} in these solutions varied from 2.50 to 6.45 mg/dm^3. The experiment was conducted in two variants: one by introducing raw sewage into the column without any pre-treatment, and the second with initially preoxidizing industrial wastewater with NaOCl and removing residual chlorine with $NaHSO_3$. The action of a strong oxidant allowed to carry out all forms of mercury possibly present in raw wastewater to ionic form, i.e., available for bioreduction by bacteria present in the deposit. Such preparation of samples allowed for almost twofold reduction of mercury concentration at the outlet from the bioreactor, from about 200 $\mu g/dm^3$ to values below 100 $\mu g/dm^3$. The results of these experiments finally confirmed the effectiveness of the proposed integrated bioremediation technology and opened the way for its industrial application.

8 Research of the Integrated Process in an Industrial Scale

The promising results of the bioremediation process investigation in laboratory scale allowed to implement the technology in an industrial scale. The full-scale installation was designed in accordance with the conclusions from the previously described studies and applied at the Chlorine Products Division of Chemical Plant in Tarnów-Mościce, Poland (ZAT).

The most important scientific objectives of the research carried out at that time were as follows:

- practical application of the mercury bioremediation method in wastewater from the Polish chlor-alkali plant using the amalgam method under real process industrial conditions;
- to identify optimal conditions for the mercury bioremediation process in an integrated installation including a large-scale bioreactor with fixed activated carbon bed and to test the effectiveness of this technology under conditions of an industrial plant over a long period of time;
- comparison, under Polish conditions, of the effectiveness of the technology originally developed by GBF (Wagner-Doebler et al. 2000; Wagner-Doebler 2003) with the integrated method, combining the process of mercury sorption and bioreduction in one apparatus;
- comparison of the effectiveness of the mercury bioreduction process in a laboratory and industrial scale and determination of the impact of a scale increase on the integrated process of mercury sorption and bioremediation from wastewater;
- comparison of the effectiveness and costs of the method of mercury removal from sewage by the hydrazine reduction, used so far in Polish chemical plants, with the integrated biological method.

Before industrial implementation of the integrated mercury bioremediation method an extensive monitoring studies were carried out in order to identify the average composition of the wastewater from the electrolysis hall of the ZAT Chlorine Products Division and to analyze the range of variation in pH and concentration of various forms of mercury, chlorine, and chlorides during the normal production activity of the chemical plant. It turned out that due to the specific nature of the plant and the installation itself (old, largely depleted equipment working for several decades, subject to relatively frequent failures), the composition of wastewater is subject to significant daily fluctuations in the concentration of mercury, chlorine, chlorides, as well as fluctuations in the pH value of sewage. In addition, it was found that some of the total mercury amount in the raw wastewater may be in metallic and/or bound form that prevents the effective action of reducing microorganisms (Gluszcz et al. 2009). This required the design and use of an additional piece of the installation for the pre-treatment of raw wastewater before feeding it to the biological module. Therefore, the complete sewage bioremediation system at ZAT in Tarnów had to consist of two main parts:

- the sewage pre-treatment plant,
- a compact module for the biological detoxication of mercury in wastewater.

9 Installation for Pre-treatment of Raw Sewage

The results of the conducted monitoring studies indicated that the installation for the pre-treatment of sewage must fulfill the following tasks:

- oxidation of all forms of mercury to Hg(II) ionic form,
- stabilization of ionic mercury by lowering the pH of the solution from approx. 12.0–13.0 to approx. pH = 3.0,
- the removal of free chlorine, which could be deadly toxic to microorganisms,
- oxygenation of sewage in order to deliver oxygen required by microorganisms,
- filtration of precipitated sludge and other fine nonsoluble impurities,
- in winter conditions— ensuring proper temperature of a sewage.

The technological diagram of the sewage pre-treatment node is shown in Fig. 8 and the view of the main elements of the installation is shown in Figs. 9 and 10.

10 Module for Biological Reduction of Mercury

Due to the conditions of location and operation of the installation at the ZAT site in Tarnów, the biological module was built in a compact form, in a typical steel transport container, which made it possible to move it to another location, if necessary, without disassembly. The schematic diagram of the biological module is shown in Fig. 11.

Wastewater after pre-treatment is pumped from the intermediate storage tank to the biological module. At the inlet to the container, a bypass is discharged from the main pipeline, which feeds a small stream of wastewater to the first point of wastewater analysis. This is where the pH, free chlorine concentration in the liquid phase, and the inlet mercury concentration are measured continuously. Mercury concentration is measured online using a continuous industrial mercury analyzer PA-1 (Mercury Instruments, Germany). The main purpose of measuring the parameters of wastewater directly at the inlet to the biological module is to protect the bioreactor against emergency exceeding of their acceptable values. The sewage supply to the container is automatically cut off in the case if:

- pH goes out of the range of 2.5–3.5;
- the concentration of chlorine exceeds 1 mg/dm^3;

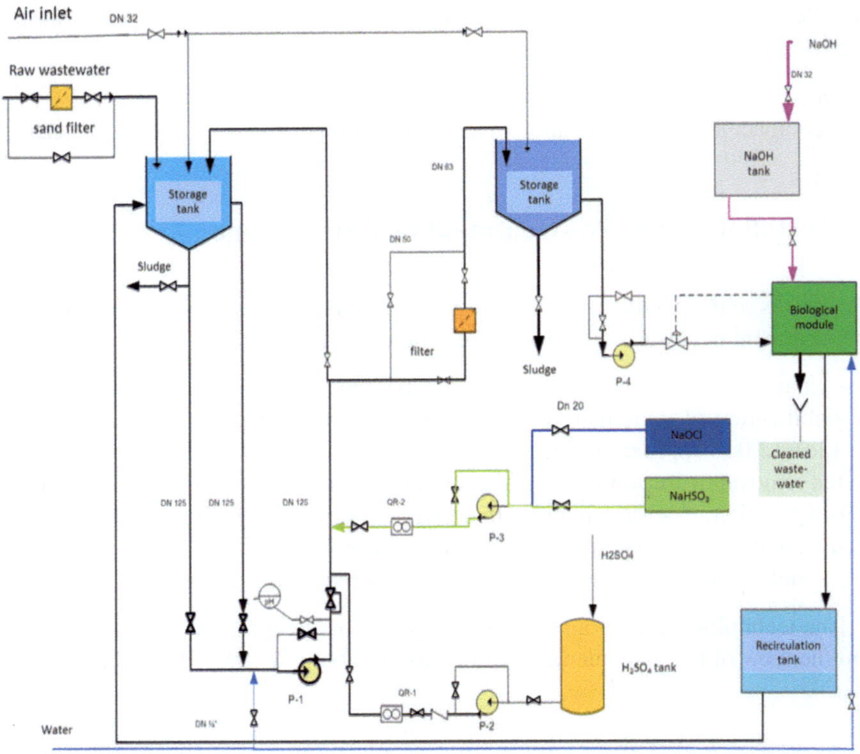

Fig. 8 Scheme of a node for the pre-treatment of raw wastewater

Fig. 9 Fragment of the sewage pre-treatment plant

– the concentration of mercury exceeds 10 mg/dm^3.

If these limits are maintained, the wastewater stream flows into the neutralizer tank. In the neutralizer tank with the capacity of approx. 1.5 m^3, the pH is adjusted from the initial value of approx. pH = 3 to pH = 7 required at the inlet to the bioreactor.

Fig. 10 View of the sewage pre-treatment plant prepared for the winter season

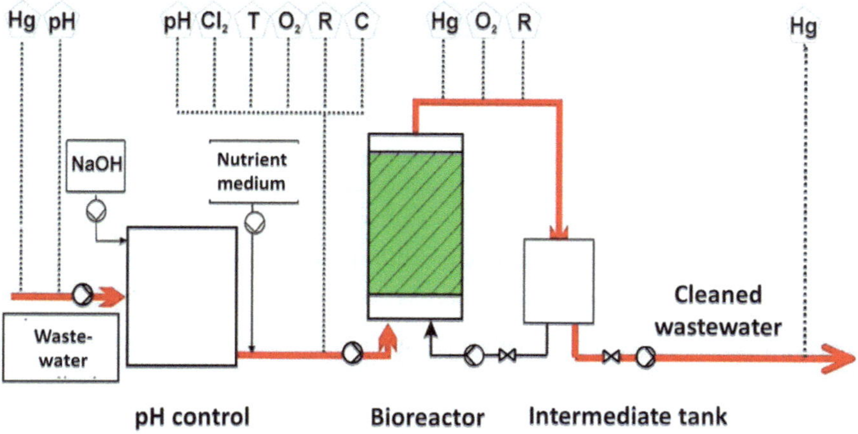

Fig. 11 Schematic diagram of the biological wastewater treatment module

In the case of a pipeline supplying wastewater from the neutralizer to the bioreactor, nutrient solution for microorganisms is also dosed. The wastewater then flows through the temperature, redox potential, and dissolved oxygen measurement system and flows into the bioreactor at the bottom. A diagram showing the structure of the bioreactor bed with dimensions is presented in Fig. 12.

The volume of the carbon deposit in the bioreactor is about 1 m^3, so with the assumed flow of wastewater within the limits of 1–2 m^3/hour, the flow rate expressed in the number of volumes of the deposit exchanged in the hour (BV/h) may vary within the limits of 1–2 BV/h. The flow rate should not change rapidly due to the risk of leaching drops of metallic mercury from the deposit or of washing out the biomass immobilized in the carbon layer. Furthermore, any change in the flow rate brings about the necessity for microorganisms to adapt to new conditions, which temporarily reduces the efficiency of the mercury bioreduction process.

In order to demonstrate the increased efficiency of the integrated mercury sorption + bioreduction process in relation to simple bioreduction, for the 6 months the bioreactor was filled with a porous pumice bed without mercury sorption properties and

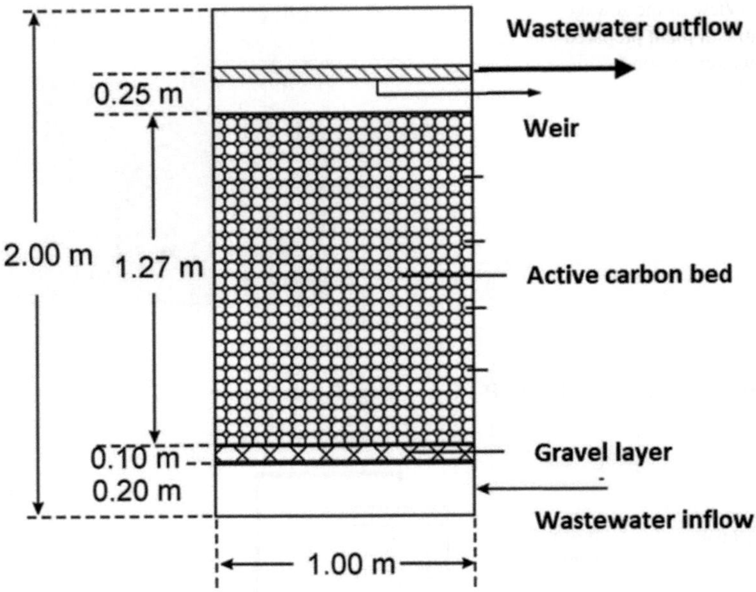

Fig. 12 Detailed schematic diagram of the integrated bioreactor fixed-bed structure

then the pumice bed was replaced with appropriately selected activated carbon. The results of the research carried out in these two phases of the installation's operation are presented below.

In order to continuously maintain the operating parameters of the process within the range acceptable for microorganisms, the biological module is equipped with an automatic measuring and control system, operating under the supervision of a computer program, specially written for the needs of the installation in the ASIX system. The control program allows continuous monitoring of all important process parameters, as well as remote monitoring and possible modification of setpoint limits, i.e., control of the installation via the internet. In addition, in the event of an emergency, the control program automatically cuts off the bioreactor from the rest of the plant if the limit values for the controlled parameters are exceeded, thus protecting the microorganisms from lethal conditions and allowing the plant to continue operating immediately after the cause of the malfunction has been removed. The view of a fragment of the control panel of the program controlling the operation of the installation is shown in Fig. 13.

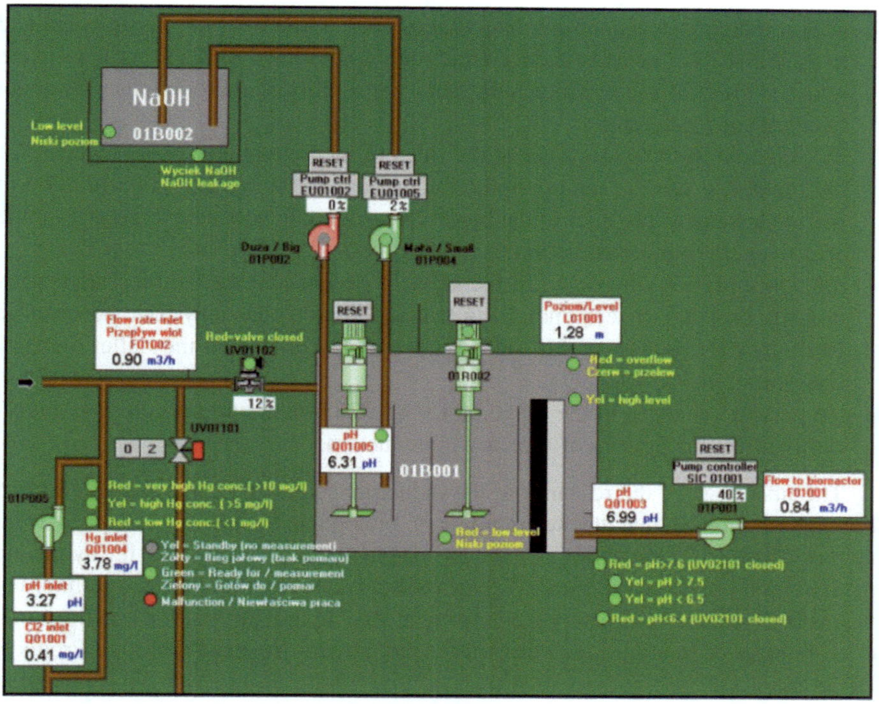

Fig. 13 Fragment of the control panel in the program for monitoring the operation of the installation

11 Evaluation of the Effectiveness of the Integrated Installation

As already mentioned, at the beginning of the operation of the industrial installation the bioreactor was filled with a pumice material without mercury sorption properties. The following conclusions could be drawn from this stage of the operation of the installation:

- the average mercury concentration at the inlet to the bioreactor in the discussed period was 3.51 mg/dm^3, and at the outlet—302 μg/dm^3; the average level of mercury removal from sewage was 91.4%;
- fluctuations of the mercury concentration at the bioreactor inlet had a decisive influence on the changes in the outlet concentration of Hg from the bioreactor;
- short-term (within 1–2 days) increases in the concentration of mercury in sewage inflowing to the bioreactor to 8 mg/dm^3 caused a temporary increase in the concentration of mercury at the outlet up to about 800 μg/dm^3; the simultaneous increase in the concentration of dissolved oxygen in the liquid within the bioreactor indicated a decrease in the activity of microorganisms after such a shock disturbance;

- after a mercury concentration disturbance ceased, the microorganisms regained their ability to effectively reduce ionic mercury within 2–3 days and the system returned to equilibrium, which indicated a high ability of microorganisms to adapt to changing conditions;
- oxygenation of the wastewater at the stage of its pretreatment was sufficient for the biological process, and periodic fluctuations in the oxygen concentration in the wastewater at the inlet to the bioreactor did not have a visible impact on the effectiveness of bioremediation;
- the concentration of oxygen in the effluent stream leaving the bioreactor periodically dropped to zero, indicating that anaerobic zones may have occurred in the bioreactor; however, this did not have a clear effect on the outlet mercury concentration;
- temperature changes in the range of 15–45 °C (winter-summer) did not significantly affect the process of mercury bioreduction.

The first stage of the research in the industrial scale, i.e., using a pumice-filled bioreactor, showed that the technology of biological treatment of sewage from mercury enabled the reduction of mercury load to a degree similar as in the hydrazine method used previously in the ZAT. Due to its advantages, bioreduction technology was able to replace completely the existing at ZAT hydrazine method and thus took over the task of treating the entire wastewater stream from the Chlorine Products Division.

For the overall assessment of the effect of the operation of the mercury sewage treatment plant in the second stage of the study, i.e., with the integrated bioreactor, a balance was made between the mercury load entering the bioreactor with the sewage stream and leaving the bioreactor along with the treated sewage. As Table 4 shows, the total load of mercury brought into the bioreactor during the period in question was 26,564 g, of which 96.3% was retained in the deposit and the remaining part left the bioreactor. The average mercury concentration at the inlet to the bioreactor was 3.18 mg/dm^3, and at the outlet from the bioreactor 121 μg/dm^3.

The analysis of the obtained results showed that by using active carbon in a bioreactor instead of pumice stones, a significant improvement in the system's efficiency was achieved; at similar values of mercury concentration in raw sewage, the degree of mercury charge reduction increased from 91.4 to 96.3%, and the average mercury concentration at the outlet from the bioreactor dropped by more than a half, from 302 μg/dm^3 to 121 μg/dm^3, which should be considered a satisfactory value. According to the environmental protection regulations in Poland, the mercury content of sewage discharged into the environment should not exceed 100 μg/dm^3. At the ZAT plant, wastewater treated by the biological module was still delivered to the main factory wastewater treatment plant, where the final stage of a treatment took place. At the same time, it was found that the stream of wastewater leaving the bioreactor may contain cells of microorganisms with adsorbed mercury. It was therefore assumed that the use of a simple sand filter would allow the retention of such mercury-bound sediments and thus further increase the efficiency of the method.

Table 4 Mercury balance in wastewater treated in the integrated installation

Month	Mercury charge at the inlet [g]	Mercury charge at the outlet [g]	Reduction of mercury load [%]
January	2449,14	41,23	98,3
February	1793,45	10,97	99,4
March	2381,48	50,73	97,9
April	2930,32	99,26	96,6
May	1543,20	68,26	95,6
June	1987,30	110,6	94,4
July	2548,91	124,51	95,1
August	2834,34	124,8801	95,6
September	2634,12	117,3274	95,5
October	2678,32	113,5675	95,8
November	2783,56	110,2401	96,0
Total	26,564,14	971,58	96,3

12 Comparison of Biological Mercury Neutralization Method with the Hydrazine Technology Previously Used at ZAT

The costs associated with the implementation and operation of both installations, as shown in Table 5, are an important aspect of the comparison.

The analysis of the above data shows that the investment cost of a hydrazine installation was more than twice as high as the investment cost of a biological apparatus. The operating costs of the mercury bioremediation plant are also lower, by approximately PLN 15,000 per year. However, the most important advantage of mercury bioreduction technology compared to the hydrazine plant is that it does not require the use of any toxic substances. In addition, there is no problem with the large amount of mercury-polluted solid waste (spent activated carbon). Due to the nature of the process, mercury in the bioreactor is separated in the form of microdroplets and mechanically retained in a highly porous bed, mainly in its bottom part. As a result, the total capacity of the deposit with respect to metallic mercury is incomparably higher than in the case of pure adsorption and the need to replace the bioreactor deposit is much less frequent than in the case of hydrazine technology. After more than a year of operation of the integrated installation, it was not necessary to replace the deposit and the increase in flow resistance through the bioreactor was insignificant. In addition, it is possible to reuse regenerated carbon as a carrier for microorganisms. In the course of operation of a bioreduction plant there is less solid waste contaminated with mercury produced. Therefore, the method based on the ionic mercury biotransformation process can be considered safer for the environment in all aspects.

Table 5 Summary of the investment and operating costs of the compared installations (PLN-Polish zloty)

	Unit	Hydrazine installation	Mercury bioreduction integrated installation
Investment cost:	[PLN]	1 150 000	480 000
Cost of consumables:			
activated carbon	[PLN/year]	25 000	–
hydrazine (N_2H_4)		40 750	–
Electricity		11 003	19 902
Sulfuric acid (H_2SO_4)		–	367
sodium hypochlorite (NaOCl)		–	7 776
sodium sulphite ($NaHSO_3$)		–	8 400
sodium lye (NaOH)		–	720
Nutrient solution		–	24 480
Total operating cost		76 753	61 645

The summary of the operation of the integrated installation in the considered period of 1 year made it possible to draw the following conclusions:

- the installation with the bioreactor of 1 m^3 volume filled with activated carbon worked with high efficiency, removing mercury from the entire stream of approximately 40 m^3/day of wastewater from the ZAT's Chlorine Products Division, achieving a significantly higher degree of mercury removal than the bioreactor with a pumice filling;
- the integration of the sorption and bioreduction processes in the bioreactor resulted in a higher stability of the deposit, i.e., lower fluctuations of the outlet mercury concentration after the inlet concentration disturbances;
- the use of a low-cost sand filter to retain biomass and small amounts of sediments from the bioreactor further reduced mercury concentrations in treated wastewater to less than 50 $\mu g/dm^3$;
- in the bioreactor filled with activated carbon, the oxygen concentration in the liquid dropped to zero more frequently, which indicated a higher activity of biomass growing on carbon than in a pumice-stone bed;
- the rate of nutrient solution dosing had a great influence on the activity of microorganisms, which resulted in significant changes in the mercury concentration at the outlet from the installation and fluctuations in the degree of a liquid phase oxygenation in the bioreactor;
- the rate of nutrient dosing for microorganisms must be carefully controlled to avoid excessive biomass growth inside the bioreactor;

- temperature fluctuations in the range of 15–45 °C in different seasons of the year had no significant influence on the process of mercury bioremediation in the integrated bioreactor;
- during the long-term performance of the bioreactor, the composition of the population of microorganisms in relation to the inoculum applied at the beginning of the research was changed; the dominant role was taken over by strains better adapted to the actual conditions in the bioreactor and not present in the original inoculum.
- The integrated mercury bioremediation method developed and tested in this study can be used not only for the treatment of wastewater from the chlor-alkali industry. This technology may be used in all cases where the initial concentration of mercury in a liquid phase is between 1–15 mg Hg/dm^3 and the final content should not exceed 100 $\mu gHg/dm^3$, i.e., the acceptable level for effluent discharged to open waters. Potentially practical applications include, e.g., the treatment of wastewater from flue gas scrubbers in power plants, the treatment of mine water, especially in plants using the mercury leaching method (precious metal mines), rainwater in areas heavily contaminated with mercury (e.g., areas of different chemical plants using mercury in the production process), leachate from landfills of mercury-containing sludge.

In conclusion, the most important scientific and practical achievements resulted from the development and industrial implementation of the integrated technology for the bioremediation of mercury in industrial wastewater are as follows:

- effective enlargement of the scale of the process from laboratory to industrial scale on the basis of extensive experimental studies;
- design of the complete installation for the specific industrial chlor-alkali plant and its construction and long-term operation;
- demonstration of the usefulness and increased effectiveness of the integrated technology by replacing the existing hydrazine method used previously in Zakłady Azotowe w Tarnowie;
- achievement of a positive economic and ecological effect as a result of the operation of the new method/ installation;
- development and experimental verification of the mathematical model of the integrated process of mercury bioremediation in the full-scale industrial bioreactor (Gluszcz et al. 2011);
- the possibility to use the developed technology in other cases of water contamination with toxic mercury, e.g., to treat wastewater from flue gas scrubbers in power plants, to treat mine water, rainwater in post-industrial areas heavily polluted with mercury, leachate from landfills for mercury-containing waste.

References

Abbas SH, Ismail IM, Mostafa TM, Sulaymon AH (2014) Biosorption of Heavy Metals—a Review. J Chem Sci Technol 3(4):74–102

Boening DW (2000) Ecolocical effects, transport and fate of mercury. Chemosphere 40:1335

Diamantopoluou I, Skodras G, Sakellaropoulos GP (2010) Sorption of mercury by activated carbon in the presence of flue gas components. Fuel Proc Technol 2(91):158–163

Dunn RF, El-Halwagi M (2003) Process integration technology review: background and applications in the chemical process industry. J Chem Technol Biotechnol 9(78):1011–1021

Gluszcz P, Fuerch K, Ledakowicz S (2013) Mercury in the chlor-alkali electrolysis industry. In: Wagner-Doebler I (ed) Bioremediation of Mercury—current research and industrial applications. Caister Academic Press, Norfolk, UK, pp 97–118

Gluszcz P, Ledakowicz S, Wagner-Doebler I, Janiszewska I, Boruta D (2009) Integrated industrial sewage treatment plant for mercury contaminated wastewater. Chem Ind 88(12):1352–1358 (in Polish)

Gluszcz P, Ledakowicz S, Zakrzewska K, Deckwer W-D (2005) Modification of the microbiological method for mercury remediation from industrial wastewater. J Biotechnol 118:S163

Gluszcz P, Petera J, Ledakowicz S (2011) Mathematical modeling of the integrated proces of mercury bioremediation in the industrial bioreactor. Bioprocess and Biosystems Eng 34(3):275–285

Głuszcz P, Wagner-Doebler I, Zakrzewska K, Ledakowicz S (2007) Integrated process of ionic mercury bioreduction and adsorption from wastewater. Chem Proc Eng 28(4):909–920

Gluszcz P, Zakrzewska K, Ledakowicz S (2004) Sorption of mercury from aqueous solutions on active carbon—phase balance and kinetics. Chem Process Eng 25(3):889–894

Gluszcz P, Zakrzewska K, Ledakowicz S, Deckwer W-D, Wagner-Doebler I (2006) Adsorption of mercury from aqueous solutions in a fixed-bed sorption column. Proc 17th Int Congr Chem Process Eng CHISA 2:625–632

Gochfeld M (2003) Cases of mercury exposure, bioavailability and absorption. Ecotoxicol Environ Safety 56:174–179

Graydon JW, Zhang X, Kirk DW, Jia CQ (2009) Sorption and stability of mercury on activated carbon for emission control. J Haz Mat 2–3(168):978–982

Hua K, Xueliu X, Zhiping L, Fang D, Bao R, Yi J (2020) Effective removal of Mercury Ions in aqueous solutions: a review. Curr Nanosci 3(16):363–375

Hyman M (2004) The impact of mercury on human health and the environment. Altern Ther Health Med 6(10):70–75

Kalitventzeff B, Marechal F, Closon H (2001) Better solutions for process sustainability through better insight in process energy integration. Appl Thermal Eng 21:1349–1368

Klemes JJ, Varbanov PS, Kravanja Z (2013) Recent developments in process integration. Chem Eng Res Design 10(91):2037–2053

Ledakowicz S, Deckwer W-D (1993) Mercury removal from aqueous solutions by biotransformation. Biotechnology 3:99–107

Rajerdran P, Muthukrishnan J, Gunasekaran P (2003) Microbes in heavy metal remediation. Indian J Experim Biol 41:1–11

Rashid S, Shah IA, Supe-Tulcan R, Rashid W, Sillanpaa M (2022) Contamination, exposure and health risk assessment of Hg in Pakistan: a review

Shah MP (2020) Microbial bioremediation & biodegradation. Springer

Shah MP (2021) Removal of emerging contaminants through microbial processes. Springer

Shahid M, Khalid S, Bibi I, Bundschuh J, Niazi NK, Dumat C (2020) A critical review of mercury speciation, bioavailability, toxicity and detoxification in soil-plant environment: ecotoxicology and health risk assessment. Sci Total Environ 711, Art 134749

Sysalova J, Kucera J, Drtinova B, Cervenka R, Zverina O, Komarek J, Kamenik J (2017) Mercury species in formerly contaminated soils and released soil gases. Sci Total Environ 584–585:1032–1039

Tekere M (2020) Biological strategies for heavy metal remediation. In: Inamuddin, Ahamed MI, Lichtfouse E, Asiri AM (eds) Methods for bioremediation of water and wastewater pollution. Environmental chemistry for a sustainable world, vol 51. Springer, Cham

Wagner-Doebler I (2003) Pilot plant for bioremediation of mercury-containing industrial wastewater. Appl Microbiol Biotechnol 62:124–133

Wagner-Doebler I, Canstein H, Li Y, Timmis KN, Deckwer W-D (2000) Removal of Mercury from chemical wastewater by microoganisms in technical scale. Environ Sci Technol 34(21):4628–4634

Zhao G, Wu X, Tan X, Wang X (2010) Sorption of heavy metal ions from aqueous solutions: a review. Open Coll Sci J 4:19–31

Phytoremediation of Metals and Radionuclides: An Emerging Technology Toward Environment Restoration

Abhishek Dadhich, Lakshika Sharma, Mamta Dhiman, and Madan Mohan Sharma

1 Introduction

Metal and radionuclide contamination is a major issue in this century's world. These elements' pollution poses a serious threat to the ecosystem, prompting researchers to classify them as "emerging pollutants" (Yu et al. 2014). Heavy metals and radionuclides have proliferated as a consequence of globalization, advanced agricultural methods, and increased urbanization practices in the area, resulting in toxicological effects on living creatures and contamination of air, water, and soil (Wijnhoven et al. 2007). Heavy metals enter the environment as a result of a single high-level exposure and show the aggregate effect as a result of low and high concentrations of exposure. Though, when present in the ecosystem, it can survive in a hazardous form for an extended length of time, producing nutritional, ecological, and environmental concerns (Das and Osborne 2018). Bioremediation, the use of biological agents to clean up the atmosphere and ecosystem has gained a lot of attention (Suresh and Ravi Shankar 2004). Plant-based remediation (phytoremediation) is a method of removing heavy metal and radioactive pollution from the soil that is both ecologically friendly and cost-effective. (Entry et al. 1997; Zhu and Shaw 2000). The ability of plants to collect concentrates and degrade contaminants from the air, water, and soil is used in this technique (Salt et al. 1995). Depending on the plant species, characteristics of the soil, and pollutants present, five different types of phytoremediation techniques have been applied: phytoextraction, plant biomass accumulated high concentration of metal and radionuclides; rhizofiltration, remediate contamination from aqueous waste streams with the help of plant roots phytovolatilization, which involves a plant's capacity to convert a to a less hazardous form of chemical compound; phytodetoxification, which involves a plant's detoxifying ability to convert a chemical species to

A. Dadhich · L. Sharma · M. Dhiman · M. M. Sharma (✉)
Department of Biosciences, Manipal University Jaipur, Jaipur, Rajasthan, India
e-mail: madanmohan.sharma@jaipur.manipal.edu

M. P. Shah (ed.), *Microbial Technologies in Industrial Wastewater Treatment*,
https://doi.org/10.1007/978-981-99-2435-6_15

a less toxic form. Phytostabilization is the chemical and physical immobilization of pollutants by plants at a specific place, limiting their migration from one side to the other. Hyperaccumulators and solar-powered pumping stations are two terms that would be used to characterize plants (Cunningham et al. 1995). Comprises homeostatic systems that maintain proper essential metal ion concentrations, while reducing harm from non-essential metal ion exposure in various cellular compartments.

Heavy metals and radionuclides are being removed from the ecosystem are also done by conventional methods in industries such as precipitation, oxidation, filtration, and adsorption. However, there are new remedial technologies that use organisms. Researchers have recently studied standard phytoremediation to revolutionary phytoremediation technologies. The present chapter examines the development and commercialization of such novel developing technologies for the removal of radionuclides and heavy metals from the environment.

2 Heavy Metal Particles Present in the Soil

Enhancement of anthropogenic and industrial activities like electroplating, smelting, mining, combustion, and agriculture, have a role in enhancing the deposition of harmful heavy metal ion concentrations such as arsenic (As), lead (Pb), mercury (Hg), cadmium (Cd), nickel (Ni), chromium (Cr), aluminium (Al) arsenic (As), and lead (Pb) (Singh et al. 2012). Even though some of the heavy metals such as zinc (Zn) iron (Fe), copper (Cu), selenium (Se), are essential in small concentrations, but their higher concentration of accumulation becomes very toxic in the environment (Awa and Hadibarata 2020; Ashraf et al. 2019). The type of the metal, characteristics (such as pH, clay, and organic matter content), and precipitation and adsorption–desorption processes all influence the metal ion concentration in soil solution (Masson et al. 2010; Naidu et al. 2003). Agricultural operations contribute significantly to heavy metal pollution from anthropogenic sources. Fertilizer, both organic and inorganic, is a primary source of heavy metal deposition. Heavy metals such as Zn, Cr, Ni, Pb, and Cd are present in fungicides in various proportions (Nagajyoti et al. 2010). For six decades, lead arsenate insecticides were employed in orchards in Canada, resulting in heavy metal poisoning in the soil. According to Paul's (2017) assessment, groundwater and soil damage in India stemmed from the abandonment of heavy fertilizers and pesticides containing heavy metals. The concentration range in soil (from 1 mg/kg to 100,000 mg/kg) is determined by the metal deposition process. Metals in soil occur as separate particles that are also coupled with inorganic metal complexes (oxides, hydroxides, phosphates, or carbonates), bounded silicate metal, exchangeable and non-exchangeable ions, and other soil components (Fig. 1).

Fig. 1 Different anthropogenic sources of heavy metal present in the environment

3 Threats of Metals in Soil

Plants and animals absorbed heavy metals as trace elements at the limit of less than 10 ppm, depending on a variety of factors. These metals often play a significant role in physiological and biochemical aspects. Oxidation–reduction reaction in the body is constituted by these metals due to their cofactor activity in the enzymatic process. Copper acts as a cofactor in peroxidases, enzyme activity of cytochrome c oxidase, and enzyme activity of cytochrome oxidase. Also, play a key role in hemoglobin formation enzyme activity. Copper cycling between the C (II) and C (I) locations is inconsistent, making it hazardous to living things and causing Wilson's disease in humans. Metals' interactions with various cell components led cellular functions being disrupted. In humans, they have the potential to harm DNA, change cellular morphology, and possibly cause cancer (Tchounwou et al. 2012). Table 1 summarizes different heavy metals' toxicity effects, symptoms, entrance routes, and pronounced health repercussions in living species, particularly humans, are discussed.

4 Metal Hyperaccumulator Plant Species and their Role

The ability of various plant species to absorb heavy metals from polluted soils in their biomass varies (Sinha et al. 2002). Hyperaccumulators plants can deposit more than 0.1 percent Pb, Co, Cr, or more than 1% Zn, Mn, or Ni in the natural environment of plant biomass (Brooks et al. 1979, 1980; Baker and Brooks 1989). There are over 720 plant species belonging to the family Euphorbiaceae, Brassicaceae,

Table 1 Different types of heavy metals, their entry routes, and toxic effect on human health with their permissible limits

Sr.No	Metal	Sources	Route of entry	Symptoms	Health effect	Permis-sible level (mg/l)	Reference
1	Nickel	Batteries, electroplating	ingesting, inhaling, and collecting through the skin	Dermatitis; nickel carbonyl: myocarditis, encephalopathy, Dermatitis; nickel carbonyl: myocarditis, encephalopathy	A respiratory disorder, kidney problems, gastrointestinal distress, dermatitis, cancer	0.2	Yan et al. (2020)
2	Cadmium	Paints and pigments, Plastic stabilizers, Electroplating, Phosphate fertilizers	gastrointestinal tract (GIT)	Pneumonitis, Proteinuria, osteomalacia, lung cancer	Oxidative stress, cancer	0.06	Singh et al. (2011)
3	Chromium	Steel industries, Tanneries, Fly ash	inhalation, ingestion, and dermal absorption	acute renal failure, gastrointestinal hemorrhage, hemolysis	A respiratory disorder, cancer	0.05	Alluri et al. (2007); Shah (2020); Singh et al. (2011)
4	Copper	Smelting, pesticides, fertilizers, and ore refinery	blood cells	Irritation/hemorrhage of the gastrointestinal tract, hemolysis, blue vomitus, and multiorgan failure syndrome	Gastric and neurological disorders	0.1	Singh et al. (2011)

(continued)

Table 1 (continued)

Sr.No	Metal	Sources	Route of entry	Symptoms	Health effect	Permis-sible level (mg/l)	Reference
5	Mercury (Hg)	Burning of fossil fuels, mining, smelting, combustion of solid waste, fertilizers, industrial effluent, use in electrical switches, and fluorescent bulbs Mercury arc lamps, municipal garbage incineration, mercury emissions from batteries, thermometers, and mercury amalgams	Absorption through the dermal, inhalation, and ingestion	GI pain, vomiting, diuresis, anemia, hypovolemic shock, gingivitis, tachycardia, goiter, high urine Hg	Nervous system disruption, brain function impairment, DNA and chromosomal damage, allergic reactions, fatigue, and headaches, as well as poor reproductive impacts	0.01	hhttp://www.atsdr.cdc.gov/toxprofiles/tp.asp?id=115&tid=24
6	Arsenic (As)	Pesticides, mining, gold, lead, copper, and nickel smelting, iron and steel production, coal combustion, and tobacco smoke	Ingestion and inhaling	Weakness, hepatomegaly, melanosis, arrhythmias, and peripheral neuropathies are all symptoms of mucosal injury, hypovolemic shock, fever, and sloughing	Defects in the womb, Lung, skin, liver, bladder, kidneys, and gastrointestinal damage are all carcinogens. Death, severe vomiting, and diarrhea	0.02	http://www.atsdr.cdc.gov/toxprofiles/tp.asp?id=22&tid=3

(continued)

Table 1 (continued)

Sr.No	Metal	Sources	Route of entry	Symptoms	Health effect	Permis-sible level (mg/l)	Reference
7	Lead (Pb)	Pesticides, gold, lead, copper, and nickel mining, smelting, iron and steel production, coal combustion, and tobacco smoke	Ingestion and inhalation	Weakness, hepatomegaly, melanosis, arrhythmias, peripheral neuropathy, and peripheral vascular disease are all symptoms of mucosal injury, hypovolemic shock, fever, and sloughing	Birth defects, Carcinogen: lung, skin, liver, bladder, Kidneys, Gastrointestinal damage, Severe vomiting, diarrhea, death	0.1	http://www3.epa.gov/ttnemc01/prelim/otm31appC; http://www.atsdr.cdc.gov/toxprofiles/tp.asp?id=96&tid=22

Asteraceae, and Rubiaceae are so far known as hyperaccumulators (Reeves et al. 2018) (Table 2). Southeast Asia had seen the highest progress in identifying hyperaccumulator plant species or families (mostly China), distinct genera species Pteris and Pityrogramma are well-known arsenic hyperaccumulators; Noccaea and Sedum are Cd and Zn hyperaccumulators, respectively, Arabidopsis, on the other hand, is a Zn hyperaccumulator (Clemens 2017; Sterckeman et al. 2017). Cd, Pb, Ti, and Zn hyperaccumulators were identified in the relevant genera of Noccaea, Sedum, Iberis, Arabidopsis, and Biscutella in a similar investigation in Europe (Reeves et al. 2018; Sterckeman et al. 2017; Wulff et al. 2013; Shah 2021a). Metals accumulation is a well-known defense mechanism pathogens and herbivores are both targets of this strategy (Hörger et al. 2013). In comparison to other plant varieties, hyperaccumulator plants construct complexes and compartmentate heavy metal in a somewhat distinct and precise way (Leitenmaier and Küpper 2013). For example, phytochelatins (PCs) are formed by cadmium, but in hyperaccumulator plants, they are weak oxygen binds them to ligand sulfur in despite of sulfur ligand (Verbruggen et al. 2009). This is a characteristic feature of a variety of other hyperaccumulation plants and their precise heavy metals (Leitenmaier and Küpper 2013). Various investigations backed up the hypothesis of "elemental defense," claiming that heavy metal hyperaccumulation protects plants from viruses and other natural rivals. Heavy metals accumulated organic protective compounds in a well-synchronized manner, therefore, increasing defense of the plant (Rascio and Navari-Izzo 2011).

5 Radionuclide Components

In the environment radionuclides, contaminants that originate from natural procedures such as erosion, weathering, and volcanic activity, isotopes of thorium (Th), radium (Ra), radon (Rn), plumbum (Pb), uranium (U), and polonium (Po), make up naturally occurring radioactive materials (NORM). Technologically enhanced naturally occurring radioactive materials (TENORM) are also found in the environment as a result of anthropogenic activity, in addition to NORM. Nuclear weapons testing, mineral mining and milling, nuclear power plant operation, nuclear waste disposal, nuclear accidents, agricultural fertilizers, fossil fuel combustion, and the manufacture of radionuclides components that can be used in medical or research are all examples of anthropogenic and industrial activities (Gupta et al. 2016; Hu et al. 2010; Jagetiya et al. 2014) such as 131I radionuclide is used for the treatment of thyroid cancer although 14C is used in agricultural research to identify the changing of the substances (Schlumberger et al. 2012; Valldor et al. 2015). From the nuclear reactors, more than 200 radionuclides are produced and a large number of them take decays to slow in a few eras (Crowley 1997). Despite the fact that a huge number of radionuclides which are anthropogenic, such as 133Xe, 131I, 134Cs, 137Cs, and 90Sr, were released into the atmosphere as a result of the Chernobyl and Fukushima disasters. They resulted after the accumulation of significant radioactive contamination of the

Table 2 A list of a few hyperaccumulator and heavy metals they can remediate with different concentrations

Sr.No	Heavy metal (HM)	Hyperaccumulator plant species	Plant family	Types of metal transport	Concent-ration (mg/kg)	Reference
1	Nickel	*Berkheya codii* *Pentacalia spp.* *Senecia spp.* *Alyssium spp.* *Bornmuellera spp.* *Thlaspi spp.* *Psychotria coronata*	Asteraceae Asteraceae Asteraceae Brassicaceae Brassicaceae Brassicaceae Rubiaceae	YSL3: Yellow-stripelike family members and ZRT-IRT-like (ZIP) transporters (poorly selective)	11,600 16,600 11,000 1280–29,400 11,400–31,200 2000–31,000	Van der Pas and Ingle (2019); Nishida et al. (2015); Schaaf et al. (2006)
2	Zinc	*Thlaspicaerulescence* *Thlaspi rotundifolium* *Dichopetalum gelonioides*	Brassicaceae Brassicaceae Brassicaceae	ZIP: Zinc-regulated transporter Ironregulated transporter Proteins (ZIP4: Major; IRT1: Minor; ZIP3/6/9/1 and HMA4: P-type metal ATPase; FRD3; Multidrug and toxin efflux family; YSL3/7: Yellowstripe-like family transporter	43,710 18,500 30,000	Gendre et al. (2007); Filatov et al. (2006); Hammond et al. (2006); Talke et al. (2006)
3	Cadmium	*Thlaspi caerulescens*	Brassicaceae	Zinc-regulated transporter (ZIP) is a zinc-regulated transporter. Iron-dependent transporter HMA4: P-type metal ATPase and Proteins	2,130	Rascio and Navari-Izzo (2011); Tian et al. (2011)

(continued)

Table 2 (continued)

Sr.No	Heavy metal (HM)	Hyperaccumulator plant species	Plant family	Types of metal transport	Concent-ration (mg/kg)	Reference
4	Selenium	*Astragalus racemosus* *Lecithis ollaria*	Leguminosae Lecithidiaceae	Sulfate/Selanate transporters SULTR1;2 (major) and SULTR1;1 (minor) and SULTR2;1: Sulfate/Selanate transporter	1,49,200 18,200	Pilon-Smits (2017); Lima et al. (2018)
5	Arsenic		Pteridaceae Brassicaceae	TIP4;1: Tonoplast intrinsic protein (High affinity AsIII transporter) and Phosphate/Arsenate transporter (High-affinity AsV transporter)	8331 3647 2900	Souri et al. (2017); Karimi and Souri (2015); He et al. (2016)
6	Cobalt	*Haumaniastum robertii* *Aeollanthus subacaulis* *Crotolaria cobalticola*	Lamiaceae Lamiaceae Fabaceae	IRT1 transporters, CDF) family proteins, zinc–iron permease (ZIP), and (AtHMA3)	10,232 4,300 30,100	Morel et al. (2009)
7	Manganese	*Maystenus bureaviana* *Maystenus sebertiana* *Macadania Neurophylla*	Celastraceae Celastraceae Proteaceae	MtZIP6 and MtZIP7, SMF1 and PMR1	19,230 22,500 55,200	Reeves et al. (2018)
8	Lead	*Minuartia verna* *Agrostis tenuis* *Festuca ovina*	Caryophyllaceae Poaceae Poaceae	Heavy metal transporting ATPases (HMAs), HMA3, P1B-ATPase	20,000 13,490 1,750	Hanikenne and Baurain (2014)

terrestrial and marine ecosystems (Yoschenko et al. 2018). Radionuclides accumulated in soil particles and also entered the aquatic ecosystem eventually causing toxic health effects to a living organism.

6 Heavy Metal and Radionuclide Remediation Technologies

6.1 Phytoextraction

Bioaccumulation of heavy metals and radionuclides by plants is a well-known mechanism today; higher plants in large numbers show different strategies when exposed to metals and radionuclides. From the metal and radionuclide, contaminated ecosystem plants take up toxic pollutants in their biomass if they are present at the toxic site, thus helping in environmental restoration. Parmar et al. (2015), reviewed phytoextraction technique as discussed the remediation in which higher plants absorb pollutants from the surface water, groundwater, or soil. This technology is mostly used to remediate contaminates metals polluted (Cd, Nickel (Ni), Cu, Zn, Pb), but it can also be utilized to remediate contaminates polluted by other metals such as Se, As, and radionuclides. Hyperaccumulator plant species that maintain a high accumulation quality of precise metals in their biomass are favored by phytoextraction (0.01–1% depending on the metal, dry weight) (Parmar et al. 2015). Phytroextraction is also known as phytoaccumulation and hyperaccumulation according to reports (Xiao et al. 2017). Metals, salts, and radionuclides are removed from the soil using cationic pumps and absorbed water from plants. Several hyperaccumulator plant species around metal mining sites thrived in heavily contaminated soils, thus selecting the right plant for absorbing and transporting pollutants in a variety of harmful situations is critical (Saxena et al. 2020). Although soils can be heavily contaminated in some circumstances, removing pollutants using this strategy will save time in a variety of ways.

6.2 Phytostablization

Phytostabilization is an in-situ approach in which hazardous materials are mixed with the lignin to create non-toxic chemicals from the root cell wall, reducing the presence of toxins in natural and agricultural ecosystems (Khalid et al. 2017; Mahar et al. 2016). The direct mechanism of root exudate produces precipitated metals in insoluble forms, which then permeate the soil medium. The major aim associated with this technology is to avoid pollution concentrations and suppress their distribution in the soil (Favas et al. 2014; Ali et al. 2013; Fan et al. 2017). Haumaniastrum, Eragrostis, Ascoliepis, Gladiolus, and Alyssum are among the hyperaccumulator plant species used in this method (Khalid et al. 2017). Many plant species can assist in phytoimmobolization due to their proclivity for releasing a large number of chelating compounds (Lebrun et al. 2018). Contaminants are immobilized by these chemicals, which restrict their uptake and reduce their mobility in soil. As a result, plants with the potential to stabilize play an important role in contaminated agricultural production sites and vegetation restoration (Eskander and Saleh 2017; Saha et al. 2017). Inorganic contaminants from contaminated sites such as arsenic (As), copper (Cu), lead (Pb), zinc (Zn), cadmium (Cd), chromium (Cr), and other metals in their remnant sediment can also be remedied via phytostabilization (Yadav et al. 2018).

6.3 Rhizodegradation

Rhizodegradation technology involved the breakdown of the organic component as a result of soil microbial activity by the root or rhizosphere part of the plant (Echereme 2018). Phytostimulation or rhizofiltration is a bioremediation technology that uses a specific plant species as well as bacteria and fungi associated with roots to remove contaminants (Ali et al. 2013; Khalid et al. 2017). These plant species roots are symbiotically associated with mycorrhizal fungi and considerably more useful for the biochemical accessibility and production of nutrients such as phosphorus (P), potassium (K), sulfur (S), Ni, nitrogen (N), cobalt (Co), Cu, zinc (Zn), and calcium (Ca) over the broad network system (Sarwar et al. 2017). Although nutrients are distributed spatially, rhizosphere and microbial associations are varied, although species of the genus Pseudomonas indicate connections associated with the root (Ali et al. 2013; Singh and Singh 2016; Salem et al. 2018). Plant death must be taken into account, and various agronomic strategies must be used to reduce it. Plant death should be considered, and agronomic approaches should be employed to reduce it by planting the plant during the rising phase in an unpolluted soil sample, for improved phytoremediation enhancement and effectiveness.

6.4 Phytovolatilization

Phytovolatilization technology entails the absorption of certain metals and radionuclides by green plants, followed by their release by vaporization. Some metal ions are taken up by the roots, converted into non-toxic compounds, and then emitted into the environment such as arsenic (As), mercury (Hg), and selenium (Se) (Limmer and Burken 2016). The phytovolatilization process, on the other hand, has the drawback of releasing dangerous metals into the environment, which can then be deposited as precipitation or form other toxic compounds (Nikolić and Stevović 2015). *Stanleya pinnata, Astragalus bisulcatus, Liriodendron tulipifera, Brassica napus, Arabidopsis thaliana, or Nicotiana tabacum* are among the plants used in this method. Heavy metals such as Se and Hg can also be remedied using this method (Ali et al. 2013; Mani and Kumar 2014) (Fig. 2).

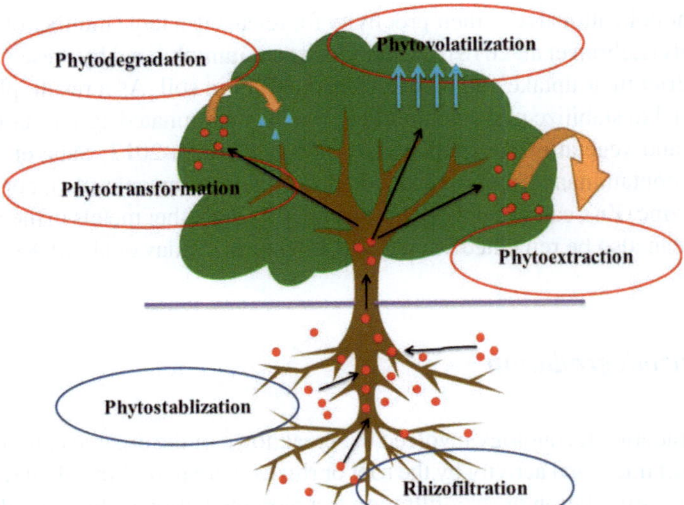

Fig. 2 Overview of different phytoremediation techniques to remediate heavy metals and radionuclide contaminants

7 Phytoremediation Using Chelating Agents

Chelating agents are another typical approach for increasing heavy metal availability in plant biomass. These agents are added to the soil to generate a chelate complex, which is made up of water-soluble, heavy metals that are available in a more mobile form that plants may easily absorb (Wuana and Okieimen 2011). Chelating chemicals can help to inhibit heavy metal precipitation and absorption in the soil, allowing heavy metals to be released more easily from soil components and thereby, enhancing heavy metal bioavailability (Salt et al. 1995; Ali et al. 2013). For chelate-assisted phytoremediation, different types of chelating agents are employed in practise. Synthetic chelating chemicals including ethylene diamine tetraacetic acid (EDTA), ethylene glycol tetraacetic acid (AGTA), and diethylene triamine penta acetic acid (DTPA) can boost heavy metal availability and absorption in plant biomass (Gupta et al. 2008; Sarwar et al. 2017). However, the impacts of these chelating compounds in the soil are limited due to their low biodegradability for a longer length of time resulted in metal leaching and negative environmental repercussions (Smolińska and Król 2012; Lee and Sung 2014). Organic chelating compounds including citric acid, malic acid, acetic acid, and oxalic acid, on the other hand, have been found to produce heavy metal compounds and increase heavy metal bioavailability in plants (Sarwar et al. 2017). Organic chelators are abundant in nature and easily degradable in soil, posing a smaller environmental risk than synthetic chelators (Souza et al. 2013); as a result, using organic chelating agents to generate chelate-assisted phytoremediation complexes will be more promising.

8 The Commercial and Global Value of Phytoremediation

The global market figure for the phytoremediation was reviewed by Glass 1999. Depending on their nutritional needs, plants take photosynthetic energy from the earth and store it in plant biomass. Due to their chemical similarities to other nutrient ions, contaminants that are required or non-essential trace elements can penetrate higher plants when present in large quantities. The roots can take up As043- or Cd2 + via the uptake systems for P043- or Fe2 + /Ca2 + , respectively (Clemens 2001).

Phytoextraction aims to maximize the bioavailability by utilizing the nutrient procurement mechanism of plants in demand, heavy metals, and other pollutants in the top portion of ground tissues. The upper portion of the plant's removing contaminant subsequently from the site by harvesting the biomass in a restricted number of following growth seasons. Plant waste can be recycled for metal smelting or dumped in specialized landfills. When a phytoextraction plant is cultivated on a pullulated site, it must accumulate significant amounts of various trace elements in the shoot, produce a high rate of biomass, and develop a large root system. By reviewing a comparative analysis of phytoremediation marketplaces (Glass 1999), the author stated that the expected phytoremediation market in 1999 was higher than the estimates in 1998. The increase can be attributed to an increase in the number of organizations offering their services, primarily in the dealing engineering industry, as well as an increase in the technology's acceptance. As a result, phytoremediation is a promising alternative to standard remediation methods. Phytoremediation is estimated to be worth \$5–\$40 for every ton of pullulated soil (Farraji et al. 2016; Wan et al. 2015). The total cost of phytoremediation of soil contaminated with heavy metals like arsenic, cadmium, and lead was \$37.7/m3, which is much cheaper than the expenses of alternative remediation procedures like solidification (\$87–\$190/m3) (Halder 2022), extraction (\$240–\$290/m3) (Halder 2022), (Halder 2022), or vitrification (\$75–\$425/ton) (Farraji et al. 2016). Despite its many benefits, phytoremediation is still not a widely used method around the world. The information provided by the United States, on the other hand, is inconclusive. According to the Environmental Protection Agency (EPA), phytoremediation has proven to be effective in a range of locations across the country.

9 Advantages and Limitations of Phytoremediation

See Table 3.

Table 3 Advantages and limitations of phytoremediation

Sr.NO	Advantage	Limitation	Reference
1	Complete studies carried out for swine wastewater treatment by *Eichhornia crassipes* and for groundwater cleanup, a deep-rooted hybrid Populus sp. is used	Contamination dissolved in groundwater is not appropriate for the case of aquatic phytoremediation	Chien et al. (2015); Quinn et al. (2001); Van Den Bos (2002)
2	Although the success of large-scale phytoremediation in a natural setting is much lower than in-vitro studies, the benefits of this remediation technology continue to attract attention and demand plans for future environmental cleaning	The key feature of footprint green technology is controlling and disposing of polluted plants using phytoremediation technology	Gomes et al. (2013); Ahalya et al. (2003); Ghosh and Singh (2005); Lasat (2000)
3	Minimal equipment required with low investment cost (constitutes substantial savings)	Contaminant cleanup that is incomplete and has a long-term low performance	Filippis (2015)
4	This treatment removes contaminants from the soil, water, or groundwater in-situ	Mostly used on the uppermost layer of the soil and mine tailings	Dhanwal (2017)
5	An effective useful method for breaking down various organic pollutants	Need additional nutrient supply for the treatment and usually slower than the other physicochemical treatments	Fatima (2017)
6	Waste production is less in secondary generation	Contaminants have the potential to enter the food chain and food web of the ecosystem	Mustafa (2021)
7	Cost efficacy, and the option of harvesting the plants' biomass accumulated or absorbed the pollutants like heavy metals and radionuclides for recycling	Mats of plants anchorages hazardous insects such as mosquitoes	Ahmad et al. (2017)

10 Conclusion

At present time, radionuclides and heavy metals are the emerging pollutants in the ecosystem having toxic properties which affect living organisms consisting mainly of human beings. Exposure to heavy metals and radionuclides has increased considerably as a result of industrial and human sources, despite the fact that natural sources contribute to environmental contamination. Immediate action is required to limit these operations being carried out in order to preserve humanity from metal and radionuclide pollution, as well as to work toward remediating existing toxins that cannot be removed from the ecosystem. These pollutants can be interchanged into a less hazardous form, trapped in rehabilitative living organisms such as microbes and plants, or utilized by plants and bacteria in their metabolic processes. Currently, genetic engineering tools and modern biotechnological methods are used to increase the remedial features of microorganisms. Despite microbes, plants show potential for phytoextraction of metal contaminants from their biomass, and filtered the metals from the root water system (rizofiltration), or waste site stabilization through erosion control and evapotranspiration (phytostabilization). After some time of the plants growing stage, it's permitted either harvested or cremated to recycle the metals. This process can be used as many times as necessary to reduce soil contamination levels to acceptable levels.

Ploughing, dredging, capping, soil flushing, pneumatic fracturing, electrokinetics, chemical conversions, soil/sediment, vitrification, and excavation are examples of traditional metal and radionuclide remediation processes are immense As a result, phytoremediation has a lot of capacity to cope with toxicants as well as recovery of useful components. Phytoremediation, as summarized in this chapter, is a green, long-term solution for removing metal and radionuclide contaminants from the environment while also protecting living organisms from their toxic effects.

References

Ahalya N, Ramachandra T, Kanamadi R (2003) Biosorption of heavy metals. Res J Chem Environ 7:71–79

Ahmad S, Pandey A, Kothari R, Pathak VV, Tyagi VV (2017) Closed photobioreactors: construction material and influencing parameters at the commercial scale. In: Photobioreactors advancement application and research. NOVA publication, pp 149–162

Ali H, Khan E, Sajad MA (2013) Phytoremediation of heavy metals concepts and applications. Chemosphere 91:869–881

Alluri HK, Ronda SR, Settalluri VS, Bondili JS, Suryanarayana V, Venkateshwar P (2007) Biosorption: an eco-friendly alternative for heavy metal removal. Afr J Biotechnol (25)

Ashraf S, Ali Q, Zahir ZA, Ashraf S, Asghar HN (2019) Phytoremediation: environmentally sustainable way for reclamation of heavy metal polluted soils. Ecotoxicol Environ Saf 174:714–727

ATSDR. Toxicological profile for arsenic (2007a) http://www.atsdr.cdc.gov/toxprofiles/tp.asp?id=22&tid=3. Accessed on 22 Sept 2015)

ATSDR. Toxicological profile for lead (2007b) http://www.atsdr.cdc.gov/toxprofiles/tp.asp?id=96&tid=22. Accessed on 22 Sept 2015

ATSDR. Toxicological profile for mercury http://www.atsdr.cdc.gov/toxprofiles/tp.asp?id=115&tid=24. Accessed on 22 Sept 2015

Awa SH, Hadibarata T (2020) Removal of heavy metals in contaminated soil by phytoremediation mechanism: a review. Water Air Soil Pollut 231:47. https://doi.org/10.1007/s11270-020-4426-0

Baker AJM, Brooks RR (1989) Terrestrial higher plants which hyperaccumulate metallic elements - a review of their distribution, ecology and phytochemistry. Biorecovery 1:81–126

Brooks RR, Morrison RS, Reeves RD, Dudley TR, Akman Y (1979) Hyperaccumulation of nickel by *Alyssum linnaeua*(Cruciferae). Proc SocLond BiolSci 203:387–403

Brooks RR, Reeves RD, Morrison RS, Malaisse F (1980) Hyperaccumulation of copper and cobalt—a review. Bull Soc Roy Bot Belg 113:166–172

Chien C, Yang Z, Cao W, Tu Y, Kao C (2015) Application of an aquatic plant ecosystem for swine wastewater polishment: a full-scale study. Desalin Water Treat 1–10

Clemens S (2001) Molecular mechanisms of plant metal tolerance and homeostasis. Planta 212:475–486

Clemens S (2017) How metal hyperaccumulating plants can advance Zn biofortification. Plant Soil 411:111–120

Crowley KD (1997) Nuclear waste disposal: the technical challenges. Phys Today 50:32–39. https://doi.org/10.1063/1.881764

Cunningham SO, Berti WR, Huang JW (1995) Phytoremediation of contaminated soils. Trends Biotechnol 13:393–397

Das A, Osborne JW (2018) Bioremediation of heavy metals. In: Environmental chemistry for a sustainable world 11. Springer, New York, p 9

Dhanwal P, Kumar A, Dudeja S, Chhokar V, Beniwal V (2017). Recent advances in phytoremediation technology. Advances in environmental biotechnology 227–241.

Echereme CB, Igboabuchi NA, Izundu AI (2018) Phytoremediation of heavy metals and persistent organic pollutants (POPs) a review. IJSRM Human 10(4):107–125

Entry JA, Watrud LS, Manasse RS, Vance NC (1997) Phytoremediation and reclamation of soils contaminated with radionuclides. Phytoremediation of soil and water contaminants, chap 22. A C S, New York

EPA. An overview of Airborne Metal Regulations, Exposure Limits, Health Effects and Contemporary Research (Appendix C) (2010) http://www3.epa.gov/ttnemc01/prelim/otm31appC.pdf. Accessed on 2 Oct 2015

Eskander S, Saleh H (2017) Phytoremediation: an overview. Environ Sci Eng, Soil Pollut Phytoremediation 11:124–161

Fan Y, Li H, Xue Z, Zhang Q, Cheng F (2017) Accumulation characteristics and potential risk of heavy metals in soil-vegetable system under greenhouse cultivation condition in northern China. Ecol Eng 102:367–373

Farraji H, Zaman NQ, Tajuddin RM, Faraji H (2016) Advantages and disadvantages of phytoremediation: A concise review. Int J Environ Tech Sci 2:69–75

Fatima K, Imran A, Naveed M, Afzal M (2017) Plant-bacteria synergism: An innovative approach for the remediation of crude oil-contaminated soils. Soil Environ 36(2):93–113

Favas PJ, Pratas J, Varun M, D'souza R, Paul MS (2014) Phytoremediation of soils contaminated with metals and metalloids at mining areas: potential of native flora. Environ Risk Assess Soil Contam 3:485–516

Filatov V, Dowdle J, Smirnoff N, Ford lloyd BR, Newbury HJ, Macnair MR (2006) Comparison of gene expression in segregating families identifies genes and genomic regions involved in a novel adaptation, zinc hyperaccumulation. Mol Ecol 15(10):3045–3059

Filippis LFD Hakeem K, Sabir M, Ozturk M, Mermut AR (2015) Role of phytoremediation in radioactive waste treatment: soil remediation and plants: prospects and challenges. Academic Press Elsevier, New York, pp 207–254

Gendre D, Czernic P, Conéjéro G, Pianelli K, Briat JF, Lebrun M, Mari S. TcYSL3 (2007) A member of the YSL gene family from the hyper-accumulator Thlaspicaerulescens, encodes a nicotianamine-Ni/Fe transporter. Plant J 49(1):1–5

Ghosh M, Singh S (2005) A review on phytoremediation of heavy metals and utilization of it's by products. Asian J Energy Environ 6:18

Glass DJ (1999) US and international markets for phytoremediation, 1999–2000. DJ Glass Associates Inc., Needham, MA, USA, p 266

Gomes HI, Dias-Ferreira C, Ribeiro AB (2013) Overview of in situ and ex situ remediation technologies for PCB contaminated soils and sediments and obstacles for fullscale application. Sci Total Environ 445:237–260

Gupta DK, Srivastava A, Singh VP (2008) EDTA enhances lead uptake and facilitates phytoremediation by vetiver grass. J Environ Biol 29:903–906

Gupta DK, Chatterjee S, Datta S, Voronina AV, Walther C (2016) Radionuclides accumulation and transport in plants. In: Gunther FA, de Voogt P (Eds) Rev Environ Contam Toxicol, vol 241. Springer, Cham, pp 139–160. https://doi.org/10.1007/398_2016_7

Halder S, Anirban A (2022) Removal of environmental pollutants (Lead, Chromium And Cadmium) using root and leaf tissues of Indian mustard, rice and wheat plants. BioRxiv 02

Hammond JP, Bowen HC, White PJ, Mills V, Pyke KA, Baker AJ (2006) A comparison of the Thlaspicaerulescens and Thlaspiarvense shoot transcriptomes. New Phytol 170:239–260

Hanikenne M, Baurain (2014) Origin and evolution of metal P-type ATPases in Plantae (Archaeplastida). Front Plant Sci 4:544.https://doi.org/10.3389/fpls.2013.00544

He Z, Yan H, Chen Y, Shen H, Xu WZ, H, (2016) An aquaporin PvTIP4;1 from Pterisvittata may mediate arsenite uptake. New Phytol 209:746–761

Hörger AC, Fones HN, Preston G (2013) The current status of the elemental defense hypothesis in relation to pathogens. Front Plant Sci 16(4):395

Hu QH, Weng JQ, Wang JS (2010) Sources of anthropogenic radionuclides in the environment: a review. J Environ Radioact 101:426–437. https://doi.org/10.1016/j.jenvrad.2008.08.004

Jagetiya B, Sharma A, Soni A, Khatik UK (2014) Phytoremediation of radionuclides: a report on the state of the art. In: Gupta DK, Walther C (Eds) Radionuclide contamination and remediation through plants. Springer, Cham pp 1–31. https://doi.org/10.1007/978-3-319-07665-2_1

Karimi N, Souri Z (2015) Effect of phosphorus on arsenic accumulation and detoxification in arsenic hyperaccumulator, Isatiscappadocica. J Plant Growth Regul 34:8–95

Khalid S, Shahid M, Niazi NK, Murtaza B, Bibi I, Dumat C (2017) A comparison of technologies for remediation of heavy metal contaminated soils. J Geochem Explor 182:247–268

Kumar SS, Kadier A, Malyan SK, Ahmad A, Bishnoi NR (2017) Phytoremediation and rhizoremediation: uptake, mobilization and sequestration of heavy metals by plants. Plant-microbe interactions in agro-ecological perspectives, pp 367–394

Lasat M (2000) Phytoextraction of metals from contaminated soil: a review of plant/soil/metal interaction and assessment of pertinent agronomic issues. J Hazard Subst Res 2:1–25

Lebrun M, Miard F, Hattab-Hambli N, Bourgerie S, Morabito D (2018) Assisted phytoremediation of a multi-contaminated industrial soil using biochar and garden soil amendments associated with Salix alba or Salix viminalis: abilities to stabilize As, Pb, and Cu. Water Air Soil Pollut 229:163

Lee J, Sung K (2014) Effects of chelates on soil microbial properties, plant growth and heavy metal accumulation in plants. Ecol Eng 73:386–394. https://doi.org/10.1016/j.ecoleng.2014.09.053

Leitenmaier B, Küpper H (2013) Compartmentation and complexation of metals in hyperaccumulator plants. Front Plant Sci 20(4):374

Lima LW, Pilon-Smits EA (1862) Schiavon M (2018) Mechanisms of selenium hyperaccumulation in plants: a survey of molecular, biochemical and ecological cues. BiochimicaetBiophysicaActa (BBA) 11:2343–2353

Limmer M, Burken J (2016) Phytovolatilization of organic contaminants Environmental. Science Technology 50:6632–6643

Mahar A, Wang P, Ali A, Awasthi MK, Lahori AH, Wang Q Li R, Zhang Z (2016) Challenges and opportunities in the phytoremediation of heavy metals contaminated soils. Ecotoxicol Environ Saf 126:111–121

Mani D, Kumar C (2014) Biotechnological advances in bioremediation of heavy metals contaminated ecosystems: an overview with special reference to phytoremediation. Int J Environ Sci Technol 11:843–872

Masson P, Dalix T, Bussière S (2010) Determination of major and trace elements in plant samples by inductively coupled plasma-mass spectrometry. Commun Soil Sci Plant Anal 41:231–243

Morel M, Crouzet J, Gravot A, Auroy P, Leonhardt N, Vavasseur A, Richaud P (2009) AtHMA3, a P1B-ATPase allowing Cd/Zn/Co/Pb vacuolar storage in Arabidopsis. Plant Physiol 149:894–904

Mustafa HM, Hayder G (2021) Recent studies on applications of aquatic weed plants in phytoremediation of wastewater: a review article. Ain Shams Eng J 12(1):355–365

Nagajyoti PC, Lee KD, Sreekanth TVM (2010) Heavy metals, occurrence and toxicity for plants: a review. Environ Chem Lett 8:199–216

Naidu R, Oliver D, McConnell S (2003) Heavy metal phytotoxicity in soils. In: Fifth national workshop on the assessment of site contamination, pp 235–241

Nikolić M, Stevović S (2015) Family asteraceae as a sustainable planning tool in phytoremediation and its relevance in urban areas. Urban Urban Green 14:782–789

Nishida S, Kato A, Tsuzuki C, Yoshida J, Mizuno T (2015) Induction of nickel accumulation in response to zinc deficiency in Arabidopsis thaliana. Int J Mol Sci 16(5):9420–9430

Paul D (2017) Research on heavy metal pollution of river Ganga: a review. Ann Agrar Sci 15:278–286

Parmar S, Singh V (2015) Phytoremediation approaches for heavy metal pollution: a review. J Plant Sci Res 2:135–147

Pilon-Smits EAH, Smits, E, Winkel L, Lin, ZQ (2017) Mechanisms of plant selenium hyperaccumulation. In: Pilon-Smits E, Winkel L, Lin ZQ (Eds) Selenium in plants, vol 11. Springer, Cham

Quinn J, Negri M, Hinchman R, Moos L, Wozniak J, Gatliff E (2001) Predicting the effect of deep-rooted hybrid poplars on the groundwater flow system at a large-scale phytoremediation site. Int J Phytoremediat 3:41–60

Rascio N, Navari-Izzo F (2011) Heavy metal hyperaccumulating plants: how and why do they do it? And what makes them so interesting? Plant Sci 1;180(2):169–181

Reeves RD, Baker AJ, Jaffré T, Erskine PD, Echevarria G, van der Ent A (2018) A global database for plants that hyperaccumulate metal and metalloid trace elements. New Phytol Apr 218(2):407–411

Saha JK, Selladurai R, Coumar, MV, Dotaniya M, Kundu S, Patra AK (2017) Remediation and management of polluted sites soil pollution an emerging threat to agriculture. Springer, Singapore

Salem HM, Abdel-Salam A, Abdel-Salam MA, Seleiman MF (2018) Phytoremediation of metal and metalloids from contaminated soil. Springer, Singapore, pp 249–262

Salt DE, Blaylock M, Kumar NPBA, Dushenkov V, Ensley BD (1995b) Phytoremediation: a novel strategy for the removal of toxic metals from the environment using plants. Nat Biotechno 13:468–474. https://doi.org/10.1038/nbt0595-468

Sarwar N, Imran M, Shaheen MR, Ishaque W, Kamran MA, Matloob A (2017) Phytoremediation strategies for soils contaminated with heavy metals: modifications and future perspectives. Chemosphere 171:710–721. https://doi.org/10.1016/j.chemosphere.2016.12.116

Saxena G, Purchase D, Mulla SI, Saratale GD, Bharagava RN (2020) Phytoremediation of heavy metal-contaminated sites: eco-environmental concerns, field studies, sustainability issues, and future prospects. Rev Environ Contam Toxicol 249:71–131

Schaaf G, Honsbein A, Meda AR, Kirchner S, Wipf D, von Wirén N (2006) AtIREG2 encodes a tonoplast transport protein involved in iron-dependent nickel detoxification in Arabidopsis thaliana roots. J Biol Chem 281(35):25532–25540

Schlumberger M, Catargi B, Borget I, Deandreis D, Zerdoud S, Bridji B, Bardet S, Leenhardt L, Bastie D, Schvartz (2012) Strategies of radioiodine ablation in patients with low-risk thyroid cancer. N Engl J Med. 366: 1663–1673.https://doi.org/10.1056/NEJMoa1108586
Shah Maulin P (2020) Microbial bioremediation & biodegradation. Springer
Shah Maulin P (2021a) Removal of refractory pollutants from wastewater treatment plants. CRC Press
Shah Maulin P (2021b) Removal of emerging contaminants through microbial processes. Springer
Shah V, Daverey A (2020) Phytoremediation: A multidisciplinary approach to clean up heavy metal contaminated soil. Environ Technol Innov 18:100774
Singh S, Singh A (2016) Phytoremediation: a sustainable approach for restoration of metal contaminated sites. Int J Sci Res 5:2171–2174
Singh R, Gautam N, Mishra A, Gupta R (2011) Heavy metals and living systems: an overview. Indian J Pharmacol 43:246–253
Sinha S, Saxena R, Singh S (2002) Comparative studies on accumulation of Cr frommetal solution and tannery effluent under repeated metal exposure by aquaticplants: Its toxic effects. Environ Monito Assess 80(1):17–31
Smolińska B, Król K (2012) Leaching of mercury during phytoextraction assisted by EDTA, KI and citric acid. J Chem Technol Biotechnol 87:1360–1365. https://doi.org/10.1002/jctb.3826
Souri Z, Karimi N, Sandalio LM (2017) Arsenic hyperaccumulation strategies: an overview. Front Cell Dev Biol 5:67
Souza LA, Piotto FA, Nogueirol RC, Azevedo RA (2013) Use of non-hyperaccumulator plant species for the phytoextraction of heavy metals using chelating agents. SciAgric 70:290–295. https://doi.org/10.1016/j.chemosphere.2008.11.007
Sterckeman T, Cazes Y, Gonneau C, Sirguey C (2017) Phenotyping 60 populations of Noccaea-caerulescens provides a broader knowledge of variation in traits of interest for phytoextraction. Plant Soil 418(1):523–540
Suresh B, Ravishankar GA (2004) Phytoremediation—a novel and promising approach for environmental clean-up. Cric Rev Biotechnol 24(2–3):97–124
Talke I, Hanikenne M, Krämer U (2006) Zinc dependent global transcriptional control, transcriptional de-regulation and higher gene copy number for genes in metal homeostasis of the hyperaccumulator Arabidopsis halleri. Plant Physiol 142:148–167
Tchounwou PB, Yedjou CG, Patlolla AK, Sutton DJ (2012) Heavy metal toxicity and the environment. Mol, Clin Environ Toxicol 2012:133–164
Tian S, Lu L, Labavitch J, Yang X, He Z, Hu H (2011) Cellular sequestration of cadmium in the hyperaccumulator plant species Sedum alfredii. Plant Physiol 157:1914–1925
Valldor P, Miethling-Graff R, Martens R, Tebbe CC (2015) Fate of the insecticidal Cry1Ab protein of GM crops in two agricultural soils as revealed by 14C-tracer studies. Appl Microbiol Biotechnol 99:7333–7341. https://doi.org/10.1007/s00253-015-6655-5
Van Den Bos A (2002) Phytoremediation of volatile organic compounds in groundwater: Case studies in plume control. Draft report prepared for the US EPA Technology Innovation Office under a National Network for Environmental Management Studies Fellowship
Van der Pas L, Ingle RA (2019) Towards an understanding of the molecular basis of nickel hyperaccumulation in plants. Plants 8(1):11
Verbruggen N, Hermans C, Schat H (2009) Mechanisms to cope with arsenic or cadmium excess in plants. Curr Opin Plant Biol 12(3):364–72
Wan X, Lei M, Chen L, Cost T (2015) Benefit calculation of phytoremediation technology for heavy-metalcontaminated soil. Sci Total Environ 563:796–802
Wijnhoven S, Leuven RSEW, Velde GVD, Jungheim G, Koelemij EI, Vries FTD, Eijsackers HJP, Smits AJM (2007) Heavy-metal concentrations in small mammals from a diffusely polluted floodplain: importance of species- and location-specific characteristics. Arch Environ ContamToxicol 52:603–613

Wuana RA, Okieimen FE (2011) Heavy metals in contaminated soils: a review of sources, chemistry, risks and best available strategies for remediation. Isrn Ecol 2011:402647. https://doi.org/10.5402/2011/402647

Wulff AS, Hollingsworth PM, Ahrends A, Jaffré T, Veillon JM, L'Huillier L, Fogliani B (2013) Conservation priorities in a biodiversity hotspot: analysis of narrow endemic plant species in New Caledonia. PLoS one 18;8(9):e73371

Xiao R, Wang S, Li R, Wang JJ, Zhang Z (2017) Soil heavy metal contamination and health risks associated with artisanal gold mining in Tongguan, Shaanxi, China. Ecotoxicol Environ Saf 141:17–24

Yadav KK, Gupta N, Kumar A, Reece LM, Singh N, Rezania S, Khan SA (2018) Mechanistic understanding and holistic approach of phytoremediation: A review on application and future prospects. Ecol Eng 120:274–298

Yan A, Wang Y, Tan SN, MohdYusof ML, Ghosh S, Chen Z (2020) Phytoremediation: a promising approach for revegetation of heavy metal-polluted land. Front Plant Sci 11:359

Yan L, Van Le Q, Sonne C, Yang Y, Yang H, Gu H, Ma NL, Lam SS, Peng W (2021) Phytoremediation of radionuclides in soil, sediments and water. J Hazard Mater 5(407):124771

Yoschenko V, Ohkubo T, Kashparov V (2018) Radioactive contaminated forests in Fukushima and Chernobyl. J Forest Res 23:3–14. https://doi.org/10.1080/13416979.2017.1356681

Yu H, Ni SJ, He ZW, Zhang CJ, Nan X, Kong B, Weng ZY (2014) Analysis of the spatial relationship between heavy metals in soil and human activities based on landscape geochemical interpretation. J Geochem Explor 146:136–148

Zhu YG, Shaw G (2000) Soil contamination with radionuclides and potential remediation. Chemosphere 41:121–128. https://doi.org/10.1016/s0045-6535(99)00398-7

Combination of Membrane-Based Biochar for Ammonium Removal from Domestic wastewater—A Review

Khac-Uan Do, Thanh-Son Bui, and Ngoc-Thuy Vu

1 Introduction

In an aqueous solution, ammonia exists in both anionic and non-ionic forms. Ammonia under the gas phase is highly soluble in aqueous solutions and combines with water to form ammonium and hydroxide in a certain proportion that upon on environmental variables, including pH and temperature (Bock 2016; Heckelman et al. 1996).

$$NH_{3(aq)} + H_2O \leftrightarrow NH_4^+ + OH^- \quad (1)$$

Ammonia can be attributed to natural processes and human activities. It is reported that 80% of produced ammonia is utilized in the field of agriculture as synthetic fertilizer, either for direct use or converted to various solid and liquid N fertilizers. The remaining 20% is utilized for other purposes, i.e., industrial applications or explosives (Giddey et al. 2013). In fact, any natural or industrial processes containing nitrogen organic matter available for breakdown can be considered a potential source of ammonia. Ammonia is regarded as one of the most widely utilized industrial chemicals. Accordingly, it can be utilized to produce fertilizers, fibers, plastics, explosives, cleaning fluids, etc. Ammonia and ammonium compounds can also be found in food as leavening agents and composting facilities. Basically, ammonia is released into the aquatic environment from non-point sources (i.e., runoff from urban and agricultural) and man-made point sources (i.e., sewage treatment plants) (Puckett 1994).

K.-U. Do (✉) · N.-T. Vu
School of Environmental Science and Technology, Hanoi University of Science and Technology, Hanoi, Vietnam
e-mail: uan.dokhac@hust.edu.vn

T.-S. Bui
School of Chemical Engineering, The University of New South Wales, Sydney, Australia

M. P. Shah (ed.), *Microbial Technologies in Industrial Wastewater Treatment*,
https://doi.org/10.1007/978-981-99-2435-6_16

In domestic wastewater, high content of nitrogen is derived from human fecal, urine, and personal care products. It is estimated that total nitrogen in domestic wastewater lies in the range of 20–70 mg/L. Nitrogen mainly exists in the following forms: organic nitrogen, nitrate, nitrite, and ammonia/ammonium. Among ammonia, normally accounts for up to 60–70%, while organic nitrogen only makes up 30–40% of domestic wastewater. More importantly, the higher content of ammonia was found in anaerobic wastewater treatment effluent, comprising permeates of anaerobic membrane bioreactors because of the lack of nitrogen oxidation (Deng 2014).

Basically, domestic wastewater is composed of up to 70% ammonia. Unfortunately, ammonia has both chronic and acute impacts on aquatic life. An increasing content of ammonia discharge into water is a serious threaten toward ecological systems as ammonia is a contributing factor to the accelerated eutrophication of water bodies and the depletion of dissolved oxygen (Du et al. 2005). Notably, toxic effects of ammonia toward fish and other aquatic animals have been shown through gill/liver/kidney damage, and reduced blood oxygen-carrying capacity (Huff et al. 2013).

Besides the detrimental effects on aquatic life, it is crucial to mention the potential impacts derived from nitrogen in domestic wastewater on human beings. Nitrate and nitrite usually occur in trace quantities in surface water. They may reach high levels due to agricultural runoff, human and animal waste, and particular effluents from sewage treatment works. The toxicity of nitrate to human beings is originated from its ability to be reduced to nitrite. The main biological impact of nitrite on human beings lies in its contribution to the oxidation of haemoglobin to methaemoglobin, which lacks the capacity to carry oxygen to the tissues. This results in death in children because of oxygen starvation (or, known as blue-baby disease) (Organization 2003).

It is obviously that an excess amount of ammonia in wastewater can bring to aquatic life and human being as well. In that context, ammonia removal is necessary for wastewater treatment plants. To date, the most widely used methods for removing ammonia from wastewater are air stripping (Limoli et al. 2016, Shah MP., 2020), biological nitrification–denitrification (Wu et al. 2008), ion exchange (Jorgensen and Weatherley 2003), adsorption method (Bernal and Lopez-Real 1993), and breakpoint chlorination (Zaghouane-Boudiaf and Boutahala 2011). The aim of this work is to provide a brief discussion of the advantages and disadvantages of each method. In particular, ammonium removal by biochar was reviewed. Especially, a novel approach of biochar added to the biological and membrane biological reactor was evaluated. It would be an alternative to enhance the ammonium removal in the system.

2 Ammonia Treatment Technologies

2.1 *Air Stripping Process*

Air stripping of ammonia is the physical separation process that is based on the principle of mass transfer, in which ammonia molecular is stripped from the liquid form to the gas phase.

$$NH_4^+ + OH^- \leftrightarrow NH_{3(aq)} + H_2O \tag{2}$$

This method involves the transfer of significant amounts of air over the wastewater stream, resulting in the formation of the partial pressure of the ammonia gas within the wastewater stream. Air stripping can be considered as a low-cost method for ammonium ion treatment in wastewater {Ahmad 2014 #58}. There are some serious disadvantages hindering the widespread use of this process, of which adjustment of pH value is one of the difficulties in this process. Another commonly encountered challenge in air stripping is the operating temperature. Air stripping tower could not work in subfreezing weather because of icing. Hence, temperature control is essential, but it also requires the additional cost. Last, air pollution and noise remain problems associated with the air stripping method.

2.2 *Ion Exchange Process*

The ion exchange process refers to a chemical reaction between ions in a solid phase and ions in an aqueous phase. In the process, the wastewater is filtered by multimedia filtration, followed by carbon filtration, and then by ion exchange utilizing a fixed-bed column containing natural zeolites. By utilizing natural zeolites the efficiency of ammonia treatment can be achieved from 94 to 97%, which is higher than that of air stripping (about 90%) (Çelik et al. 2001). However, the ion exchange requires more additional costs compared to air stripping. It is recommended that the use of ion exchange for ammonia treatment in wastewater should be considered toward concentrations of around 70 mg/L (Chan 2003).

2.3 *Breakpoint Chlorination Process*

The process of ammonia ion treatment from wastewater by the addition of chlorine is known as breakpoint chlorination. The applied chlorine dose is measured as chlorine.

$$2NH_4^+ + 3Cl_2 \rightarrow N_2 + 6HCl + 2H^+ \tag{3}$$

It is reported that approximately 95–99% of the ammonia might be oxidized to nitrogen gas. However, several undesirable by-products (i.e., dichloramine, trichloramine, or nitrate ions) were formed. The main drawbacks of using chlorination consist of the greater presents of dissolved solids in wastewater, and the additional costs regarding operation and maintenance. More importantly, the health problems associated with trihalomethanes that are formed in chlorination remain a serious issue when applying this method (Chan 2003).

2.4 *Biological Nitrification–denitrification Process*

The traditional technique of removing ammonia from aqueous solution is based on biological treatment. However, the widespread use of this method is generally hindered by several factors, such as pH change, toxic shock, low temperature, and low dissolved oxygen (Seruga et al. 2019). The conventional biological ammonia removal system comprises two parts: nitrification and denitrification. To nitrification occur, it is mandatory to provide a lot of energy for aeration, whereas denitrification requires a sufficient amount of organic carbon. In case of insufficient organic carbon (less than required) in wastewater, it is very important to add an external carbon source to the anoxic reactor to promote the growth rate of biomass (Chan 2003). In that context, removal of the ammonia via the precipitation of struvite ($MgNH_4PO_4.\ 6H_2O$) is an alternative. However, this technique is heavily depending on several factors (i.e., temperature, pH, the certain content of calcium, magnesium, and phosphorus). With a high concentration of NH_4^+ ions in aqueous solution, it is essential to add an additional amount of PO_4^{3-} or Mg^{2+} and maintain a weakly alkaline environment to achieve the crystallization of struvite (Huang et al. 2014).

2.5 *Adsorption Process*

Among ammonia removal methods, the adsorption method is worthy of consideration as a potential alternative technique to eliminate ammonia ions in wastewater due to its promising properties like simplicity, effectiveness, economic efficiency, and feasibility in comparison to mentioned methods above. Obviously, the selection of suitable adsorbent material plays an irreplaceable role in the overall efficiency of the adsorption process. To date, the common adsorbents that have been utilized for ammonia adsorption can be listed as follows: activated carbon, charcoal, limestone, zeolite, and coal fly ash. However, the widespread use of these materials is seriously restricted by secondary treatment requirements, which leads to an increase in operating costs.

Recently, biochar has attracted intensive interest from the scientific community in environmental engineering. Owing to the large surface area, highly porous structures that contain oxygen-containing functional groups on the surface, and abundant raw materials, biochar can be considered as an effective and low-cost adsorbent (Inyang et al. 2011, Shah MP., 2021). Notably, the biochar-based adsorption process does not require secondary treatment (Jayakumar et al. 2021). To date, many efforts have been made to optimize the biochar-based adsorption process toward heavy metals and organic matters. However, the studies focusing on the application of biochar in the removal of ammonium from aqueous solutions are still in the early stage and seldom reported.

3 Application of Biochar for Ammonium Removal

3.1 Charateristics of Biochar

Biochar is a carbon-rich solid material, which is generally obtained via the thermochemical conversion of biomass at temperatures usually lower than 700 °C in an oxygen-limited or inert environment. Biomass is originated from wood chips, leaves, crop stalks, animal manure, carcasses, sludge (Tan et al. 2017). Biochar is defined as a combination of two types: charcoal and char, excluding products derived from fossil fuel, and they are produced via incomplete combustion (smoldering or charring) of carbon-containing organic matter such as plants and trees. Biochar preparation methods comprise pyrolysis, hydrothermal carbonization, gasification, and other methods. Pyrolysis is a process of thermally decomposing raw material at high temperatures (300–650 C) under anoxic conditions (Kambo and Dutta 2015). Pyrolysis produces the three major products, comprising carbon-containing solid product as biochar, volatile products which are converted to liquid matters via condensation, and the non-condensable gases, such as H_2, CH_4, CO, and CO_2 (Mohan et al. 2006). Depending upon the reaction time, temperature, and heating rate the pyrolysis process can be categorized into: slow, fast, and flash pyrolysis. Among, slow pyrolysis can be considered a main process for biochar production due to the higher solid yield (ranged from 25 to 35%), the highest value compared to those of other pyrolysis processes (Mohan et al. 2006). In the slow pyrolysis process, biomass is gradually heated to temperatures ranged from 300 to 650 C with low heating rates (10–30 C/min) during long residence time (up to hours) (Onay and Kockar 2003). The process parameters, such as operating temperature, heating rates, reaction time, and moisture content of raw material play an essential role influencing physicochemical properties and the final yield of biochar. Accordingly, it is reported that when the whole process run at a slow heating rate and low operating temperature, the percentage yield of obtained biochar are significantly improved (Oliveira et al. 2017; Onay 2007).

Hydrothermal carbonization is a thermochemical conversion process in which raw materials are converted into carbon-rich solid products as hydrochar. Hydrothermal carbonization is carried out at temperature ranged from 180 to 260 C in a Teflon-lined autoclave in which feedstocks are immersed in liquid, followed by heating in a confined system under pressure (2–6 MPa) for minutes to hours (Hoekman et al.

2013; Mumme et al. 2011). Hence, this process is not influenced by the high moisture content of raw materials, and this advantage eliminates the predrying requirement of wet feedstocks, which indirectly saves a significant amount of required energy and investment cost. Hydrothermal carbonization produces three main products: solid products (hydrochar), liquid products (bio-oil), and gases (mainly, CO_2) (Weber and Quicker 2018). Hydrochar is the preferred product in the hydrothermal carbonization process with the final yield in the range of 40–70% (Yan et al. 2010). Gasification is the process of incomplete oxidation of biomass at a high operating temperature ranging from 600 to 1200 C for short residence time (10–20 s) (McKendry). The primary product of gasification is a mixture of gases (CO_2, CO, and H_2), a potential fuel. Gasification only produces a low yield of biochar, less than 10%, which is due to the complete conversion of most of the organic matter to ash and gases (Brewer et al. 2009).

It is imperative to determine the characteristics of biochar as they play an irreplaceable role in defining the scale of application in environmental remediation. As discussed above, each thermochemical pretreatment method possesses distinct process parameters and operating conditions. This leads to the formation of corresponding products which own distinct physicochemical properties. The main element compositions of biochar are carbon, oxygen, hydrogen, and nitrogen, in which carbon content normally accounts for higher than 50%, except for manure- and sludge-derived biochar. The alkyl groups and aromatic structures constituted by these elements are the key constituents of biochar (Glaser et al. 1998). When the operating temperature of pretreatment methods increases, the ratio of H/C or O/C content in biochar tends to reduce, and this implies enhanced aromaticity and carbonization degrees, except for polarity (Yang et al. 2018). The existence of N is mainly depending on the type of raw materials and there is no obvious correlation between the content of N and operating temperature. The contents of trace elements in biochar derived from animals, comprising calcium, phosphorus, magnesium, and potassium, are higher than those in plant-based biochar. Notably, organic composition in the feedstocks easily decomposed into volatile gas during pretreatment at high temperatures, resulting in the formation of disordered crystal plates through cross-linking of carbon atoms (Chen et al. 2018). This leads to the formation of a number of pores of various sizes on the surface of final products as biochar. When the operating temperature promptly increases, the quantity of large hole and microporous also rise quickly, resulting in the rapid increase in the specific surface area of biochar. It is also essential to mention oxygen-containing functional groups (i.e., carboxyl, carbonyl, and phenol hydroxyl) that exist on the surface of biochar. The number of acidic polar functional groups on the biochar surface enhance its polarity, which in turn directly promotes the adsorption capacity toward polar material (Chen et al. 2018). These functional groups make biochar an ideal candidate of superior adsorbents to treat heavy metals in wastewater. An increase in pyrolysis temperature during pretreatment methods is not favorable to the formation of functional groups (Qambrani et al. 2017). Recently, (Xu et al. 2014) reported a significant content of magnesium found in biochar produced from pig manure. However, there is no presence of these elements even in detectable quantities in wheat straw-based biochar.

It was reported that the properties such as surface area, pH, or surface charge of biochar synthesized from different feedstocks were considerably different. Similarly, the pore diameter and surface area of the biochar were significantly different and heavily depended on the initial compositions of biomass, which in turn influences the adsorption capacity toward heavy metals. (Liu and Zhang 2009) pointed out that pinewood-derived biochar possessed twice the Pb adsorption capacity, compared to that of rice husk biochar.

3.2 Ammonium Removal by Biochar Adsorption

The potential utilization of biochar for the elimination of heavy metals and organic compounds from wastewater has received intensive attention in the field of environmental remediation during the past decade. Various functional groups that exist in cellulose, hemicelluloses, proteins, and sugars in agricultural residue feedstocks can be physically activated either by pyrolysis or steam and CO_2 treatment, and this promotes the adsorbtion capacity of biochar toward pollutants (Inyang et al. 2011). As we mentioned above, the amount of functional groups in the biochar play an irreplacable role in the removal of contaminants via adsorbate removal mechanism. In short, high carbon content and void structures, along with high degree of porosity, large surface area, and strong affinity for non-polar substances make biochar serve as promising sorbent material in wastewater treatment technology (Qambrani et al. 2017; Yenisoy-Karakaş et al. 2004). Recent reports by (Cao et al. 2009) and (Lu et al. 2012) pointed out biochar was much more effective than commercial activated carbon for heavy metal adsorption. Development of biochar-based wastewater treatment techniques could be considered a low-cost route due to the wide variety and abundance of natural feedstocks for biochar production. (Lehmann 2007) reported that the cost of agricultural by-products-derived biochar production is only USD \$4 per gigajoule, which is mainly dependent on the required equipment and heating system.

Biochar could effectively absorb ammonia in aqueous phase (Zhang et al. 2012). In fact, post-treatment of biochars bring positive impacts on ammonium adsorption capacity. (Xu et al. 2018) reported magnesium-oxides-modified biochar (Mg-biochar) produced from wood waste exhibited an impressive removal capacity of ammonium, achieving 47.5 mg/g, which is higher than those of other carbon-based absorbents. The struvite precipitation on the surface of wood waste-derived biochar was the predominant mechanism in the removal of ammonium. (Wang et al. 2016) reported an remarkably enhanced ammonium adsorption capacity in oxidized maple wood-derived biochar, in comparison to pristine maple wood-based biochar. (Li et al. 2017) prepared MgO-impregnated porous biochar via an integrated adsorption-pyrolysis method for ammonium absorption from an aqueous solution. Obtained results revealed that 20% Mg-biochar possessed the highest ammonium adsorption ability, being higher than 22 mg/g. The authors described that the factors consisting of struvite crystallization, electrostatic attraction, and π–π interactions are the main driving force which contributed to the enhanced adsorption of ammonium. Furthermore, pyrolysis temperatures are another factor that affects adsorption ability of biochar toward ammonium. Interestingly, Yang and his colleagues (Yang et al. 2018) found that lower pyrolysis temperatures resulted in a higher ammonium adsorption capacity. Accordingly, pine sawdust-derived biochars at 300 C exhibited the best ammonia adsorption ability (5.38 mg/g), which is due to electrostatic interaction and chemical bonding between ammonium and functional groups on the surface of biochar. The higher H/C and O/C ratios of synthesized biochar at 300 C (0.78 and 0.32, respectively) pointed out the greater presence of functional groups on the surface of biochar as compared to those of the sample at 550 C.

4 Combination of Biochar and Membrane System for Ammonium Removal

4.1 Addition of Biochar into Membrane System

Membrane filtration technology has been applied to wastewater treatment widely in recent years (Manikandan et al. 2022). Membrane could be used in both aerobic and anaerobic wastewater treatment technologies (Simonič 2021). During operation, membrane fouling normally happened due to the deposition of the organic and inorganic foulants and/or biofoulants. They could be attached with the membrane surface. Besides, they could be inside membrane pores. In fact, fouling is still a big challenge of membrane application. Membrane fouling could be controlled by several strategies, such as, physical way (using vibrating membranes; gas scouring; ultrasonic waves), chemical method (using acid or alkalis to remove foulants). Besides, the addition of chemicals or adsorbents could also be used to reduce the membrane fouling by removing the foulants in the soluble form (Lee et al. 2021).

So far, granular activated carbon has been added to the membrane system. It could play as a carrier for bacterial growth. Besides, it could be scoured with the air in the reactor, resulting in reduction of foulants,which were attached to the membrane surface (Schumann et al. 2020). Recently, biochar produced from agricultural wastes was used widely in wastewater treatment. Biochar has been a low-cost absorbent. Biochar could be used for adsorption of pollutants independently as a physicochemical method. The adsorption affinity of ammonium on biochar depended on several factors. For example, the availability of functional groups in biochar surface could enhance the adsorption of ammonium. In addition, the multilayer adsorption could be happened in biochar because of the hydrogen bonding and polar interactions between the functional groups with ammonium. Besides, biochar could be also used by combining it with biological processes. In particular, it would be interesting to add biochar into a membrane bioreactor system. Figure 1 presents a schematic diagram of the biological system with addition of biochar. Biochar could be added to a conventional activated sludge. It could also be added to a membrane bioreactor

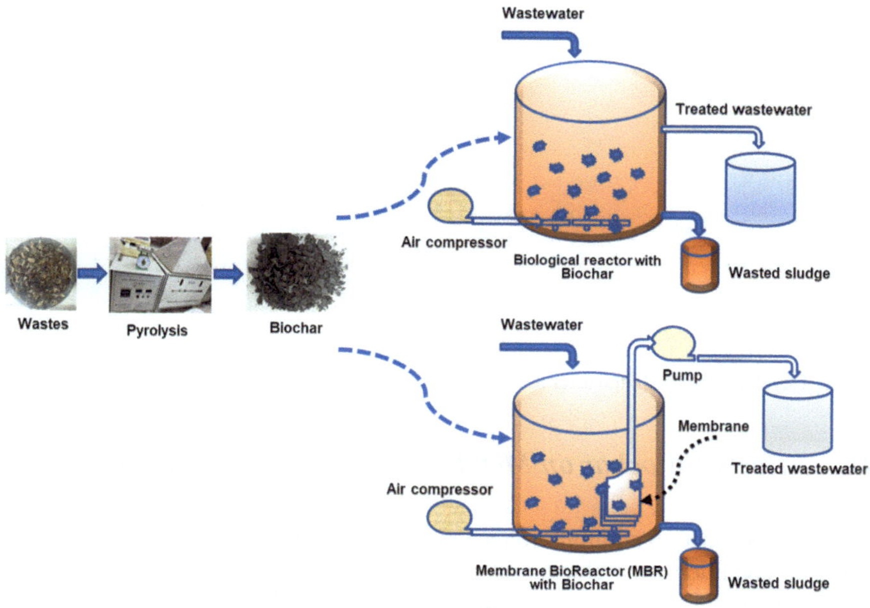

Fig. 1 A development of biochar addition in the biological and membrane bioreactor systems treating ammonium in wastewater

system. Several membranes such as MF (microfiltration membrane), UF (ultrafiltration membrane) could be used in the system. The membrane was characterized by the membrane type (hollow fiber or flat sheet membrane); production material (PES, PVDF); pore sizes, and flux (Manikandan et al. 2022). By this way, biochar could enhance the performance efficiency of the system. In addition, it could help to control the membrane fouling during operation.

Adding biochar into a biological system could help to create the granular sludge. As seen in Fig. 2, the granular sludge was formed after a week. It was developed well after three months. This shows that the biochar has played as a suitable carrier for bacteria growth. Biochar has been used as an adsorbent in a membrane system (Schumann et al. 2020). Membrane system combined with biochar showed a good rate of reversible membrane fouling. Biochar could adsorb small organic compounds. Therefore, it could help to reduce the rate of the irreversible membrane fouling. In addition, biochar was characterized by high hydrophilicity. It contains several polar functional groups so that it could adsord other pollutants, such as ammonium in the reactor.

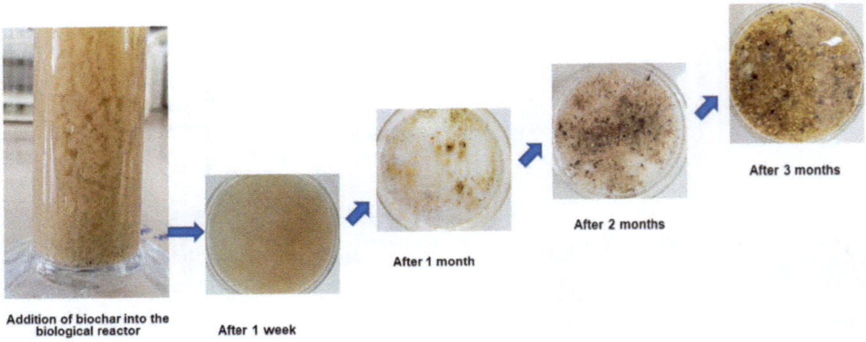

Fig. 2 Development of sludge after the addition of biochar into the biological reactor

4.2 *Effect of Biochar on Membrane System Performances*

Biochar was added to ultrafiltration membrane system to remove humic acid. Under unstirred conditions, the biochar addition did not show good ability in in term of membrane fouling reduction. However, under the stirred conditions, biochar could enhance the humic acid removal. As a result, the membrane flux was increased and the membrane fouling was reduced significantly (Shankar et al. 2017). The presence of humic acid in water and wastewater could be one of the factors affecting the membrane fouling significantly. Besides, the divalent ions in waters and wastewaters (such as Ca^{2+} ion) was reported as a factor affecting the humic acid fouling in the membrane. This could be due to the functional groups in humic acid interacted with Ca^{2+}. As a results, a humic acid fouling layer structure could be formed and deposited onto the surface of the membrane. Finally, it could affect the membrane fouling. Therefore, removal of humic acid is necessary to control membrane fouling. Addition of biochar into the membrane system would be a good solution. Biochar could be acted as an adsorbent, which could adsort the humic acid at a relative high adsorption capacity (Miao et al. 2018).

Biochar was incorporated with a mixed matrix membrane as an adsorbent to remove phosphate from wastewater. In this case, the adsorption characteristics of the membrane could be examined by using the cross-flow filtration tests. With this matric membrane–biochar, the adsorption process of phosphorus removal could be expressed by several mechanisms such as precipitation, ionic exchange, ligand exchange, electrostatic interaction, and bonding of hydrogen. Biochar enhanced phosphate removal significantly, however, membrane fouling during operation was not mentioned (Mohammadi et al. 2021). Biochar was combined with a ceramic membrane system to treat textile wastewater. In this work, biochar was produced from market vegetable waste. Textile wastewater was filtered by a ceramic ultrafiltration membrane to reduce the organic, suspended, and dissolved substances before the introduction of adsorption with biochar. A maximum adsorption capacity of biochar could be about 300 mg/g. It could enhance the organic compound removal up to 80%. However, one of the disadvantages of this system is that the membrane fouling could be happened. This would decline the membrane flux which results in the reduction of the permeate for the adsorption process (Santra et al. 2020). Therefore, the addition of biochar into the ceramic membrane system would be a better approach for textile wastewater treatment.

Novel membrane with biochar was fabricated for wastewater treatment. In this work, biochar was produced from wood under the pyrolysis temperature of 300 C and 700 C. Then, the produced biochar was blended with polymer (polyvinylidene fluoride) at dosages 10%, 30%s and 50% to fablicates a novel biochar membranes. This biochar membrane could enhance the adsorption capacity greatly, upto 187 mg/g. In addition, based on the sieving effect of membrane, it could retain bacteria effectively. More important, biochar played an important role in minimization of membrane fouling (Ghaffar et al. 2018). The biochar membrane could give an alternative solution for wastewater treatment and membrane fouling control. Addition of biochar as an adsorbent could affect the membrane fouling. Biochar could play a function in adsorption process. It also could help to minimize the sludge attached on the membrane surface, resulting in t reduction of membrane fouling. Therefore, this aspect should be evaluated in detail in the integrated system (Wang et al. 2018; Yin et al. 2021). Physiochemical characteristics of biochar such as size, pore size, negative charge, and hydrophilic and hydrophobic should be examined. Besides, the properties of the membrane such as material, pore size, surface charge, hydrophilic, and hydrophobic, should be analyzed (Chu et al. 2017).

Figure 3 illustrates the removal of ammonium in the biological membrane system combined with biochar. Ammonium can be removed by the adsorbent. Ammonium in wastewater could be also adsorbed onto the membrane pores which could cause membrane fouling. In addition, the adsorbent, in some cases, could play as a foulant. It could block the pores of membrane resulted in reducing the flux of membrane filtration (Kim et al. 2022). Therefore, an extensive work should be carried out to examine the possibility of the addition of biochar into the membrane bioreactor use for removing ammonium in wastewater.

During operation, the flux will be declined. The obtained permeate could be used for the ammonium removal tests. This could be used to identify the treatment efficiency of the membrane bioreactor with biochar addition. Based on the obtained data, the variation of flux and ammonium removal with time could be presented in a table or in a graph.

The ammonium removal by the biochar and membrane system could be estimated by the following equation.

$$R = \frac{NH_{4-in} - NH_{4-out}}{NH_{4-in}} \times 100\%$$

In which, R is the ammonium removal efficiency; $NH_{4\text{-in}}$ is the ammonium in the influent; $NH_{4\text{-out}}$ is the ammonium in the effluent.

The flux could be estimated based on the volume concentration factor (F) as follows.

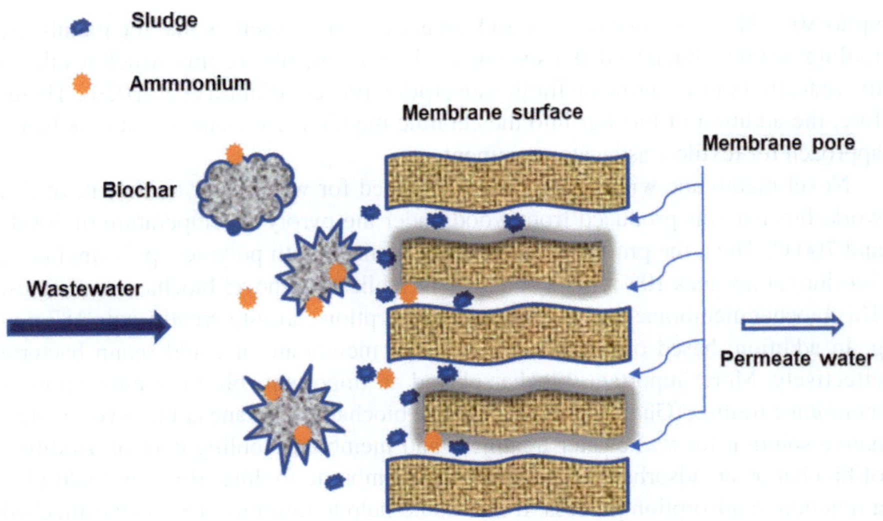

Fig. 3 Minimization of membrane fouling by biochar addition

$$F = \frac{V_F}{V_R} = 1 + \frac{V_P}{V_R}$$

In which V_F is the feed volume; V_R is the retentate volume, and V_P is the permeate volume, where $V_F = V_R + V_P$.

Membrane flux could be calculated based on Darcy's law as below (Mrazík and Kříž 2021).

$$J = \frac{\Delta P}{\eta R_m}$$

In which J is the membrane flux (L/m^2.h), ΔP is the pressure reduction through membrane (kPa), η is the water dynamic viscosity (kg/m.s), and R_m is the membrane resistance (1/m). Membrane resistance contains several factors such as cake formation on the membrane surface and concentration polarization phenomena on the membrane surface. Therefore, addition of biochar in the bioreactor could affect the membrane resistance.

It is important to identify the potential mechanisms for ammonium removal in an integrated system where the biochar was added in a membrane reactor system. Effect of operational conditions such as hydraulic retention time, amount of biochar addition, air scouring on the flux reduction, and ammonium removal in a combination of biochar and membrane bioreactor should be investigated in detail. The effect of biochar addition on fouling of membrane should be examined. The biochar could improve the adsorption of ammonium and organic compounds, resulting in a flux increase during filtration (Ahmad et al. 2014). It was reported that under the stirred conditions addition of biochar of 40 mg/L could give high flux. However, under the unstirred condition, the flux was reduced even by adding biochar of 20 mg/L. This could be due to biochar addition causing the pore blockage which results in a decrease in membrane flux (Shankar et al. 2017). On the other hand, the presence of organic compounds on the membrane surface could increase membrane fouling, resulting in flux reduction. Therefore, adding biochar could help to reduce the organic compound on the membrane surface based on the adsorption process. As a result, the addition of biochar could reduce membrane fouling. This could be an important aspect to integrate the biochar with the membrane bioreactor. It could enhance the treatment efficiency and also could help to reduce the membrane fouling.

5 Conclusions and Perspectives

Ammonium remaining in treated wastewater could cause the eutrophication in the receiving waters. Ammonium could be removed effectively by the adsorption process. Biochar could be used as an effective adsorbent for ammonium removal. It could be enhanced by a combination of biochar and biological reactors. Addition of biochar into the biological reactors, such as activated sludge system, or sequencing bath reactors could improve the ammonium removal efficiency. In those systems, the ammonium could be removed by adsorption and biological mechanisms. In these combination systems, sludge and biochar could be washed out, therefore, application of membranes in those systems could be an alternative solution. Membrane could help to increase the sludge concentration in the system. In addition, biochar will be remained in the reactor. This would be a good carrier for bacterial growth, resulting in increasing the ammonium removal (i.e., nitrification and denitrification). In addition, biochar was moved and scoured in the reactor by the aeration. This would help to reduce the foulants attached the membrane surface, resultin in the minimization of membrane fouling. Therefore, addition of biochar into the membrane bioreactor would provide a novel system for enhancing ammonium removal. This novel system could provide several advantages as below.

- Ammonium could be removed by adsorption in biochar and by biological mechanisms.
- Activated sludge and biochar were kept completely in the system at high concentration by membrane filtration.
- Biochar could serve as a carrier for bacterial growth.
- High concentration of sludge could improve the nitrification and denitrification processes.
- Moving/scouring biochar could play as a factor to remove the foulants and control membrane fouling effectively.

Finally, a combination of biochar and membrane bioreactor could provide a good solution for removing ammonium in domestic and industrial wastewater.

Acknowledgements The authors thank the Hanoi University of Science and Technology for facility supports. The financial supports from Kurita Overseas Research Grant (No. 19PVN005) were sincerely acknowledged. The addition fundings supported by Vingroup Innovation Foundation Post-Graduate Scholarship Program (VINIF.2019.TS.66) was also grateful.

References

Ahmad M et al (2014) Biochar as a sorbent for contaminant management in soil and water: A review. Chemosphere 99:19–33. https://doi.org/10.1016/j.chemosphere.2013.10.071

Bernal MP, Lopez-Real JM (1993) Natural zeolites and sepiolite as ammonium and ammonia adsorbent materials. Biores Technol 43:27–33. https://doi.org/10.1016/0960-8524(93)90078-P

Bock G (2016) Removal of high and low levels of ammonium from industrial wastewaters

Brewer CE, Schmidt-Rohr K, Satrio JA, Brown RC (2009) Characterization of biochar from fast pyrolysis and gasification systems. Environmental progress & sustainable energy: an Official Publication of the American Institute of Chemical Engineers 28:386–396

Cao X, Ma L, Gao B, Harris W (2009) Dairy-Manure derived biochar effectively sorbs lead and atrazine, Environ Sci Technol 43:3285–3291

Çelik MS, Özdemir B, Turan M, Koyuncu I, Atesok G, Sarikaya HZ (2001) Removal of ammonia by natural clay minerals using fixed and fluidised bed column reactors. Water Supply 1:81–88. https://doi.org/10.2166/ws.2001.0010

Chan TY (2003) Ammonia removal in wastewater with anaerobic ammonium oxidation process. Concordia University

Chen H, Xie A, You S (2018) A review: advances on absorption of heavy metals in the waste water by biochar. In: IOP Conference Series: Materials Science and Engineering, vol 1. IOP Publishing, p 012160

Chu KH, Shankar V, Park CM, Sohn J, Jang A, Yoon Y (2017) Evaluation of fouling mechanisms for humic acid molecules in an activated biochar-ultrafiltration hybrid system. Chem Eng J 326:240–248. https://doi.org/10.1016/j.cej.2017.05.161

Deng Q (2014) Ammonia removal and recovery from wastewater using natural zeolite: an integrated system for regeneration by air stripping followed ion exchange. University of Waterloo

Du Q, Liu S, Cao Z, Wang Y (2005) Ammonia removal from aqueous solution using natural Chinese clinoptilolite. Sep Purif Technol 44:229–234

Ghaffar A, Zhu X, Chen B (2018) Biochar composite membrane for high performance pollutant management: Fabrication, structural characteristics and synergistic mechanisms Environ Pollut 233:1013–1023. https://doi.org/10.1016/j.envpol.2017.09.099

Giddey S, Badwal S, Kulkarni A (2013) Review of electrochemical ammonia production technologies and materials. Int J Hydrogen Energy 38:14576–14594

Glaser B, Haumaier L, Guggenberger G, Zech W (1998) Black carbon in soils: the use of benzenecarboxylic acids as specific markers. Org Geochem 29:811–819

Heckelman PE, Kinneary JF, O'Neil MJ, Smith A (1996) The Merck Index: an encyclopedia of chemicals, drugs, and biologicals.

Hoekman SK, Broch A, Robbins C, Zielinska B, Felix L (2013) Hydrothermal carbonization (HTC) of selected woody and herbaceous biomass feedstocks. Biomass Convers Biorefinery 3:113–126

Huang H, Xiao D, Zhang Q, Ding L (2014) Removal of ammonia from landfill leachate by struvite precipitation with the use of low-cost phosphate and magnesium sources. J Environ Manage 145:191–198. https://doi.org/10.1016/j.jenvman.2014.06.021

Huff L, Delos C, Gallagher K, Beaman J (2013) Aquatic life ambient water quality criteria for ammonia-freshwater US Environmental Protection Agency 10

Inyang M, Gao B, Ding W, Pullammanappallil P, Zimmerman AR, Cao X (2011) Enhanced lead sorption by biochar derived from anaerobically digested sugarcane bagasse. Sep Sci Technol 46:1950–1956

Jayakumar A, Wurzer C, Soldatou S, Edwards C, Lawton LA, Mašek O (2021) New directions and challenges in engineering biologically-enhanced biochar for biological water treatment Science of the total environment:148977

Jorgensen TC, Weatherley LR (2003) Ammonia removal from wastewater by ion exchange in the presence of organic contaminants. Water Res 37:1723–1728. https://doi.org/10.1016/S0043-1354(02)00571-7

Kambo HS, Dutta A (2015) A comparative review of biochar and hydrochar in terms of production, physico-chemical properties and applications. Renew Sustain Energy Rev 45:359–378. https://doi.org/10.1016/j.rser.2015.01.050

Kim S et al. (2022) Review of adsorption–membrane hybrid systems for water and wastewater treatment. Chemosphere 286:131916. https://doi.org/10.1016/j.chemosphere.2021.131916

Lee J, Kwon D, Kim J (2021) Long-term performance evaluation of granular activated carbon fluidization and biogas sparging in anaerobic fluidized bed membrane bioreactor: Membrane fouling and micropollutant removal. Process Saf Environ Prot 154:425–432. https://doi.org/10.1016/j.psep.2021.08.024

Lehmann J (2007) Bio-energy in the black. Front Ecol Environ 5:381–387

Li R, Wang JJ, Zhou B, Zhang Z, Liu S, Lei S, Xiao R (2017) Simultaneous capture removal of phosphate, ammonium and organic substances by MgO impregnated biochar and its potential use in swine wastewater treatment. J Clean Prod 147:96–107

Limoli A, Langone M, Andreottola G (2016) Ammonia removal from raw manure digestate by means of a turbulent mixing stripping process. J Environ Manage 176:1–10. https://doi.org/10.1016/j.jenvman.2016.03.007

Liu Z, Zhang F-S (2009) Removal of lead from water using biochars prepared from hydrothermal liquefaction of biomass. J Hazard Mater 167:933–939

Lu H, Zhang W, Yang Y, Huang X, Wang S, Qiu R (2012) Relative distribution of Pb2+ sorption mechanisms by sludge-derived biochar Water research 46:854–862

Manikandan S, Subbaiya R, Saravanan M, Ponraj M, Selvam M, Pugazhendhi A (2022) A critical review of advanced nanotechnology and hybrid membrane based water recycling, reuse, and wastewater treatment processes Chemosphere 289:132867. https://doi.org/10.1016/j.chemosphere.2021.132867

McKendry P Energy production from biomass: overview of biomass. Bioresour Technol 83:55–63

Miao R et al. (2018) New insights into the humic acid fouling mechanism of ultrafiltration membranes for different Ca2+ dosage ranges: results from micro- and macro-level analyses. Water Sci Technol 77:2265–2273. https://doi.org/10.2166/wst.2018.141

Mohammadi R, Hezarjaribi M, Ramasamy DL, Sillanpää M, Pihlajamäki A (2021) Application of a novel biochar adsorbent and membrane to the selective separation of phosphate from phosphate-rich wastewaters. Chem Eng J 407:126494. https://doi.org/10.1016/j.cej.2020.126494

Mohan D, Pittman CU Jr, Steele PH (2006) Pyrolysis of Wood/biomass for Bio-Oil: a Critical Review. Energy Fuels 20:848–889

Mrazík L, Kříž P (2021) Porous medium equation in graphene oxide membrane: nonlinear dependence of permeability on pressure gradient explained. Membranes 11:665

Mumme J, Eckervogt L, Pielert J, Diakité M, Rupp F, Kern J (2011) Hydrothermal carbonization of anaerobically digested maize silage. Biores Technol 102:9255–9260

Oliveira FR, Patel AK, Jaisi DP, Adhikari S, Lu H, Khanal SK (2017) Environmental application of biochar: Current status and perspectives. Biores Technol 246:110–122

Onay O (2007) Influence of pyrolysis temperature and heating rate on the production of bio-oil and char from safflower seed by pyrolysis, using a well-swept fixed-bed reactor Fuel processing technology 88:523–531

Onay O, Kockar OM (2003) Slow, fast and flash pyrolysis of rapeseed Renewable energy 28:2417–2433

Organization WH (2003) Nitrate and nitrite in drinking-water: Background document for development of WHO Guidelines for Drinking-water Quality. World Health Organization

Puckett LJ (1994) Nonpoint and point sources of nitrogen in major watersheds of the United States vol 94. vol 4001. US Geological Survey

Qambrani NA, Rahman MM, Won S, Shim S, Ra C (2017) Biochar properties and eco-friendly applications for climate change mitigation, waste management, and wastewater treatment: A review. Renew Sustain Energy Rev 79:255–273

Santra B, Ramrakhiani L, Kar S, Ghosh S, Majumdar S (2020) Ceramic membrane-based ultrafiltration combined with adsorption by waste derived biochar fortTextile effluent treatment and management of spent biochar. J Environ Health Sci Eng 18:973–992. https://doi.org/10.1007/s40201-020-00520-w

Shah Maulin P (2020) Microbial bioremediation & biodegradation. Springer

Schumann P, Ordóñez Andrade JA, Jekel M, Ruhl AS (2020) Packing granular activated carbon into a submerged gravity-driven flat sheet membrane module for decentralized water treatment. J Water Process Eng 38:101517. https://doi.org/10.1016/j.jwpe.2020.101517

Seruga P, Krzywonos M, Pyżanowska J, Urbanowska A, Pawlak-Kruczek H, Niedźwiecki Ł (2019) Removal of ammonia from the municipal waste treatment effluents using natural minerals. Molecules 24:3633

Shah Maulin P (2021) Removal of refractory pollutants from wastewater treatment plants. CRC Press

Shankar V, Heo J, Al-Hamadani YAJ, Park CM, Chu KH, Yoon Y (2017) Evaluation of biochar-ultrafiltration membrane processes for humic acid removal under various hydrodynamic, pH, ionic strength, and pressure conditions. J Environ Manag 197:610–618. https://doi.org/10.1016/j.jenvman.2017.04.040

Simonič M (2021) Reverse osmosis treatment of wastewater for reuse as process Water—a case study. Membranes 11:976

Tan X-f et al (2017) Biochar as potential sustainable precursors for activated carbon production: Multiple applications in environmental protection and energy storage. Biores Technol 227:359–372

Wang B, Lehmann J, Hanley K, Hestrin R, Enders A (2016) Ammonium retention by oxidized biochars produced at different pyrolysis temperatures and residence times RSC Advances 6:41907–41913

Wang X, Huang D, Cheng B, Wang L (2018) New insight into the adsorption behaviour of effluent organic matter on organic-inorganic ultrafiltration membranes: a Combined QCM-D and AFM Study R Soc Open Sci 5:180586–180586. https://doi.org/10.1098/rsos.180586

Weber K, Quicker P (2018) Properties of Biochar Fuel 217:240–261

Wu Z, An Y, Wang Z, Yang S, Chen H, Zhou Z, Mai S (2008) Study on zeolite enhanced contact–adsorption regeneration–stabilization process for nitrogen removal. J Hazard Mater 156:317–326. https://doi.org/10.1016/j.jhazmat.2007.12.029

Xu D et al (2014) Cadmium adsorption on plant-and manure-derived biochar and biochar-amended sandy soils: impact of bulk and surface properties. Chemosphere 111:320–326

Xu K, Lin F, Dou X, Zheng M, Tan W, Wang C (2018) Recovery of ammonium and phosphate from urine as value-added fertilizer using wood waste biochar loaded with magnesium oxides. J Clean Prod 187:205–214

Yan W, Hastings JT, Acharjee TC, Coronella CJ, Vásquez VR (2010) Mass and energy balances of wet torrefaction of lignocellulosic biomass. Energy Fuels 24:4738–4742

Yang HI, Lou K, Rajapaksha AU, Ok YS, Anyia AO, Chang SX (2018) Adsorption of ammonium in aqueous solutions by pine sawdust and wheat straw biochars. Environ Sci Pollut Res 25:25638–25647

Yenisoy-Karakaş S, Aygün A, Güneş M, Tahtasakal E (2004) Physical and chemical characteristics of polymer-based spherical activated carbon and its ability to adsorb organics. Carbon 42:477–484

Yin S et al. (2021) An investigation of the membrane fouling behaviors in constant flux mode IOP Conference Series: Earth and Environmental Science 647:012203. https://doi.org/10.1088/1755-1315/647/1/012203
Zaghouane-Boudiaf H, Boutahala M (2011) Kinetic analysis of 2,4,5-trichlorophenol adsorption onto acid-activated montmorillonite from aqueous solution. Int J Miner Process 100:72–78. https://doi.org/10.1016/j.minpro.2011.04.011
Zhang M, Gao B, Yao Y, Xue Y, Inyang M (2012) Synthesis of porous MgO-biochar nanocomposites for removal of phosphate and nitrate from aqueous solutions. Chem Eng J 210:26–32

Microbial Remediation of Synthetic Microfiber Contaminated Wastewater

Sunanda Mishra and Alok Prasad Das

1 Introduction

Synthetic fibers, a consequence of man mimicking nature, find its existence from raw materials like petrochemicals. Some are profoundly tough and durable, while some dry pretty quickly than others or maybe more absorptive and easier to dye. Microfibers, arising from clothing, laundering, and textile industry, have been identified as significant environmental pollutants in water bodies (Mishra et al. 2020; Singh et al. 2020). Although textile industries have tremendous economic impact and generate a lot of employment, however, a thriving textile industry is also a potent source of environmental pollution, as a variety of toxins produced are dumped straight into the surrounding environment without any further treatment (Mishra et al. 2019a). The potential sources of microfiber in the environment are depicted in Fig. **1**. Synthetic microfiber is a form of microplastic with a diameter of less than 5 mm. Microfiber of synthetic textiles is released from clothing during manufacturing, use, and disposal steps. Although clothing has been an integral part of the human population for centuries, pollution has become problematic since the past decade due to high consumption of synthetic textiles (Ramasamy and Subramanian 2021; Mishra et al. 2019b; Das and Ghosh 2017; Shah MP 2020). In recent years, textile contamination has become a major environmental concern; synthetic fibers account for 15% of worldwide plastic output and can be fragmented and degraded to produce synthetic microfiber (Avio et al. 2020, Boucher and Friot 2017). According to the results of the global synthetic fiber consumption survey, synthetic clothing usage has expanded dramatically, accounting for roughly 65% of total fiber use. Textile industry effluents are one of the most significant sources of liquid effluent pollutants, with research

S. Mishra
Department of Botany, Bhubaneswar, India

A. P. Das (✉)
Department of Life Sciences, RD Women's University, Bhubaneswar, India
e-mail: alok1503@gmail.com

M. P. Shah (ed.), *Microbial Technologies in Industrial Wastewater Treatment*,
https://doi.org/10.1007/978-981-99-2435-6_17

estimating that roughly 280,000 tons of textile dyes are emitted each year (Saratale et al. 2017). Textile manufacturing uses a lot of water and produces a lot of effluent, which contains a lot of additives, detergents, suspended solids, carcinogenic amine, and aldehyde groups (Biswal et al. 2021b). It has been estimated that 10–15% of total dyes are discharged into the water bodies during textile processing, producing severe water pollution, however, this varies depending on stages of processing, synthetic dyes' types, fabric material, and processing. The functioning parameters used in the transformation of fiber to textile fabric, as well as the concentration of chemical agents (Kalyani et al. 2009). Textile effluents have long been a serious environmental problem because they affect the aquatic ecosystem, soil fertility, plant growth, and plant productivity (Kumar and Das 2017). Inhalation of airborne microfibers and intake of microfibers found in popular foods also have negative effects on human health (Das et al. 2015a, b).

In the fabric industry, dyes and pigments play a critical role in value addition, appearance, and consumer satisfaction (Das and Singh 2011). Synthetic dyes have been shown to have a rapid increase in application in the textile industry due to their low cost of synthesis and great stability. In addition, when compared to natural dyes, a

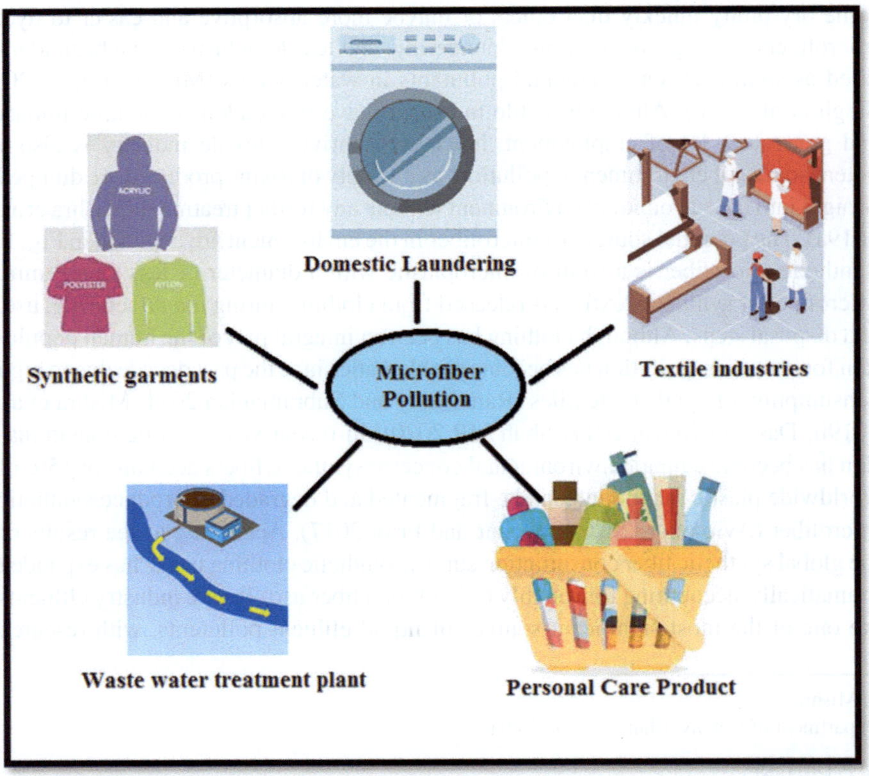

Fig. 1 Potential sources of Microfiber into the environment

wide range of hues may be synthesized. As a result, a massive flow of polluted sewage was discovered in many industries. Heavy metals, which are typically a constituent of textile effluent in free ionic or complex compound form, constitute an important component of dyes. When these heavy metals enter water bodies, they settle down in the sediment and start building up in the bodies of aquatic creatures, producing major health effects (Biswal et al. 2021a; Mishra and Das 2021). These toxins made their way into the food chain, posing a health danger to plants, animals, and humans. The issue with textile pollutants is not the presence of synthetic dyes, but their resistance to degradation and stability (Kumar and Das 2016; Bhattacharjee et al. 2021). There is rising concern about the negative impact of textile effluents on aquatic biota and humans (Mishra et al. 2021). Synthetic fibers are so small that collecting them from nature is nearly impossible. A solution must be devised to reduce the number of fibers released into the environment and to clean polluted areas in an environmentally acceptable manner. Ultraviolet treatment, hydrolysis, methanolysis and biodegradation are some of the environmental, chemical, and biological degrading methods that have been developed. Among all of these processes, biodegradation is thought to be the most environmentally friendly and cost-effective process to control textile pollution.

This paper provides a thorough review of types of synthetic fabrics and their growing applications, describes the factors affecting the release of microfibers into the environment, presents a clear idea of microfiber pollution associated risks on environmental and human health, possible mechanisms, and process of microbial degradation of microfibers.

2 Types of Synthetic Fibers

In current era, most clothing materials are made of artificial fabrics as compared to natural fabrics. Natural fabrics were mostly used for clothing since early times. But these days, we use synthetic fibers for clothing and other purposes on a regular basis. Fibers which are manmade and are produced from chemical substances are called synthetic fibers (Napper and Thompson 2016; Das et al. 2014). Mostly, they are attained from polymers of small molecules, i.e., raw materials such as natural gas, coal, petroleum, and petrochemicals (Das and Mishra 2008, 2010). Generally, there are some steps that convert fiber into fabric. The first step of this process is the extraction of sources. Then, the fibers are twisted together firmly to produce yarn with the help of weaving method and rollers (Das 2016; Auta et al. 2017). And as a result, fabrics are obtained and further, they are treated with different types of chemicals, bleach, dyes, and printing to produce varieties of attires. There are some synthetic fibers, which dominate the market they are polyester, nylon, acrylic, spandex, rayon.

3 Polyester

The term polyester is generally used in polyethylene terephthalate (PET). This fabric is considered as the most popular textile all over the world due to its impressive durability, cheap price, and variations of fabric. It is a synthetic fabric that is usually derived from coal and petroleum. They are basically composed of petroleum derivative, i.e., ethylene and polymer within the ester functional group. This polyester is governing the market, starting from a party wear to a formal dress, polyester fabrics are endless. It is also used to make different types of garments like dresses, sarees, kurtas, shirts, suits, skirts, curtains, towels, rugs, blankets, etc. It is also used for making water bottles, food containers, disposable plates and cups, chairs, pillows, etc. Look anywhere around you, surely, you will find polyester around you. Fossil fuels are used to derive the basic constituents which are used in the manufacturing of polyester. As fossil fuel is a natural and limited resource, rapid use of it is knocking a negative impact on the environment. Further, it is also producing various toxins in the procedure of transforming ethylene into PET and by releasing microfibers as by-products. They are non-degradable in nature and release dyes in the environment which are harmful to the environment and living beings. Global destructive influence of polyester is constantly increasing since this fabric material has made its way into the consumer wardrobe. Overall, polyester damages the environment at each phase starting from production to degradation.

4 Nylon

Nylon is an important type of synthetic fabric. It is a long-chain polymer of carbon monomer called adipic acid and hexamethylene diamine that is obtained from crude oil. In making polymer familiar as nylon, diamine acid is required to enter into a reaction with adipic acid and it has different forms named as polyamide, PA, PA6, PA12, PA66, PA69, PA1212, etc., but PA66 is the type of substance which is termed as nylon salt in crystallized structure. Then, the substance is heated to produce a molten substance which is extruded through a spinneret and immediately hardens. Hence, resulting fibers to fabrics. The invention of nylon is considered to have a negative environmental impact. Excessive expanse of energy is essential to make nylon fabric, and a number of waste materials are also generated during the industrial process. A huge amount of water is used to cool nylon fabrics, and that water repeatedly carries pollutants into the water surface. In the production of the secondary constituent part of nylon fabric, i.e., adipic acid, that releases nitrous oxide in the atmosphere, and it is the worse air-polluting gas. As they are synthetic, they are not biodegradable and last for hundreds of years in the environment.

5 Acrylic

Acrylic is a synthetic fabric which is made from polyacrylonitrile named polymer formed by acrylonitrile or vinyl cyanide. This fabric is manufactured by reacting specific petroleum or coal-based chemical compounds with an array of monomers. It is frequently considered an artificial wool due to its heat retention potentials. It is the least breathable form of textile among all the fabrics. It is often used to create fake fur and fleece, tracksuits, hoodies, boot lining, sweatshirts, carpets (Harrison et al. 2018), etc. Nevertheless, there are major concerns about acrylic that it may be carcinogenic so it is advisable to avoid this fabric. Its environmental impact is mostly negative because, during production, various toxic substances are used which effect workers' health as they are uneducated and are unaware of its consequences even though safety measures are taken it fails. It is non-biodegradable and causes damaging effects on both nature and living beings (Ghosh and Das 2018). Well-nigh, it is impossible to recycle acrylic fabric, which means there is no operative means of disposal of acrylic fabrics once it is produced. It simply remains in the environment for a longer period. Sometimes, during the wash, small microfibers discharge and enter the water bodies resulting polluting the water.

6 Spandex

Spandex is a synthetic fabric. This fabric is also known as elastane or Lycra. It is also referred to as polyether-polyurea copolymer fabric that is made up of synthetic polymers called polyurethane having elastic properties. Synthetic materials like polyester, nylon, etc., are directly derived from natural resources like coal and petroleum, but this material is completely made from chemicals that are synthesized in test centers. Mostly, spandex is not used directly as fabric, a small amount of this material is used in garments. It is intermingled with numerous types of fabric to add stretch starting from jeans to socks. These are used mostly in yoga pants, skinny jeans, underwears, gym wears, etc. Spandex is made from repetition monomer chains that are held together with the support of an acid. They have the ability to resist heat, which means that, especially heat-sensitive fabrics like polyester and nylon get better when combined with spandex fabric. In the environment, the impact of this fabric is not as noticeable as the negative impact of other synthetic fabrics. But, it is also not friendly to the environment as it is a non-biodegradable material.

7 Rayon

This fabric is generally soft and durable and considered as the world's most beloved fabric. This fabric mostly resembles cotton, but they are cheap and durable compared to cotton. This fabric is used for different apparels like shirts, tops, dresses, kurtis, etc. It is an artificial fabric commonly known as viscose fabric due to its semi-liquid form. After cooling down the viscose changes into another form which, is called as rayon. They are semi-synthetic because they are produced from wood pulp. As it goes through a lengthy procedure, it is considered semi-synthetic. The cellulose is treated with sodium hydroxide and carbon disulfide, which results in sodium cellulose xanthate's viscous solution, which is dipped into a sulfuric acid and sodium sulfate solution. After conversion of filaments into yarn, it is desulfurized, then bleached after it is washed and dried. The practice of toxic chemicals is necessary in this process and it is nearly impossible to produce rayon without using sodium hydroxide, which contaminates water bodies and air. Plant and animal life are wounded by the manufacture of this fabric, by polluting the ecosystem. In addition, the production of rayon depletes forests at a rapid rate.

8 Impact on Environment and Ecosystem

With the never-ending debate over renewable energy vs. fossil fuel, some environmental scientists warn us about the harmful effect it can impose on our environment, such as deadly carbon emission, ozone layer depletion, and global warming (Lant et al. 2020; Das and Swain 2013; Ghosh et al. 2018a, b). The textile sector is responsible for more environmental deterioration than any other industry, accounting for roughly 17% of worldwide carbon emissions. Polyester, nylon, rayon, and acrylic are the most common textile materials. These synthetic fibers are less expensive to manufacture, which could be beneficial to the profit and the economy (Mohanty et al. 2018; Shah MP 2021). This is also causing significant environmental damage. According to reports, the apparel industry is responsible for about 25% of global industrial water contamination. Synthetic fibers contaminate water bodies in other ways as well, from waste products originating in industrial units to artificial colors which ultimately find their way into sewers and waterways, making the environment more hazardous and deadly. Polyester and nylon, being non-biodegradable and unsustainable affect the environment negatively. Nylon produces nitrous oxide, a gas that is more harmful to the ozone layer. Dye is another substance commonly used in the garment industry. Unused colors are then washed into rivers and waterways, causing pollution. Microfibers may contain adsorbed poisons on their surface in addition to harmful compounds in their formulation. Microplastics have been discovered to carry a wide range of noxious metals and organic pollutants like PAHs, DDT, PCBs (Gaylard et al. 2021), hence creating an elevated threat to aquatic biota.

Apparels made from synthetic materials such as polyester and nylon, according to NBC's MACH, are major contributors to microplastic contamination, which can culminate in the ocean and seafood we eat. According to IUCN report, microplastics along with microfibers are assumed to account for nearly 1.7 million tons of the total 8 million tons of plastic pollutants that end up in the ocean every single year (Mishra et al. 2019b). Microplastic entering the water is believed to be 37% made up of fibers from synthetic garments. Both marine species and human health are suffering as a result of synthetic fibers. Synthetic nylon has been identified in the digestive tracts of most fishes, according to many investigations. Many aquatic animals, such as zooplankton, polychaetes, crabs, and deep-sea invertebrates, have been seen to consume synthetic microfiber (Alnajar 2021; Avio et al. 2020) microplastic and was observed to be transported from algae to zooplankton and to the goldfish in a scientific investigation by (Smith et al. 2018; Mohanty et al. 2017a, b). Seabirds have also been discovered dead, with the cause of death being the consumption of synthetic fibers mistaken for food (Zhao et al. 2021). Synthetic fibers are growing long-term hazards that must be addressed head-on, or our planets will be polluted beyond repair. When fibers like microplastic come into contact with food production, they create an issue. Because these fibers are so microscopic, cleaning them out of streams can be difficult. Microfibers wind up in drinking water and fertilizer, which means, our food is being poisoned not only via the digestive tracts of shellfish in our plate but also through the manures and irrigation systems used in crop fields.

9 Microbial Degradation of Synthetic Fibers

Degradation of synthetic fibers by microorganisms involves transformation of pollutants into metabolites that can be included by microscopic organisms as their carbon source in metabolic pathways and biological systems (De Falco et al. 2018; Das et al. 2012; Yang et al. 2019). Synthetic polymers are promising resources of energy due to their higher carbon content. Hindering or accelerating this process, physical and chemical properties like molecular structure, composition, melting point, surface structure, chain mobility, crystallinity, molecular weight, density act as regulatory parameters in the biodegradation process. Complete biodegradation of synthetic polymers may require the involvement of more than one microbe where one will break the polymeric chain and others will incorporate smaller monomeric fragments by further breaking the previously broken pieces. Various types of synthetic textiles are degraded differently under several conditions by diverse groups of microorganisms (Ghosh and Das 2015; Gago et al. 2018).

Biodegradation is the process of converting textile contaminants into compounds that bacteria can use in their metabolic systems and pathways. This means that organisms exploit the fibers as a source of energy and nutrients, converting them to harmless natural chemicals like water, carbon dioxide, and methane. The biodegradability of synthetic fibers is influenced by its physical and chemical qualities (Amaral-Zettler et al. 2020). In bioremediation, many parameters such as surface roughness, molecular weight, density, crystalinity, and melting point are relevant. Synthetic materials can be biodegraded by microbes with the utmost flexibility. Microorganisms try to adsorb onto the fiber structure, connect to the surface, and adapt to the environment so that they can begin to develop. Because of their high carbon content, synthetic fibers are a promising energy source. The adsorption to a surface is influenced by its chemical and physical structure as well as its electrical charge. From a variety of microbes, one that breaks the polymer and others that may break these smaller polymers and utilize the expelled monomers in their metabolism the biodegradation process is generally affordable, the operating expenses are cheap, and the complete mineralization end products are nontoxic. The majority of the decline happens when the bacteria are actively growing.

During degradation of polymers, in the presence of microorganisms, macromolecular polymeric chain is broken down, hence starting the biodegradation process. Depending upon the types of microbes involved and polymer characteristics as mentioned above, this process is greatly influenced. Non-enzymatic degradation process involves random hydrolytic chain cleavage of ester linkage giving rise to small molecules. Higher degradation temperature i.e., 40–60°C is found to be an enhancer of this process. Digestion of lactic acid oligomers by microbes is initiated when the molecular weight becomes 10,000 and subsequently, CO_2, CH_4 and water are released which becomes part of natural cycles. In enzyme catalyzed degradation process, mainly two processes are involved: (i) depolymerization and (ii) mineralization (Siracusa 2019). Hydrolysis and oxidation together with extracellular microbial enzymes causing internal linkage breakage (endo-action) and terminal polymer molecules breakage (exo-action) lead to depolymerization. Mineralization happens inside the cell responsible for degradation. It can take place in both aerobic condition (CO_2 and H_2O are released) and anaerobic condition (CH_4, CO_2, and H_2O are released) (Singh and Sharma, 2008). To make plastic degradable, reduction in dimension and increase in surface area is essential. Biodegradation of synthetic polyesters such as poly lactic acid (PLA), poly L lactide (PLLA), polycaprolactone (PCL), poly-(butylene succinate) (PBS), poly-butylene succinate-co-adipate (PBSA) by optimizing depolymerase enzyme production was reported by Sriyapayi 2018 where yeast extract combined with $(NH_4)_2(SO)_4$ as nitrogen sources played an important role of a persuader for depolymerase activity in thermophilic bacteria. The enzymes belonged to serine hydrolase superfamily and the bacterial species involved were *Actinomadura* sp. S14, *Actinomadura* sp.TF1, *Streptomyces* sp. APL3, *Laceyella* sp. TP4. In a biotic environmental degradation, degradation of crystalline form takes place after the degradation of amorphous form. Asmita et al. (2015); Katarzyna and Grażyna (2010) studied biodegradation of polyester where they identified a novel bacterium *Ideonella sakaiensis* 201-F6 capable of producing PETase that released Mono(2-hydroxyethyl) terephthalic acid (MHET) as a most important product and TPA, bis (2-hydroxyethyl) terephthalic acid (BHET) as by products upon incubation with PET film. PETase was slightly thermolabile, yet was substantially more active than other PET hydrolyzing enzymes studied in this experiment. The organism binds to PET surface and starts secreting the enzyme to induce biodegradation but the exact binding mechanism is unknown as PET lacks surface binding motifs and its 3D structure is still unknown. It indicates polyester is mostly inert to biodegradation except some of its biodegradable form.

10 Biodegradation of Nylon

Nylon-6 is known to be poorly biodegraded (Tokiwa et al. 2009), but fortunately, the reports of its biodegradation has been successfully reported by few. Mainly, thermophilic bacteria *Geobacillus thermocatenulatus* (Tomita et al. 2003) and many marine bacterial and fungal strains (Sudhakar et al. 2007) have been reported to associated with nylon degradation. Enzymatic hydrolysis of amine bonds is associated with biodegradation of Nylon-12, followed by formation of 12-amino dodicanoic acid (Mayaada et al. 2016). Proceeding in degradation of 12-amino dodicanoic acid by oxidation would result in formation of carboxyl and other degradation compounds (Prabhakar et al. 2019; Gautam et al. 2007). Studies by Tomita et al. 2003 isolated a bacterium that could efficiently biodegrade Nylon-6 and Nylon-12, but not Nylon-66. Nylon-6 and Nylon-66 both were found to be maximally degraded by bacterial strain *Bacillus cereus* out of other three strains considered in biodegradation study by (Sudhakar et al. 2007). Degradation of Nylon-6 by *Anoxybacillus rupiensis* used chemical groups such as N–H, C = O, C-H for carbon and nitrogen requirements in its nutrition (Mayaada et al. 2016). Fungal involvement in Nylon-66 membrane biodegradation under lignolytic condition was reported by Deguchi et al. (Deguchi et al. 1997) revealed formation of chemical componds ending with CHO, NHCHO, CH_3,$CONH_2$ from oxidative process.

11 Biodegradation of Cellulosic Fibers

Cotton as a cellulose-based material is an instantly available source of carbon and are susceptible to biodegradation by microorganisms whereas rayon is a semi-synthetic cellulosic fabric having biodegradability property nearly equivalent to cotton. In an aquatic biodegradation study, cotton and rayon were degraded about 75% and 63%, respectively, after 243 days of experimental time frame (Zambrano et al. 2019). Another experiment conducted by Vildan et al. by taking some cellulosic textiles along with other synthetic fabric, showed decrease in intensity of functional groups and nearly 90% weight loss of cotton in just four months. Modal and viscose were almost completely biodegraded. Breakage in OH bonds, methyl and methylene of cellulose was due to microbial attack on cellulosic chain (Vildan et al. 2019). Degradation of dyed cotton study by Anjali et al. (2013) found attachment of five fungal species from families of Mucoraceae, Aspergillaceae, Mycosphaerellaceae with the predominance of Rhizopus species. Cellulase is an important enzyme in the degradation of cellulose-based material. Conversion of cellulose to glucose is a two-step reaction as studied in Trichodermaviridae where in first step involves break down of glycosidic linkage in to cellobiose by b-1,4-glucanase with a b-1,4 bond (a glucose dimer). Subsequently, in second step, this b-1,4-glycosidic bond is broken by beta glycosidase resulting in a utilizable energy source glucose. Wool is animal based cellulosic fiber, built from low sulfur, high sulfur and high tyrosine types of keratins. Microorganisms exhibiting proteolytic and keratinolytic enzymes are capable of degrading this fiber. Till now, 299 fungal species possessing keratinolytic properties have been identified, out of which 107 have pathogenicity toward humans (B1yskal 2009). Woolen fabric decomposition follows deamination, sulfitolysis, and proteolysis. The first step proceeds by splitting of disulfide bonds, which contribute toward keratins' resistance asset. This is tailed by enzymatic protein decomposition into oligopeptides by protease enzymes and then into amino acids by the action of peptidases. The resulting amino acids release ammonia into the atmosphere through the process of oxidative deamination (Vassilenko et al. 2021). Microbial decomposition of wool can be characterized by the presence of diversely colored stains, distinctive odor (H_2S is produced in anaerobic decomposition), and disappeared stretching potential (Beata 2012).

12 Conclusion

The nature of clothing has witnessed dramatic changes since its inception. From animal skin to complex synthetic fiber blends, whatever one can think of wearing is now included in the human lifestyle. The textile industry offers one of the greatest looming production patterns concerning global environmental impact. From the use of toxic chemicals to higher consumption of fresh water and energy, and the generation of wastes and effluents in larger quantity, each of them is contributing to environmental pollution. Fast fashion, perhaps the worst facet of textile industry, thanks to population burst and consumerism, affects the earth by promoting the production of poor-quality clothes appreciated for a short period of time, which in turn adds to synthetic textile waste. From the deepest ocean trench to the highest peak in the world, there are evidences of their presence. They are ubiquitous, even in hostile and impossible environments, hence require robust research and immediate cleaning steps and strategies to check its further progression to mitigate the hazardous effect mankind is going to witness in the near future. Microorganisms capable of degradation, responsible enzymes, and mechanisms behind degradation are not so adequately explored areas yet. Sustainable fashion must be a priority and textile manufacturing must be focused on the usage of ecofriendly and biodegradable fibers such as organic cotton, organic wool, hemp, lyocell, flax, and organic dyes. A combined approach from the selection of raw materials to sewing can be a greater help in attaining environmental sustainability. One thing we have to remember is that, in order to flaunt fashion, we have to be alive first.

References

Alnajar N, Jha AN, Turner A (2021) Impacts of microplastic fibers on themarine mussel, mytilus galloprovinciallis. Chemosphere 262:128290

Avio CG, Pittura L, d'Errico G, Abel S, Amorello S, Marino G, Gorbi, S. and Regoli F (2020) Distribution and characterization of microplastic particles and textile microfibers in Adriatic food webs: General insights for biomonitoring strategies. Environmental pollution (Barking, Essex : 1987), 258, 113766. https://doi.org/10.1016/j.envpol.2019.113766

Amaral-Zettler LA, Zettler E, and Mincer TJ (2020) Ecology of the plastisphere. Nat Rev Mi5 crobiol. 18, 139–151. https://doi.org/10.1038/s41579-019-0308-0

Asmita K, Shubhamsingh T, Tejashree S (2015) Isolation of plastic degrading micro-organisms from soil samples collected at various locations in Mumbai. India. Curr. World Environ. 4:77–85

Auta HS, Emenike CU, Fauziah SH (2017) Distribution and importance of microplastics in the marine environment: a review of the sources, fate, effects, and potential solutions. Environ Int 102:165–176. https://doi.org/10.1016/j.envint.2017.02.013

Anjali, Deshmukh, Shruti Deshmukh, Varsha Zade and Vaibhao Thakare (2013)"The microbial degradation of cotton and silk dyed with natural dye:a laboratory investigation." Int J Theor and Appl Sci, 50–59

Beata, Gutarowska, Andrzej Michalski. (2012)"Microbial degradation of woven fabrics and protection against biodegradation." In wooven fabrics, 267–296. china: BoD—Books on Demand, 2012

Blyskal B (2009) "Fungi utilizing keratinous substrates." Int Biodeterior and Biodegrad, 631–653

Bhattacharjee J, Mishra S, Das AP (2021) Recent advances in sensor-based detection of toxic dyes for bioremediation application: a Review. Appl Biochem Biotechnol. https://doi.org/10.1007/s12010-021-03767-7.Advanceonlinepublication.doi:10.1007/s12010-021-03767-7

Biswal P, Pal, Ghosh SA, Das AP (2021a) Exploration of probiotic microbial biodiversity in acidic environments (curd) and their futuristic pharmaceutical applications. Geomicrobiology

Biswal P, Pal A, Das AP (2021b) Screening for probiotic potential of Lactobacillus Rhamnosus strain CRD4. Biointerface Res Appl Chem, 11(2)

Boucher J, Friot D (2017) "primary microplastics in the oceans: A global evaluastion of sources." 43 IUCN, Switzerland

De Falco F, Gullo M P, Gentile G, Di Pace E, Cocca M, Gelabert L, Brouta-Agnésa M, Rovira A, Escudero R, Villalba R, Mossotti R, Montarsolo A, Gavignano S, Tonin C, Avella M (2018) Evaluation of microplastic release caused by textile washing processes of synthetic fabrics. Environmental pollution (Barking, Essex : 1987), 236, 916–925. https://doi.org/10.1016/j.envpol.2017.10.057

Das AP, Singh S (2011) Occupational health assessment of chromite toxicity among Indian miners. Indian J Occup Environ Med 15(1):6–13

Das A, Swain S, Panda S, Pradhan N, Sukla L (2012) Reductive Acid Leaching of Low Grade Manganese Ores,". Geomaterials 2(4):70–72

Das A, Mishra S (2008) Hexavalent Chromium (VI): Health hazards and environmental pollutant. J Environ Res Dev 2:386–392

Das AP, Mishra S (2010) Biodegradation of the metallic carcinogen hexavalent chromium Cr (VI) by an indigenously isolated bacterial strain. J. Carcinogenesis. 9:6

Das AP, Swain S (2013) Algal biosorption of toxic dye Methylene blue. A potential source of food, feed, biochemicals, biofuels and biofertilizers, International conference on Algal Biorefinery, Indian Insitiute of Technology, 13 January 2013, Siksha O Anusandhan University, India

Das AP, Kumar PS, Swain S (2014) Recent advances in biosensor based endotoxin detection. Biosens Bioelectron 51:62–75

Das AP, Ghosh S, Mohanty S, Sukla LB (2015a) Consequences of manganese compounds: a review. Toxicol Environ Chem 96:981–997
Das A, Bal B, Mahapatra P (2015b). Chromogenic biosensors for pathogen detection. https://doi.org/10.1201/b18654-15
Das AP, Ghosh S (2017) Bioleaching of Manganese of mining waste materials. Materials Today Proceedings. 5:2381–2390
Das AP (2016) Biosensors: the future of diagnostics. Sensor network data communication. https://doi.org/10.4172/2090-4886.S1-e001
Deguchi T, Kakezawa M, Nishida T (1997) "Nylon biodegradation by lignin-degrading fungi." Appl Environ Microbiol, 329–331
Ghosh S, Das AP (2018) Metagenomic insights into the microbial diversity in manganese-contaminated mine tailings and their role in biogeochemical cycling of manganese. Sci Rep 8. https://doi.org/10.1038/s41598-018-26311-w
Ghosh S, Das AP (2015) Modified titanium oxide (TiO2) nanocomposites and its array of applications: a review. Toxicol Environ Chem 97(5):491–514
Ghosh S, Bal B, Das AP (2018a) Enhancing manganese recovery from low-grade ores by using mixed culture of indigenously isolated bacterial strains. Geomicrobiol J 35(3):242–246
Ghosh S, Kumar MS, Bal B, Das AP (2018b) Application of bioengineering in revamping human health. In: Singh S (eds) Synthetic Biology. Springer, Singapore. https://doi.org/10.1007/978-981-10-8693-9_2
Gautam R, Bassi AS, Yanful EK, Cullen E (2007) "Biodegradation of automotive waste polyester polyurethane foam using Pseudomonas chlororaphis ATCC55729." Int Biodeterior Biodegrad, 245–249
Gago J, Carretero O, Filgueiras AV, Viñas L (2018) Synthetic microfibers in the marine environment: A review on their occurrence in seawater and sediments. Mar Pollut Bull 127:365–376. https://doi.org/10.1016/j.marpolbul.2017.11.070
Gaylarde CC, Baptista Neto JA, da Fonseca EM (2021) Nanoplastics in aquatic systems—are they more hazardous than microplastics?. Environmental pollution (Barking, Essex : 1987), 272, 115950. https://doi.org/10.1016/j.envpol.2020.115950
Harrison JP, Hoellein TJ, Sapp M, Tagg AS, Ju-Nam Y, Ojeda JJ (2018) Microplastic-associated biofilms: a comparison of freshwater and marine environments. In: Wagner M, Lambert S (eds) Freshwater microplastics. The handbook of environmental chemistry, 58. Springer, Cham
Katarzyna L, Grażyna L (2010) Polymer biodegradation and biodegradable. Pol J Environ Study, 255–266
Kalyani DC, Telke AA, Dhanve RS, Jadhav JP (2009) Ecofriendly biodegradation and detoxification of Reactive Red 2 textile dye by newly isolated Pseudomonas sp. SUK1. J Hazard Mater, 163(2–3), 735–742. https://doi.org/10.1016/j.jhazmat.2008.07.020
Kumar MS, Das AP (2017) Emerging nanotechnology based strategies for diagnosis and therapeutics of urinary tract infections: A review. Adv Coll Interface Sci 249:53–65
Kumar MS, Das AP (2016) Molecular identification of multi drug resistant bacteria from urinary tract infected urine samples. Microb Pathog 98:37–44
Lant NJ, Hayward AS, Peththawadu M, Sheridan KJ, Dean JR (2020) Microfiber release from real soiled consumer laundry and the impact of fabric care products and washing conditions. PLoS ONE 15(6):e0233332
Mishra S, Rout PK, Das AP (2019a) Solar Photovoltaic Panels as next generation waste: A Review. Biointerface Research in Applied Chemistry. 9(6):4539–4546
Mishra S, Rath CC, Das AP (2019b) Marine microfiber pollution: A review on presentstatus and future challenges. Mar Pollut Bull 140: 188–197.https://doi.org/10.1016/j.marpolbul.2019.01.039.
Mishra S, Singh RP, Rath CC, Das AP (2020) Syntheticmicrofibers: Source, transport and their remediation.J Water Process Eng 38:101612

Mishra S, Das AP (2021) Current treatment technologies for removal of microplastic and microfiber pollutants from wastewater. Wastewater treatment: cutting edge molecular tools, techniques and applied aspects. 237–251
Mishra S, Rout PK, Das AP (2022) Emerging Microfiber Pollution and Its Remediation. Environmental pollution and remediation, environmental and microbial biotechnology, https://doi.org/10.1007/978-981-15-5499-5_9
Mohanty S, Ghosh S, Bal B, Das AP (2018) A review of biotechnology processes applied for manganese recovery from wastes. Reviews in environmental science and bio/technology, 17https://doi.org/10.1007/s11157-018-9482-1
Mohanty S, Ghosh S, Nayak S, Das AP (2017a) Bioleaching of manganese by Aspergillus sp. isolated from mining deposits. Chemosphere 172:302–309
Mohanty S, Ghosh S, Nayak S, Das AP (2017b) Isolation, identification and screening of manganese solubilizing fungi from low-grade manganese ore deposits. Geomicrobiol J 34(4):309–316
Mayaada S, Mahdi, Rasha S, Ameen and Hiba K (2016) Ibrahim. "Study on Degradation of Nylon 6 by thermophilic bacteria Anoxybacillus rupiensis Ir3 (JQ912241)." Int J Adv Res Biol Sci, 200–209.
Napper IE, Thompson RC (2016) Release of synthetic microplastic plastic fibres from domestic washing machines: Effects of fabric type and washing conditions. Mar Pollut Bull 112(1–2):39–45. https://doi.org/10.1016/j.marpolbul.2016.09.025
Prabhakar A, Mishra S, Das AP (2019) Isolation and Identification of Lead (Pb) Solubilizing bacteria from automobile waste and its potential for recovery of lead from end of life waste batteries. Geomicrobiol J 36(10):894–903. https://doi.org/10.1080/01490451.2019.1654044
Ramasamy R, Subramanian RB (2021) Synthetic textile and microfiber pollution: a review on mitigation strategies. Environ Sci Pollut Res Int 28(31):41596–41611. https://doi.org/10.1007/s11356-021-14763-z
Sudhakar M, Mukesh D, Sriyutha Murthy P, Venkatesan R (2007) Marine Microbe Mediated Biodegradation of Low andHigh Density Polyethylene. International Biodegardat Ion and Biodeterioration. https://doi.org/10.1016/j.ibiod.2007.07.011
Singh RP, Mishra S, Das AP (2020) Synthetic microfibers: Pollution toxicity andremediation, Chemosphere, https://doi.org/10.1016/j.chemosphere.2020.127199.
Shah MP (2020) Microbial bioremediation and biodegradation. Springer
Saratale RG, Kuppam C, Mudhoo A, Saratale GD, Periyasamy S, Zhen G, Koók L, Bakonyi P, Nemestóthy N, Kumar G (2017) Bioelectrochemical systems using microalgae—A concise research update. Chemosphere 177:35–43. https://doi.org/10.1016/j.chemosphere.2017.02.132
Shah MP (2021) Removal of refractory pollutants from wastewater treatment plants. CRC Press
Smith M, Love DC, Rochman CM, Neff RA (2018) Microplastics in seafood and the implications for human health. Curr Environ Health Rep 5(3):375–386. https://doi.org/10.1007/s40572-018-0206-z
Siracusa V (2019) Microbial degradation of synthetic biopolymers waste. Polymers 11(6):1066. https://doi.org/10.3390/polym11061066
Singh B, Sharma N (2008) Mechanistic implications of plastic degradation. Polym Degrad Stab 93:561–584
Tokiwa Y, Calabia BP, Ugwu CU, Aiba S (2009) Biodegradability of Plastics. Int J Mol Sci 10(9):3722–3742
Tomita K, Ikeda N, Ueno A (2003) "Isolation and characterization of a thermophilic bacterium, Geobacillus thermocatenulatus, degrading nylon 12 and nylon 66." Biotechnology letters, 1743–1746.
Vassilenko E, Watkins M, Chastain S, Mertens J, Posacka AM, Patankar S, Ross PS (2021) Domestic laundry and microfiber pollution: exploring fiber shedding from consumer apparel textiles. PLoS 16(7):e0250346. https://doi.org/10.1371/journal.pone.0250346

Vildan S, Gökberk D (2019) "Biodegradation behaviour of different textile fibres: visual, morphological, structural properties and soil analyses." Fibres and textiles in Eastern Europe, 100–111

Yang L, Qiao F, Lei K, Li H, Kang Y, Cui S, An L (2019) Microfiber release from different fabrics during washing. Environmental pollution (Barking, Essex : 1987), 249, 136–143

Zambrano MC, Pawlak JJ, Daystar J, Ankeny M, Cheng JJ, Venditti RA (2019) Microfibers generated from the laundering of cotton, rayon and polyester based fabrics and their aquatic biodegradation. Mar Pollut Bull 142:394–407. https://doi.org/10.1016/j.marpolbul.2019.02.062

Zhao Y, Qiao R, Zhang S, Wang G (2021) Metabolomic profiling reveals the intestinal toxicity of different length of microplastic fibers on zebrafish (Danio rerio). J Hazard Mater 403:123663. https://doi.org/10.1016/j.jhazmat.2020.123663

MIX
Papier aus verantwortungsvollen Quellen
Paper from responsible sources
FSC® C105338

If you have any concerns about our products,
you can contact us on
ProductSafety@springernature.com

In case Publisher is established outside the EU,
the EU authorized representative is:
Springer Nature Customer Service Center GmbH
Europaplatz 3, 69115 Heidelberg, Germany

Printed by Libri Plureos GmbH
in Hamburg, Germany